Process Plant Piping

This book is designed as a complete guide to manufacturing, installation, inspection, testing and commissioning of process plant piping. It provides exhaustive coverage of the entire piping spool fabrication, including receiving material inspection at site, material traceability, installation of spools at site, inspection, testing and pre-commissioning activities. In a nutshell, it serves as a complete guide to piping fabrication and erection. In addition, typical formats for use in piping fabrication for effective implementation of QA/QC requirements, inspection and test plans, and typical procedures for all types of testing are included.

Features:

- Provides an overview of development of piping documentation in process plant design with a number of illustrations
- Gives exposure to various codes used in piping and pipelines within its jurisdiction
- Quick reference guide to various applicable sections of ASME B 31.3 provided
- Coverage of entire construction contractors' scope of work with regard to plant piping
- Written with special emphasis on practical aspects of construction and final documentation of plant piping for later modifications/investigations

This book is aimed at mechanical, process and plant construction engineers/supervisors, specifically as a guide to all novices in the above disciplines.

Process Plant Piping
Practical Guide to Fabrication, Installation, Inspection, Testing, and Commissioning

Sunil Pullarcot

CRC Press
Taylor & Francis Group
Boca Raton London New York

CRC Press is an imprint of the
Taylor & Francis Group, an **informa** business

Designed cover image: © Shutterstock

First edition published 2023
by CRC Press
6000 Broken Sound Parkway NW, Suite 300, Boca Raton, FL 33487-2742

and by CRC Press
4 Park Square, Milton Park, Abingdon, Oxon, OX14 4RN

CRC Press is an imprint of Taylor & Francis Group, LLC

© 2023 Sunil Pullarcot

Library of Congress Cataloging-in-Publication Data
Names: Pullarcot, Sunil, 1957- author.
Title: Process plant piping : practical guide to fabrication, installation,
inspection, testing and commissioning / Sunil Pullarcot.
Description: First edition. | Boca Raton, FL : CRC Press, 2023. |
Includes bibliographical references and index. |
Identifiers: LCCN 2022036340 (print) | LCCN 2022036341 (ebook) |
ISBN 9781032357072 (hbk) | ISBN 9781032357089 (pbk) |
ISBN 9781003328124 (ebk)
Subjects: LCSH: Chemical plants–Piping.
Classification: LCC TP159.P5 P85 2023 (print) | LCC TP159.P5 (ebook) |
DDC 660/.28–dc23/eng/20221006
LC record available at https://lccn.loc.gov/2022036340
LC ebook record available at https://lccn.loc.gov/2022036341

ISBN: 9781032357072 (hbk)
ISBN: 9781032357089 (pbk)
ISBN: 9781003328124 (ebk)

DOI: 10.1201/9781003328124

Typeset in Times
by codeMantra

Contents

About the Author

Sunil Pullarcot is a postgraduate in mechanical engineering having more than 38 years of industrial experience, one-third of which is in the oil and gas industry.

During his career spanning over 3½ decades, he had worked in heavy fabrication and construction industry in a multitude of disciplines like QA/QC, engineering (static and rotating equipment), project management, construction management and commercial and that too in different roles representing manufacturer/contractor, consultant and finally as owner.

Based on the vivid experience he gathered over the years, he established himself as an accomplished trainer on topics related to pressure vessels, heat exchangers, storage tanks, plant piping, NDT and welding. He is in the panel as training consultant to NExT, a Schlumberger Company, IBC Academy UK and an approved ASME instructor.

In his endeavour to pass on the valuable and unique experience he gained over 3½ decades to the forthcoming generations, Mr. Pullarcot has so far authored two books in his domain of expertise, namely, *Practical Guide to Pressure Vessel Manufacturing*, published in 2002 by Marcel Dekker, and *Above Ground Storage Tanks: Practical Guide to Construction, Inspection, and Testing* published by CRC Press in 2015.

Apart from the books, he published a few articles in renowned journals like *Hydrocarbon Processing, Inspectioneering* and *World Pipelines* published from the USA and UK in related topics.

The book *Process Plant Piping: Practical Guide to Fabrication, Installation, Inspection, Testing, and Commissioning* is his latest publication dealing with entire construction activities involved in process plant piping, with all finer details, which often are not documented or considered relevant, especially in view of beginners in this field.

Mr. Sunil Pullarcot is now retired from his last job as specialist with Kuwait Oil Company and is settled back in India.

Preface

Piping is the interconnection used in process plants for transferring process fluid from one equipment (static/rotating) to another, to complete chemical process involved.

Many books are available in the market dealing with many aspects of piping including portions of construction and associated services, whereas one covering all aspects related to construction of piping was not available in the market, especially for reference and guidance to fresh engineers and technicians joining this line of business.

The purpose of this book is to serve as a single point of resource providing basic information on all aspects of piping and detailed information on all aspects pertaining to manufacture, inspection, installation, testing and commissioning of a process plant. In a nutshell, the book deals in detail with all aspects of construction after engineering until completion and handing over for commissioning. However, this does not mean that the book eludes basic engineering involved in piping. To make the coverage complete and comprehensive, a brief about development of piping documents, starting from basic concept documents to isometrics, is also included. For the same purpose, a brief about all commonly used piping elements (pipes, fittings, flanges, bolting, gaskets, valves, etc.) is also included in the book, even though such basics would be known to most of the novices in the field.

Agencies concerned with the safety of operating process plants and industry together developed standards and specifications for almost all industrial components, and process plant piping was no exception. The author believes that the predominant standard followed in construction of process plant piping across the world is the one framed by the American Society of Mechanical Engineers (ASME) designated as ASME B 31.3 – Process Piping. Therefore, the entire contents of this book are organized in line with requirements of ASME B 31.3 and in sequence of actual work progression on site.

As mentioned earlier, the book is entirely based on practical experience of the author, and solutions offered in the book are simple, straightforward and could be understood even by engineers without much experience. Since methodologies and solutions proposed in the book are well within the requirements of ASME 31.3, it can be safely applied to other piping as well, which are not covered by its jurisdiction such as water piping, etc. In addition, this book also provides logical explanation to various code requirements and is capable of throwing some light into the unexplained side of piping code, and this feature of the book makes it really different from other books already on shelves.

The book principally targets beginners in the construction industry and deals with practical tips for manufacture of piping spools, its installation, field welding/joining, inspection, testing up to commissioning of the plant, including documentation pertaining to piping construction. Many of the tips included in the book are from the author's practical improvisation of existing practices within industry, which were proved successful in projects he was involved in.

In conclusion, as in the case of previous two books by the author, *Practical Guide to Pressure Vessel Manufacturing,* published in 2002 by Marcel Dekker, and *Above Ground Storage Tanks: Practical Guide to Construction, Inspection, and Testing* published by CRC Press in 2015, this book could also be considered as an effort to bridge the awareness gap between codes/standards and the actual construction practice followed at construction sites.

At this time of publication of this book, the author fondly remembers the immense support he received from renowned contractors like Petrofac and various inspection agencies and personnel he worked with, in giving this book the present stature, by way of their valuable opinions and advice.

1 Introduction

1.1 GENERAL

A process plant (be in oil and gas production, refining, petrochemical, fertilizer chemical or any other industry) can be compared with the human body (heart, arteries, veins and other organs connected through arteries and veins). The piping system is equivalent to arteries and veins, pumps equivalent to heart and organs equivalent to other process equipment of the plant. Piping truly resembles the blood circulation system consisting of heart, arteries and veins and other organs of the human body.

A schematic of both is shown in Figures 1.1 to 1.3 to depict similarities.

As mentioned earlier, piping system in a process plant is equivalent to arteries and veins, and they carry process fluids from one equipment to the other under pressure, developed by pumps in between, which is equivalent to the heart in the human body).

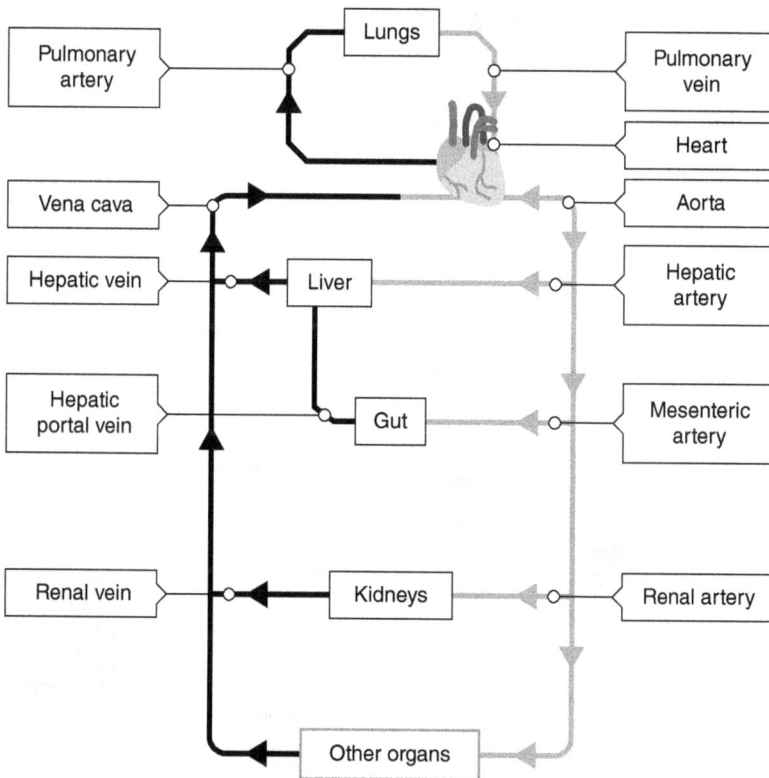

FIGURE 1.1 Blood circulation system (Ellenberger, 2010).

DOI: 10.1201/9781003328124-1

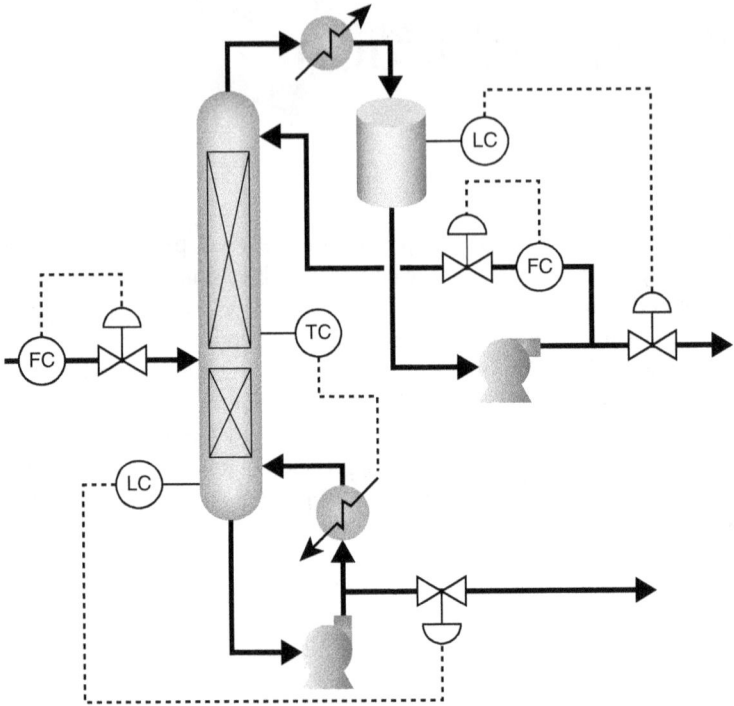

FIGURE 1.2   Typical process flow diagram of a section of process plant (1) (Nayyar, 1999).

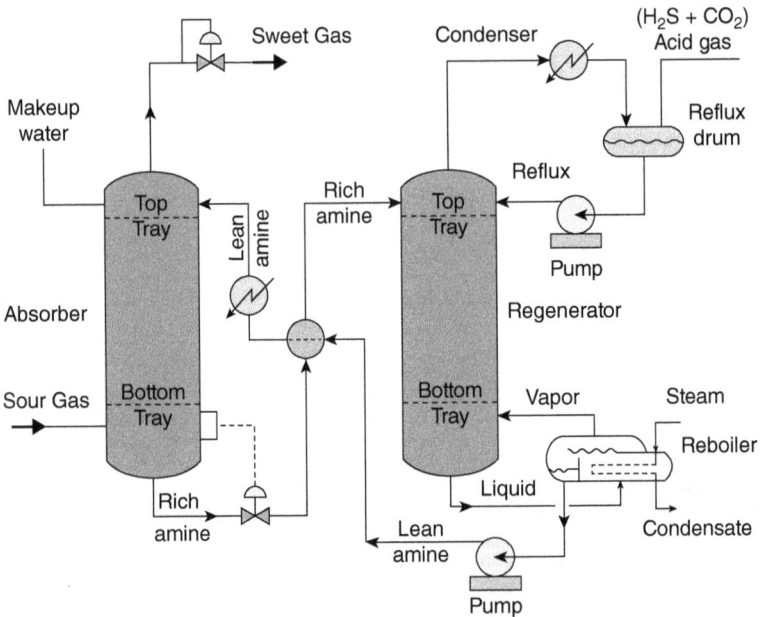

FIGURE 1.3   Process flow diagram (2) (NPTEL IITM, n.d.).

By virtue of the nature of the chemical process involved, piping system may require to carry liquids, vapours, slurries and sometimes solids as well, that too in mixed phases as required by the chemical process handled by the process plant. Process design followed by mechanical design of piping system shall take care of all these aspects. While in operation, piping systems are often subjected to extreme process conditions like high pressures, temperatures, flow velocities and flow regimes. Added to that are the corrosion and erosion aspects, which may dramatically alter material requirements. Above all are the statutory requirements imposed by governments of the locality for handling fluids that are considered toxic and/or lethal in nature.

Consideration of all the mentioned aspects during the design phase of piping makes the life of a piping design engineer highly challenging. Since piping works as the link or connection between different types of process equipment, the piping engineer shall have wide knowledge in related disciplines like static and rotating equipment engineering as well. This will enable the piping design engineer to be more effective and practical in considering a combination of scenarios that may occur during startups, running and stoppage of the plant during the design phase itself, thereby making the process design more robust.

With the advent and availability of newer and better materials, advancements in design aids, construction methods, inspection and testing methods and researches carried out on failures of the past, process and mechanical designs have undergone many evolutions over the years, and piping design has become much easier of late. However, use of the latest design aids requires a very good understanding of the fundamentals of design and its underlying concepts and considerations.

1.2 PIPING

Piping system consists of pipes, fittings, flanges, fasteners (bolts and nuts), gaskets, valves and pressure-containing portions of other piping components. It also includes pipe hangers and supports and other items necessary to prevent over-pressurization and overstressing of pressure-containing components. From this description, it is evident that pipe is one of the elements or a part of the piping system. Therefore, pipe sections when joined with fittings, valves and other mechanical equipment and properly supported by hangers and supports are called piping.

1.3 PIPING ENGINEERING

Piping engineering is highly important for safety and reliability of the process plant, including the safety of personnel operating the plant, public and environment in the vicinity of the plant. However, this matter (the way piping engineering is used in the process of plant design) is not often included in the curriculum of even mechanical engineers. Of late, many industrial training institutes have started add-on or special training programmes, of which a few are in association with renowned universities. Previously, the only way to gain expertise in this subject was through working with experienced engineers in this discipline.

The primary purpose of piping engineering is to make sure that the entire piping system involved in the process plant is:

- Specified according to requirements (including life cycle)
- Designed according to specifications and other statutory requirements
- Fabricated and installed according to the design
- Inspected, tested and commissioned as envisaged
- And finally, operated and maintained according to the recommendations (Operation and Maintenance (O&M) manual)

The design of pipes and piping elements of plant piping is simple and straightforward, and the pipes have easy-to-design components. However, when the system is assembled or installed in different configurations over pipe racks as per the piping layout design (to meet the site requirements of a specific plant connecting specific static/rotating equipment), the system gets complicated and is subjected to many different kinds of loadings. At a minimum, the following aspects shall be essentially considered while designing a piping system:

- Material selected for piping components is compatible with fluid to be transferred, that too in the temperature ranges specified and flow regimes anticipated.
- Entire piping is designed for internal pressure and the temperature ranges specified.
- Under tolerance on wall thickness of pipes and piping components during manufacture also needs to be considered while calculating the pressure thickness of piping.
- Variation in temperatures possible during operation as well as due to change in atmospheric temperatures.
- Expansion and contraction of piping system due to variations in temperature of piping system.
- Loadings that can be transferred to the intervening equipment.
- Requirements of pipe support to take care of empty weight, live weight and weight at the time of hydrostatic testing.
- Creep and fatigue effects of piping over the entire life span of the plant.
- Corrosion and erosion aspects that are likely to occur.
- Ensure that all piping elements like pipe, fittings, flanges, valves, gaskets, bolting, etc. meet the minimum design requirements of pressure, temperature and the kind of stresses it is subjected to, including those at joints.

1.4   PIPING DESIGN

The best piping design/configuration is the least expensive one, taking into consideration the initial project cost and maintenance/replacement cost over the life cycle of the plant. Considerations involved while arriving at the initial cost include but are not limited to installation cost, pressure drop, piping stresses, fatigue, supports, expansion loops, anchors and ease of operation and maintenance.

The complete process plant design consists of the following steps:

1. Conceptual layout design
2. Equipment layout design
3. Piping layout design

It may be noted that the above-mentioned steps are not really distinctive since equipment arrangement can be done along with piping layout. However, in large process plants, piping layout itself emerges as a very distinct step.

1.4.1 CONCEPTUAL LAYOUT DESIGN

This is a highly innovative activity, and hence, the right vision and concept at this point could have far-reaching implications in the economic operation of the plant during its service life. In conceptual design, essential process design requirements are established. Considering the process and its requirements, horizontal and vertical relationship between equipments is spelled out in this process. This process also includes 3D space allocation for all equipment within the process plant along with space allocation for all other ancillary services. In many cases, this may include laboratories, control rooms, workshops, storage areas, office and other amenities required for the smooth operation and maintenance of the plant. Changes in process, operating philosophy or equipment type or size may result in substantial changes in conceptual arrangement, thereby affecting all the subsequent activities, and hence the plant is expected to be developed after considering the various pros and cons on a long-term perspective.

1.4.2 EQUIPMENT LAYOUT DESIGN

This is an extension of conceptual layout in a detailed manner, wherein precise locations of equipment of the plant with main interconnections are clearly laid out. Equipment layout is the basic document for mechanical design of plant (culminating in detailed construction drawings for equipment and piping), similar to the piping and instrumentation diagram (P&ID) for process design.

1.4.3 PIPING LAYOUT DESIGN

Plant piping is often routed through pipe racks, and it passes through the entire plant driving the piping, electrical and instrumentation cable through it, controlling the operation of almost all individual equipment of the plant and for conveying its status to the control room for monitoring.

Upon completion of the layout design of equipment and pipe racks, the piping design can be undertaken to yield a very accurate material take-off (MTO). The steps involved in arriving at the MTO are provided in the following sections.

1.5   DESIGN BASIS

P&ID which was developed subsequent to the process flow diagram (PFD) in line with the project design basis (PDB) (a write-up describing the process and control system requirements) is the starting documentation for the piping design. P&ID provides all basic requirements like pipe size, schematic of the interconnecting equipment, its type, primary and secondary branching from main lines, etc. However, before finalizing the routing of the piping and its design, the design basis needs to be frozen, which essentially includes the following:

- Design codes
- Loading conditions
- Equipment requirements
- Client/project-specific requirements
- Statutory regulations (if any)
- Material specifications
- Failure modes expected

1.6   DESIGN CODES

For piping systems, applicable codes specify requirements for design, materials, fabrication, erection, inspection and testing, practically covering all stages up to pre-commissioning and commissioning activities of any plant construction project, whereas for individual piping components like pipes, fittings (elbows, tees, returns, reducer, etc.), flanges, valves, and other inline items, applicable standards referred in piping codes specify the design and construction rules for the respective components.

Compliance to applicable Piping Code is considered as the basic minimum project requirement. Statutory organizations under whose jurisdiction the project site is located also often insist on compliance to international codes of their preference. In addition, many times, such compliances are also insisted by insurance agencies while assessing the insurance tariffs for the plant. Apart from compliance to piping codes, insurers look at the consequential damages and their probabilities as well while assessing insurance tariff.

Every code at its outset defines its jurisdiction. Similarly, standards also specify the scope of application of each standard. By specifying the code or standard in contract, it makes only the mandatory parts applicable, whereas supplementary or non-mandatory requirements are applicable only when they are specifically stated as additional requirements in the contract. Because of this, while specifying requirements as to codes and standards, users shall be quite conversant with the coverage and applicability of codes in great detail, failing which the essential requirements may be missed out.

Various organizations across the world are publishing codes and standards in consultation with research organizations and representatives from related industries, often through technical committees constituted with experts drawn from various organizations interested in respective disciplines. This includes but is not limited to representatives from industry associations, manufacturers, professional groups, users, government agencies, insurance companies, etc. Technical committees so constituted are also responsible for maintaining, updating and revising the codes and standards on a periodic basis or as required. Revisions are required in line with the technological developments, research findings, experience-based feedback and also when changes in referenced codes, standards, specifications and regulations take place. Since revisions to codes and standards are published periodically, it is important that the users of these codes and standards remain informed of the release of latest editions, addenda or revisions itself, and also the changes implemented through each.

As far as plant piping is concerned, the basic code followed by most of the consultants and clients across the world is that of the American Society of Mechanical Engineers (ASME), ASME B 31 codes. ASME developed Code for Pressure Piping which covers Power Piping, Fuel Gas Piping, Process Piping, Pipeline Transportation Systems for Liquid Hydrocarbons and Other Liquids, Refrigeration Piping and Heat Transfer Components and Building Services Piping. Earlier ASME B 31 was known as American National Standards Institute (ANSI) B 31.

1.6.1 AMERICAN SOCIETY OF MECHANICAL ENGINEERS

The ASME is one of the prominent organizations across the world, instrumental in the development of codes and standards for more than a century. The first ASME Code was rolled out in 1915 (1914 edition), known as the "Rules for Construction of Steam Boilers and other Pressure Vessels", the motive force behind which was the Brockton Shoe Factory Boiler Explosion in 1905. Subsequent to the Brockton Explosion, the ASME established a Technical Committee, known as ASME Boiler & Pressure Vessel Committee, in 1911 to formulate rules for the construction of steam boilers and other pressure vessels, and this committee was instrumental in developing the first ASME Code published in 1915.

Since then, ASME has established other technical committees as well and developed codes and standards, such as ASME B 31, Code for Pressure Piping and many more.

1.6.2 EDITIONS AND ADDENDA

ASME codes are revised and published every three years, incorporating all the addenda issued in between revisions.

Code editions are published every three years and incorporate all additions and revisions made to the code during the preceding three years.

Code addenda are published in coloured sheets, which include additions and revisions to individual sections of the code periodically (twice a year as summer and winter addenda).

1.6.3   Interpretations

In addition to the above, ASME also issues written replies to queries concerning interpretation of technical aspects provided in the code. Interpretations are also published separately as part of the update service and are issued twice a year, till the next edition is published. Interpretations neither form part of the code nor the addenda.

1.6.4   Code Cases

Code cases may arise out of special requirements, which were not covered by code, for example, new materials and new construction methods. The Boiler and Pressure Vessel Committee meets regularly to consider such proposals to be incorporated as additions or revisions to existing code. Code cases are also published periodically and are issued to code holders or buyers until publication of the next edition of code.

1.7   HISTORY OF ASME CODES FOR PIPING

ASME piping codes evolved much later compared to the code for Boilers and Pressure Vessels. At the behest of ASME, the American Standards Association (ASA) initiated the project B 31 in 1926 to develop a code for piping. Consequently, the first issue of the Piping Code was published in 1935 as the American Tentative Standard Code for Pressure Piping.

Development of codes is not an easy task. It requires a pool of talents from many disciplines. Even though the piping looks similar visually, the specific requirements vary from industry to industry and process to process, that too depending on fluids handled by the piping systems.

During the period from 1942 to 1955, the code evolved into B 31.1, the American Standard Code for Pressure Piping, and later ASA started publishing various sections of the code as separate documents and published as different books for convenience, as a part of the entire B 31 Piping Code. The first of these separate books was ASA B 31.8, Transmission and Distribution Piping Systems.

Similarly, ASA B 31.3 was published as a separate book in 1959, superseding Section 3 of B 31.1 of 1955, which further evolved into Petroleum Refinery Piping Code in 1973 and thereafter to its current form as B 31.3, Process Piping.

ASA later became ANSI, and ASME adopted these codes as such under the same numbers. As such, it encompasses petroleum refinery, chemical, cryogenic, and paper processing requirements.

After adoption of ANSI codes by ASME, over the years, ASME has published different codes for piping depending on requirements for specific service piping and pipelines; a brief about each is provided in the following table:

ASME B 31.1   Power Piping

Used in electric power generating stations, industrial and institutional plants, geothermal heating systems, and central and district heating systems. B 31.1 is intended to be applied to:

- Piping for steam, water, oil, gas, air and other services
- Metallic and nonmetallic piping
- All pressures
- All temperatures above −29°C (−20°F)

B 31.1 is mandatory for piping that is attached directly to an ASME Section I boiler up to the first isolation valve, except in the case of multiple boiler installations where it is mandatory up to the second isolation valve.

ASME B 31.2   Fuel Gas Piping

Used in Fuel Gas Piping, until it was withdrawn in 1988 and responsibility for that piping was assumed by ANSI Z 223.1. It was a good reference for design of gas piping systems (from the metre to the appliance) and although it has been withdrawn, ASME makes it available as a reference.

ASME B 31.3   Process Piping

ASME B 31.3 Process Piping (once called the Chemical Plant and Petroleum Refinery Piping Code) is perhaps the code with the broadest coverage.

Principally used in process facilities such as petroleum refineries; chemical, pharmaceutical, textile, paper, semiconductor and cryogenic plants; and related processing plants and terminals. B 31.3 is intended to be applied to:

- Piping for all fluid services
- Metallic and nonmetallic piping
- All pressures
- All temperatures

Code prescribes requirements for materials and components, design, fabrication, assembly, erection, examination, inspection and testing of piping and applies to piping for all fluids including: (i) raw, intermediate and finished chemicals; (ii) petroleum products; (iii) gas, steam, air and water; (iv) fluidized solids; (v) refrigerants; and (vi) cryogenic fluids. It also includes piping which interconnects pieces or stages within a packaged equipment assembly as well.

The owner of the facility is responsible for designating the category of piping when certain fluid services (i.e., Category M (toxic), high purity, high pressure, elevated temperature or Category D) (nonflammable, nontoxic fluids at low pressure and temperature) are applicable to specific systems and for designating whether a Quality System is to be imposed.

ASME B 31.4   Pipeline Transportation Systems for Liquid Hydrocarbons and Other Liquids

Used in pipeline transporting of products which are predominately liquid between wells, plants and terminals, and within terminals, pumping, regulating and metering stations. B 31.4 is intended to be applied to:

- Piping transporting liquids such as crude oil, condensate, natural gasoline, natural gas liquids, liquefied petroleum gas, carbon dioxide, liquid alcohol, liquid anhydrous ammonia and liquid petroleum products
- Piping at pipeline terminals (marine, rail and truck), tank farms, pump stations, pressure reducing stations and metering stations, including scraper traps, strainers and prover loops
- All pressures
- Temperatures from −29°C to 121°C (−20°F to 250°F) inclusive

B 31.4 covers the design, construction, operation and maintenance of these piping systems; B 31.4 does not have requirements for auxiliary piping, such as water, air, steam and lubricating oil.

ASME B 31.5 **Refrigeration Piping and Heat Transfer Components**

Used in Refrigeration Piping and Heat Transfer Components. Piping and heat transfer components containing refrigerants and secondary coolants include water that is used as a secondary coolant. B 31.5 is intended to be applied to:

- Refrigerant and secondary coolant piping
- Heat transfer components such as condensers and evaporators
- For all pressures
- For temperatures at and above −196°C (−320°F)

ASME B 31.6 **Chemical Process Piping**

ASME B 31.6 was at one time going to be the Chemical Plant Piping Code but the 1974 draft document was so parallel to the existing B 31.3 that decision was taken by the B 31 Main Committee to fold it into B 31.3 before it was published.

ASME B 31.7 **Nuclear Power Piping**

ASME B 31.7 Nuclear Power Piping Code was withdrawn after two editions and responsibility was assumed by ASME B&PV Code, Section III, Subsections NA, NB, NC and ND.

ASME B 31.8 **Gas Transmission and Distribution Piping Systems**

ASME B 31.8 for Gas Transportation and Distribution Piping Systems is used in pipeline transportation of products, which are predominately natural gas between sources and end-use services. B 31.8 is intended to be applied to:

- Onshore and offshore pipeline facilities used for the transport of gas
- Gathering pipelines
- Gas distribution systems
- Piping at compressor, regulating and metering stations
- All pressures
- Temperatures from −29°C to 232°C (−20°F to 450°F) inclusive

Code covers design, fabrication, installation, inspection and testing of pipeline facilities used for transportation of gas and covers safety aspects of operation and maintenance of those facilities.

ASME B 31.8S **Managing System Integrity of Gas Pipelines**

This standard applies to onshore pipeline systems constructed with ferrous materials and that transport gas. Pipeline systems mean all parts of physical facilities through which gas is transported, including pipe, valves, appurtenances attached to pipe, compressor units, metering stations, regulator stations, delivery stations, holders and fabricated assemblies. This standard is specifically designed to provide the operator with the information necessary to develop and implement an effective integrity management programme utilizing proven industry practices and processes. The processes and approaches within this standard are applicable to the entire pipeline system.

ASME B 31.9 **Building Services Piping**

ASME B 31.9 for Building Services Piping is typically found in industrial, institutional, commercial and public buildings, and in multi-unit residences and is intended to be applied to:

- Piping for water and anti-freeze solutions for heating and cooling, steam and steam condensate, air, combustible liquids and other nontoxic, nonflammable fluids contained in piping not exceeding the following:
  - Dimensional limits
    - Carbon steel: NPS 42 (DN 1050) and 0.500″ (12.7 mm) wall
    - Stainless steel: NPS 24 (DN 600) and 0.500″ (12.7 mm) wall
    - Aluminium: NPS 12 (DN 300)
    - Brass and copper: NPS 12 (DN 300) and 12.125″ (308 mm) for copper tube
    - Thermoplastics: NPS 24 (DN 600)
    - Ductile iron: NPS 24 (DN 600)
    - Reinforced thermosetting resin: NPS 24 (DN 600)

○ Pressure and temperature limits, inclusive:
- Compressed air, steam and steam condensate to 1,035 kPa (150 psi) gauge
- Steam and steam condensate from ambient to 186°C (366°F)
- Other gases from ambient to −18°C to 93°C (0°F to 200°F)
- Liquids to 2,415 kPa (350 psi) gauge and from −18°C to 121°C (0°F to 250°F)
- Vacuum to 1 Bar (14.7 psi)
- Piping connected directly to ASME Section IV Heating Boilers

This code prescribes requirements for the design, materials, fabrication, installation, inspection, examination and testing of piping systems for building services and includes piping systems in the building or within the property limits.

ASME B 31.10   Cryogenic Piping

ASME B 31.10 Cryogenic Piping Systems (like B 31.6) was also in the process of development. Later the Committee recognized that the 1981 draft was very similar to ASME B 31.3 and decision was taken to include it into ASME B 31.3 Code and hence never issued as a separate document.

ASME B 31.11   Slurry Transportation Piping Systems

ASME B 31.11 Slurry Transportation Piping Systems is used in pipeline transportation of aqueous slurries between plants and terminals and within terminals and is intended to be applied to
- Piping transporting aqueous slurries of nonhazardous materials
- Piping in pumping and regulating stations
- For all pressures
- For temperatures from −29°C to 121°C (−20°F to 250°F) inclusive

ASME B 31.12   Hydrogen Piping and Pipelines

ASME B 31.12 Hydrogen Piping and Pipelines is applicable to piping in gaseous and liquid hydrogen service and to pipelines in gaseous hydrogen service. The code is applicable up to and including the joint connecting the piping to associated pressure vessels and equipment.

ASME B 31 G   Manual for Determining Remaining Strength of Corroded Pipelines

This document is intended for providing guidance in the evaluation of metal loss in pressurized pipelines and piping systems. It is applicable to all pipelines and piping systems that are part of ASME B 31 Code for Pressure Piping.

ASME B 31 Q   Pipeline Personnel Qualification

This Standard establishes requirements for developing and implementing an effective pipeline personnel qualification programme (qualification programme) utilizing a combination of technically based data, accepted industry practices and consensus-based decisions.

With regard to piping components used in plant piping, the following ASME standards are generally specified as the dimensional control standards:

Other ASME Codes Referred in Piping

ASME B 16.1	Cast Iron Pipe Flanges and Flanged Fittings
ASME B 16.3	Malleable Iron Threaded Fittings, Classes 150 and 300
ASME B 16.4	Cast Iron Threaded Fittings, Classes 125 and 250
ASME B 16.5	Pipe Flanges and Flanged Fittings
ASME B 16.9	Factory Made Wrought Steel Butt Welding Fittings
ASME B 16.10	Face to Face and End to End Dimensions of Valves
ASME B 16.11	Forged Fittings, Socket Welding and Threaded
ASME B 16.12	Cast Iron Threaded Drainage Fittings
ASME B 16.14	Ferrous Pipe Plugs, Bushings and Locknuts with Pipe Threads
ASME B 16.15	Cast Bronze Threaded Fittings Classes 125 and 250

ASME B 16.18	Cast Copper Alloy Solder Joint Pressure Fittings
ASME B 16.20	Ring Joint Gaskets and Grooves for Steel Pipe Flanges
ASME B 16.21	Nonmetallic Flat Gaskets for Pipe Flanges
ASME B 16.22	Wrought Copper and Copper Alloy Solder Joint Pressure Fittings
ASME B 16.23	Cast Copper Alloy Solder Joint Drainage Fittings – DWV
ASME B 16.24	Cast Copper Alloy Pipe Flanges and Flanged Fittings Classes 150, 300, 400, 600, 900, 1500 and 2500
ASME B 16.25	Butt Welding Ends
ASME B 16.26	Cast Copper Alloy Fittings for Flared Copper Tubes
ASME B 16.28	Wrought Steel Butt Welding Short Radius Elbows and Returns
ASME B 16.29	Wrought Copper and Wrought Copper Alloy Solder Joint Drainage Fittings – DWV
ASME B 16.32	Cast Copper Alloy Solder Joint Fittings for Solvent Drainage Systems
ASME B 16.33	Manually Operated Metallic Gas Valves for Use in Gas Piping systems up to 125 psig (sizes ½ through 2)
ASME B 16.34	Valves – Flanged, Threaded and Welding End
ASME B 16.36	Orifice Flanges
ASME B 16.37	Hydrostatic Testing of Control Valves
ASME B 16.38	Large Metallic Valves for Gas Distribution (Manually Operated, NPS 2½–12, 125 psig maximum)
ASME B 16.39	Malleable Iron Threaded Pipe Unions, Classes 1150, 250 and 300
ASME B 16.40	Manually Operated Thermoplastic Gas Shutoffs and Valves in Gas Distribution Systems
ASME B 16.42	Ductile Iron Pipe Flanges and Flanged Fittings, Classes 150 and 300
ASME B 16.47	Large Diameter Steel Flanges (NPS 26 through NPS 60)

Other ASME B 36 Piping Component Standards used in piping are the following:

ASME B 36.10	Welded and Seamless Wrought Steel Pipe
ASME B 36.19	Stainless Steel Pipe

The above is not a comprehensive list of the standards usually referred in the plant piping process. What was listed above only constitutes the commonly referred ones by the ASME, as the specific discussion hereunder is about ASME B 31.3. Piping may require many more references as to other standards as well, such as American Petroleum Institute (API), supply plumbing (SP), ANSI, and American Society for Testing and Materials (ASTM) as required.

Depending on plant location and type of facility, either an ASME Code or any other equivalent Piping Code shall be selected. Clients and consultants usually decide on which codes to be followed for the piping.

The general practice is to apply a single Piping Code to all piping systems within the plant, but sometimes this may not be appropriate. In the case of a petrochemical plant, generally the plant piping is designed as per ASME B 31.3. But by chance when some power boiler is included within the plant, the steam piping needs to be designed as per ASME B 31.1 and many other equipment may have to be designed according to BPV Code as well. Further, the transfer lines from one plant to another are designed in accordance with either B 31.4 or 31.8. Though this is the present status of codes and regulations, it can and will change in due course. Based on researches and studies taking place across the world, the codes are continuously evolving. In that process, technical committees responsible will also consider the commercial aspects, and hence could expect new additions/deletions or total restructuring of above codes as well, as seen in the past.

Most of the above codes have many similarities such as in calculation of minimum wall thickness, inspection and testing methods, philosophy, etc., whereas the specific rules could be different based on the type of facility to be designed. The allowable stresses considered in each code are different, implying a different factor of safety, which is principally based on the expected loading conditions, various possibilities and probabilities of combination loading conditions envisaged.

As the codes provide only basic minimum requirements, more rigorous analysis, precision in manufacturing, inspection and testing are often required by clients across the world, which are unique and specific to the industry concerned. These are imposed as project requirements and vary from project to project and industry to industry.

The main codes mentioned herein are basically meant for the design and construction and installation, and hence do not cover the operation and maintenance aspects of piping. However, operation and maintenance aspects are lately added as non-mandatory appendices to the codes. An example is Appendix V to ASME B 31.1, Recommended Practice for Operation and Maintenance and Modifications of Power Piping.

While selecting the applicable codes for the project, it shall be applied in entirety for a system. This implies that the use of B 31.3 for calculation of wall thickness of a piping and the use of B 31.8 for allowable stress value are not acceptable. However, the codes are not comprehensive enough to take care of all the permutations and combinations that could occur in a specific industry, for instance. When code is silent on the unique issues being faced by designer, other codes also can be researched to find an acceptable and sound way forward. In such instances, apart from other codes, books, published articles, etc. are also often referred in arriving at a solution.

Apart from the ASME Codes related to piping as mentioned earlier, the other often referred codes are the following:

- API maintenance and inspection of piping components
- Manufacturers Standardization Society (MSS) for pipe supports MSS-SP-58 and 69
- ANSI for various piping components, including valves, fittings and radiographic plugs
- ASTM materials, manufacturing inspection methods and test methods
- American Welding Society (AWS)
- American Water Works Association (AWWA)
- National Fire Protection Association (NFPA)

1.8 LOADING CONDITIONS FOR PIPING

The loading conditions that may arise in piping can be bifurcated into two groups, static and dynamic loading, with its constituents listed in the following.

1.8.1 STATIC LOADING

- Temperature – Including multiple operating temperatures and cycles
- Pressure – Standard operating pressure, upset pressure and design pressure

- Equipment Movement – Thermal movement of piping when equipment heats up or cools down
- Dead Weight – Including weight of pipe, fluid, inline components, insulations, branch line, pipe support attachment or any other attachment
- Wind – Though this is technically a dynamic condition, it is usually analysed as an equivalent static condition
- Cyclic Conditions – Arising out of batch operations, by which a pipe is filled and emptied many times per shift. Depending on process, this may need to be considered as a static fatigue such as thermal or a dynamic load condition

1.8.2 DYNAMIC LOADING

- Steam hammer created by sudden closure of valves creating pressure waves in the pipe
- Surge or pressure waves caused by opening and closing of inline valves
- Thrust created by safety valves and other pressure relief devices
- Water hammer and other issues related to two-phase flows
- Thermal shocks from rapid cooling or heating of pipe surface
- Seismic event
- Pipe whip caused by sudden fracturing of pipe often considered only in nuclear piping
- Various upset conditions created due to out-of-control chemical reactions
- Various upset conditions caused due to loss of control of devices like control valves, etc.
- Flow-induced vibrations

1.9 EQUIPMENT REQUIREMENTS

The interface between pipe and equipment is extremely important and needs to be managed properly in the entire process plant. The following aspects about the equipments falling in between piping need to be known and aligned with the piping design parameters:

- Location size and type of each nozzle on equipment
- Design conditions (pressure and temperature) consistent with that of the piping
- Safety valve set pressures shall be consistent with that of operating conditions of the piping
- Equipment nozzle movement due to temperature – accommodated by the piping flexibility
- In case nozzle loads are limited by design (usual in most cases), expansion joints may be required between piping and equipment if the piping is considered rigid.

1.10 CLIENT/PROJECT REQUIREMENTS

In addition to the international standards and specifications mentioned in the contract, many more client specifications may also be listed in the contract. Often such specifications impose stringency to the minimum requirements indicated in international standards:

- Piping needs to take care of all project-specific requirements, which are normally above code requirements.
- The above specifications are required based on typical requirements of the process, medium to be handled and other intricacies of the process.
- Designs are often rejected due to non-compliance to client requirements.
- Many times, the above requirements are insisted based on some past bad experience of the client and may be requiring changes in current practices. But as long as such requirements are included in the contract, piping engineer needs to address the same.

1.11 MATERIAL SPECIFICATIONS

Material selection is yet another specialized job in piping, which often requires the expertise of a metallurgist and essentially includes the following:

- Selection of proper materials based on fluid characteristics and their behaviour at operating conditions is a herculean task requiring the services of a metallurgist.
- Specifications for individual piping elements need to be derived from client standards and specifications.
- In case older or discontinued international standards are specified in the contract, they are to be substituted with new specifications and consequential change in client specifications as required and need to be notified to client.
- Client standard often restricts the choices provided within specified international standards and accordingly needs to be captured by the purchase or construction specifications to be developed.

1.12 FAILURE MODES

The following failure modes are often considered during piping design:

- Velocity of fluid causing erosion in piping
- Hardening of piping elements of piping systems where two-phase flow is expected
- Embrittlement of materials in low-temperature piping
- Creep damage in high-temperature piping

- In high-temperature piping using welded pipes, failures can occur due to creep degradation, high stress intensification at weld seam, etc. (stress-concentrated corrosion).

1.13 SCOPE, COVERAGE AND LAYOUT OF ASME CODE B 31.3 – PROCESS PLANT PIPING

1.13.1 GENERAL

As mentioned in Section 1.7, ASME B 31 Code for Pressure Piping consists of a number of individually published sections, under the ambit of ASME Committee B 31, Code for Pressure Piping. The individual jurisdictions of each of the codes that fall within the ambit of ASME B 31 are also briefly addressed in Section 1.7.

The selection of applicable code falls within the responsibility of the owner/consultant and shall be done judiciously considering the nature of service of the piping system and the code that is almost completely applicable to the service.

Factors to be considered in that process include the limitations or exclusions of Code Section selected; jurisdictional requirements; and applicability of other codes and standards and their jurisdiction. Once selected, all applicable mandatory requirements of the selected Code Section shall be applicable without fail. For some installations or plants, more than one Code Section may apply to different parts of the installation. Further, the owner is responsible for imposing additional requirements, above code requirements, if found essential for safe operation of piping system within the installation.

As mentioned earlier, code sets forth engineering requirements deemed necessary for safe design and construction of pressure piping and its safe operation. While safety is the basic consideration, this factor alone shall not necessarily govern final specifications for any piping installation. Since the code is not a design handbook, it cannot eliminate the need for a competent person to carry out design with proper engineering judgement of probable technical scenarios the piping may pass through during its service life.

Piping code specifies basic design principles and formulae to be used in the design of piping. These are to be supplemented as necessary with specific requirements to ensure uniform application of principles to entire system of piping and to guide in selecting appropriate piping elements. The code prohibits use of designs and practices known to be unsafe and warns about precautions as well, wherever required so.

1.13.2 SCOPE OF ASME B 31.3 PROCESS PLANT PIPING

ASME B 31.3 specifies design, materials, fabrication, erection, inspection and testing requirements for process plant piping systems. Process plants envisaged in ASME B 31.3 may belong to any of the industries like petroleum refineries; chemical, pharmaceutical, textile, paper, semiconductor and cryogenic plants; their related intermediary process plants, terminals; etc.

Similarly, ASME B 31.3 applies to piping and piping components that are used for all fluid services, not just hydrocarbon services and may include but is not limited to the following services:

- Raw, intermediate and finished chemicals
- Petroleum products
- Gas, steam, air and water
- Fluidized solids
- Refrigerants
- Cryogenic fluids

Apart from the piping interconnecting equipments within the process plant, the interconnecting piping used in packaged equipment like instrument air system, dehydration system, etc. also falls within the ambit of ASME B 31.3.

1.13.3 EXCLUSIONS FROM ASME B 31.3

Specific exclusions to the scope of ASME B 31.3 are as follows:

- Piping systems designed for internal gauge pressures at or above zero but less than 105 kPa (15 psi), provided the fluid handled is nonflammable, nontoxic and not damaging to human tissues and its design temperature is from −29°C (−20°F) through 186°C (366°F).
- Power boilers in accordance with BPV Code Section I and boiler external piping which are required to conform to B 31.1.
- Tubes, tube headers, crossovers and manifolds of fired heaters which are internal to heater enclosure.
- Pressure vessels, heat exchangers, pumps, compressors and other fluid handling or processing equipment, including internal piping and connections for external piping.

In other words, except for the first item, all others fall under the jurisdiction of other codes/standards and are excluded from the scope of ASME B 31.3.

The pictorial presentation of coverage of ASME B 31.3 (as provided in Figure 300.1.1) is provided in Figure 1.4 for clarity on scope. For defining the jurisdiction of ASME B 31.3 clearly, the joint connecting piping to equipment is considered to be within the scope of B 31.3.

1.13.4 TYPES OF FLUID SERVICES COVERED

- *Category D fluid service*: nonflammable, nontoxic and not damaging to human tissues, the design pressure does not exceed 1,050 kPa (150 psig), and the design temperature is from −29°C (−20°F) to 186°C (366°F).
- *Category M fluid service*: a fluid service in which potential for personnel exposure is judged to be significant and in which a single exposure to a very

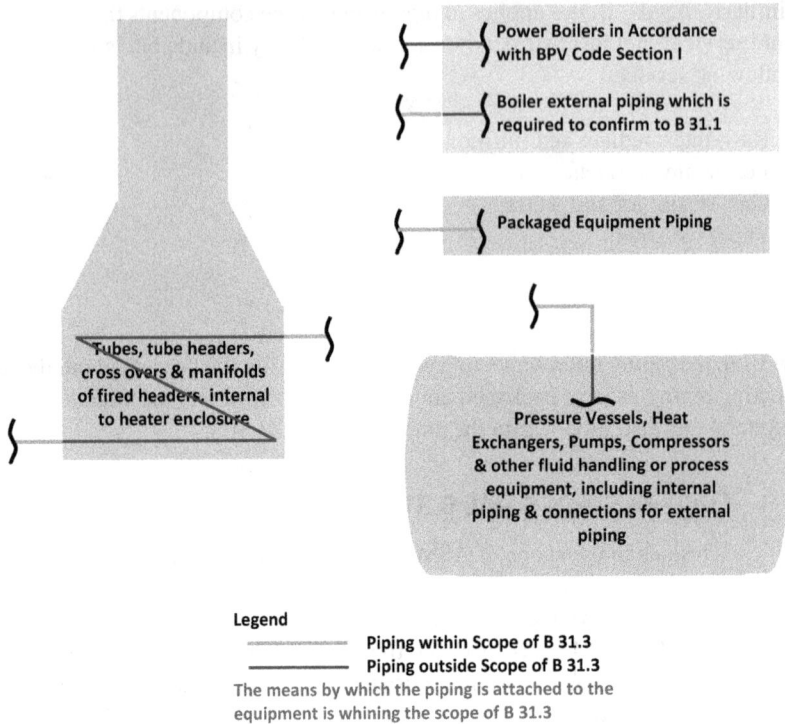

FIGURE 1.4   Depiction of scope of ASME B 31.3 (ASME, 2022).

small quantity of a toxic fluid, caused by leakage, can produce serious irreversible harm to persons on breathing or bodily contact, even when prompt restorative measures are taken.

- *Elevated temperature fluid service*: a fluid service in which piping metal temperature is sustained equal to or greater than Tcr, where Tcr = temperature 25°C (50°F) below the temperature identifying start of time-dependent properties of material of piping.
- *Normal fluid service*: a fluid service pertaining to most piping covered by this code, that is, not subject to rules for Category D, Category M, Elevated Temperature, High-Pressure or High-Purity Fluid Service.
- *High-pressure fluid service*: a fluid service for which owner specifies use of Chapter IX for piping design and construction. High pressure is considered herein to be pressure in excess of that allowed by the ASME B16.5 Class 2500 rating for the specified design temperature and material group.
- *High-purity fluid service*: a fluid service that requires alternative methods of fabrication, inspection, examination and testing not covered elsewhere in the code, with the intent to produce a controlled level of cleanness. The term thus applies to piping systems defined for other purposes as high purity, ultra-high purity, hygienic or aseptic.

2 Piping Documents Development Process

2.1 KEY DOCUMENTS

The flow chart in Figure 2.1 shows a schematic of how the key piping documents are evolved starting from the project design basis (PDB) up to purchase order for piping components. The evolution of piping drawings culminating in piping material purchase follows a definitive pattern. The common adopted document evolution pattern is described in its order in sections below.

2.2 PROJECT DESIGN BASIS

PDB is basically a descriptive document consisting of a brief description of chemical process involved in the project, with a list of static and rotating equipment. This document also elaborates on the control system requirements and essentially contains the following information:

- Overall facility configuration
- Operating mode and basic philosophy
- Redundancy in design
  - Specific minimum redundancy requirements for equipment
- Site-specific design conditions
- Air emission limitations
- Fuel gas
- Water supply
  - Raw water
  - Plant waste water
  - Storm water runoff
- Noise limits
- Sub-surface conditions
- Electrical and communication interconnections
  - Permanent electrical power
  - Standby electric power
  - Communications
- Codes, standards and specifications
- Prohibited materials

DOI: 10.1201/9781003328124-2

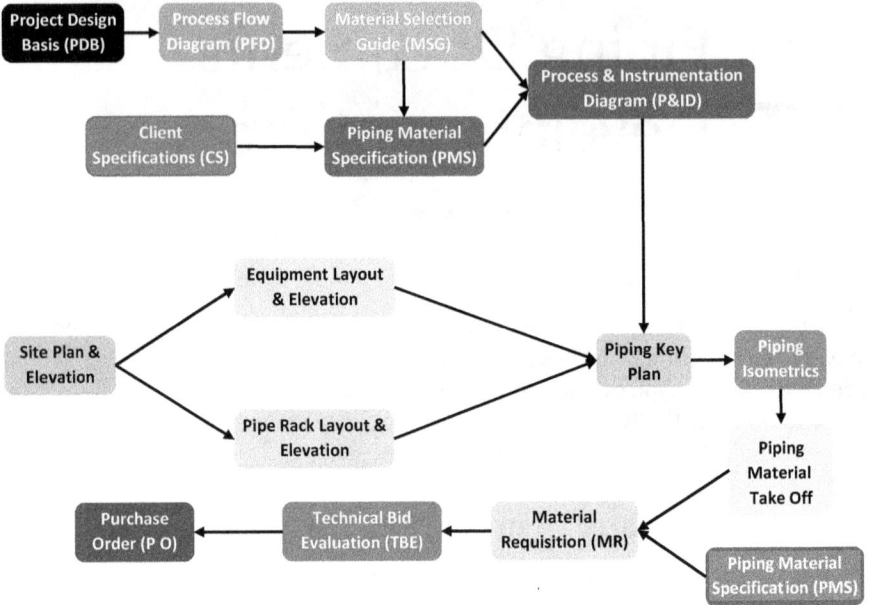

FIGURE 2.1   Piping documents development process.

2.3   BLOCK FLOW DIAGRAM

The block flow diagram (BFD) is a very simple diagram which is a condensed version of the entire process of the plant into a single A3 or A4 sheet. This is something similar to an organization chart, consisting of textboxes with embedded text and interconnected by lines with flow direction and commodity transferred between boxes. Typically, a BFD contains the following information.

- Individual equipment or package equipment represented by a single symbol, usually a rectangular text box.
- Clear labels describing the function of the process, in case of package item.
- Order of process flow arranged from left to right with due weightage for the gravity flow.
- Lines connecting equipment or packages shall have flow direction indicated by arrows.
- In case more than one line leaves an equipment, the processed commodity through each line is to be specified (Figure 2.2).

2.4   PROCESS FLOW DIAGRAMS

Process flow diagrams (PFD) contain more information than BFDs and is considered as the next level document made from the BFD.

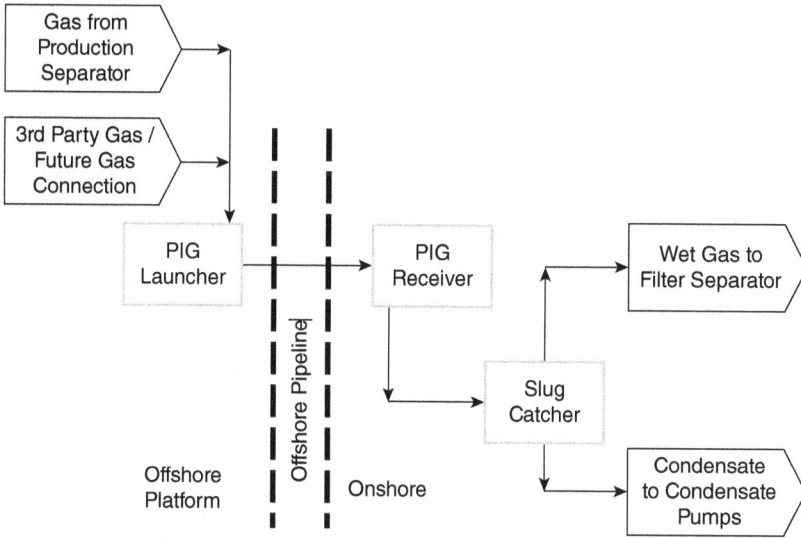

FIGURE 2.2 Typical block flow diagram (BFD) (Walker, 2009).

PFD mainly defines:

a. A schematic representation of sequence of all relevant operations occurring during a process including information considered desirable for analysis.
b. The process representing events, which occur to the material(s) to convert feedstock(s) to specified products.
c. An operation occurring when an object (or material) is intentionally changed in any of its physical or chemical characteristics, is assembled or disassembled from another object or is arranged or prepared for another operation, transportation, inspection or storage.

PFDs indicate many more details about the major equipment or packages represented in the BFD, the subsystems and flow of product between them. PFDs are often accompanied by heat and mass balance, pressure and temperature of feed and product lines from all major equipment or package equipment like pressure vessels, heat exchangers, storage tanks, rotating equipment like pumps, compressors, turbines, etc. PFDs also shall indicate main headers, points of pressure, temperature and flow control requirements.

For static equipment like pressure vessels, heat exchangers and storage tanks shown in PFD, design pressure and temperature shall be shown. Similarly for rotating equipment like pumps, capacity, pumping head and horse power of the driver are specified, whereas for compressors, the horse power or power in other units, along with pressures and temperatures before and after, is also expected to be specified in PFD.

2.4.1 Purpose of PFD

The purpose of PFD is generally as follows:

a. *Broadly describes plant design basis*: PFD shows the plant design basis indicating feedstock, product and mainstream flow rates and operating conditions.
b. *Describes scope of process*: PFD serves to identify the scope of process considered.
c. *Details of equipment configuration*: PFD shows graphically the arrangement of major equipment, process lines and main control loops.
d. *Lists required utilities*: PFD shows utilities which are used continuously in the process.

2.4.2 Contents of PFD

The PFD generally contains the following information:

a. All process lines, utilities and operating conditions essential for heat and material balance.
b. Utility flow lines and their types which are used continuously within battery limits.
c. Equipment diagrams to be arranged according to process flow, designation and equipment number.
d. Simplified control instrumentation pertaining to control valves and similar controls to be included in the process.
e. Major process analysers.
f. Operating conditions around major equipment.
g. Heat duty for all heat transfer equipment.
h. Changing process conditions along individual process flow lines, such as flow rates, operating pressure and temperature, etc.
i. All alternate process and utility lines' operating conditions.
j. Material balance table for essential streams.

The following details are generally not to be shown on PFD, except in special cases:

a. Minor process lines which are not usually used in normal operation and minor equipment, such as block valves, safety/relief valves, etc. unless otherwise specified.
b. Elevation of equipment.
c. All spare equipment.
d. Heat transfer equipment, pumps, compressor, etc., to be operated in parallel or in series shall be shown as one unit.
e. Piping information such as size, orifice plates, strainers and classification into hot or cold insulated, of jacketed piping.
f. Instrumentation not related to automatic control.
g. Instrumentation of trip system.

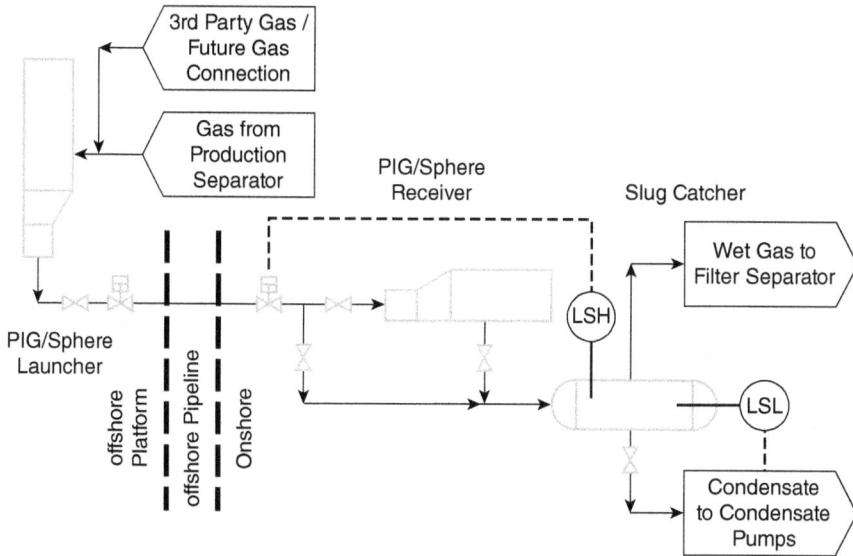

FIGURE 2.3 Typical process flow diagram (PFD) (Walker, 2009).

h. Drivers of rotating machinery except where they are important for control line of the process conditions.

i. Any dimensional information on equipment, such as internal diameter, height, length and volume. Internals of equipment shall be shown only if required for a clear understanding of the working of the equipment.

In a nutshell, a typical PFD usually provides the following information (Figure 2.3):

- Process piping
- Process flow direction
- Major equipment through symbols and package equipment as a rectangular text box
- Major bypass and recirculation/minimum flow lines
- All critical valves' process control
- Processes identified by the system name
- System ratings and operational range
- Composition of fluids
- Connection between systems

2.5 PIPING AND INSTRUMENTATION DIAGRAM

Piping and instrumentation diagram (P&ID) carries a lot of information and hence is a key document for cross-verification during construction phase of a process plant, which includes erection of equipment and piping followed by its electrics and control systems (Figure 2.4).

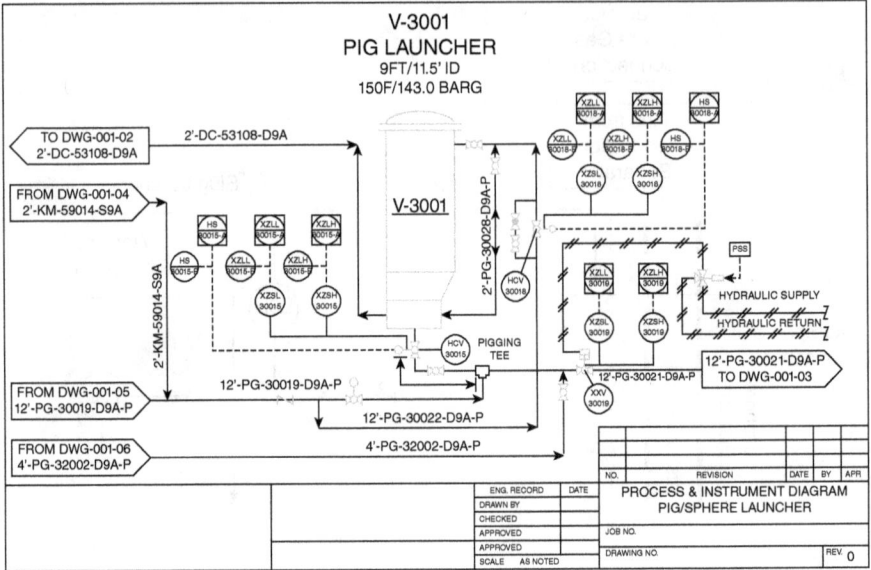

FIGURE 2.4 Typical process and instrumentation diagram (P&ID) (Walker, 2009).

P&ID is developed taking into consideration the conceptual aspects of PDB and PFD and elaborated further by incorporating the following:

- Detailed symbols
- Detailed equipment information
- Equipment order and process sequence
- Process and utility piping
- Process flow direction
- Major and minor bypass lines
- Line numbers, size and piping material specification (PMS)
- Isolation and shut-off valves
- Maintenance drains and vents
- Relief and safety valves
- Instrumentation
- Controls
- Types of process component connections
- Vendor and contractor battery limits
- Skid and package interfaces
- Hydrostatic test vents and drains
- Design requirements for hazardous operations

2.5.1 FUNCTION AND PURPOSE OF P&IDs

P&IDs are the basic graphical document representing process. For processing facilities, it is a graphic representation of:

- Key piping and instrument details
- Control and shutdown schemes
- Safety and regulatory requirements
- Basic start-up and operational information

2.5.2 When and Who Use P&IDs

P&IDs are a schematic illustration of functional relationship of piping, instrumentation and system equipment components. They are typically created by engineers who are designing a manufacturing process for any process plant.

These facilities usually require complex chemical or mechanical steps that are mapped out with P&IDs to construct a plant and also to maintain plant safety as a reference for process safety information (PSI) in process safety management (PSM). In case of any mishaps, the investigation starts with review of P&ID. P&IDs are an invaluable document in streamlining an existing process, replace a piece of equipment or make modifications or improvements to existing plant. Any alteration in process after commissioning shall be implemented only through a documented management of change (MOC) proceedings.

P&IDs are used extensively by field technicians, engineers and operators to understand the process and how the instrumentation is interconnected. They are quite useful in training new operators and maintenance technicians as well.

2.5.3 Difference between a PFD and a P&ID

Simplified or conceptual designs are called process flow diagrams (PFDs). A PFD shows fewer details than a P&ID and is usually the first step in the design process – more of a bird's-eye view.

2.5.4 Limitations of P&ID

Since P&IDs are graphic representations of processes, they have some inherent limitations. They can't be relied on as real models, because they aren't necessarily drawn to scale or are geometrically accurate.

2.5.5 P&ID and Associated Documents

Because P&IDs are schematic overview graphics, anyone who reads it need documents to clarify the details and specifications. Here are some of them:

- *Process flow diagrams*: P&IDs originate from PFDs. A PFD is a picture of the separate steps of a process in sequential order. Elements that may be included are: sequence of actions, materials or services entering or leaving the process (inputs and outputs), decisions that must be made, people who become involved, time involved at each step and/or process measurements.

- *Piping material specifications*: In PMS, one can find details about all piping materials used in the process plant, including but not limited to pipes, fittings, valves, gaskets, bolts etc.
- *Equipment and instrumentation specifications (EIS)*: Standards and details too extensive to fit into the P&ID are included in the EIS including Scope, Standards, Codes and Specifications, Definitions and Terminology, Materials of Construction, Design Basis, Mechanical/Fabrication, Guarantees, Testing and Inspection, Documentation and Shipping.
- *Functional requirement specification (FRS)*: How the plant or system operates is detailed in the FRS. It includes the functional description, communication and scope definition of the process.

2.5.6 CONTENTS OF P&ID

P&ID shall essentially contain the following information:

- Mechanical equipment with names and numbers
- All valves and their identifications
- Process piping, sizes and identification
- Miscellaneous – vents, drains, special fittings, sampling lines, reducers, increasers and swagers
- Permanent start-up and flush lines
- Flow directions
- Interconnections' reference
- Control inputs and outputs, interlock
- Seismic category
- Interfaces for class changes
- Quality level
- Annunciation inputs
- Computer control system input
- Vendor and contractor interfaces
- Identification of components and subsystems delivered by others
- Intended physical sequence of the equipment
- Equipment rating or capacity

2.5.7 SPECIFIC EXCLUSIONS

Finer details can be contained in supporting documents. P&ID shall not be too much crowded and for that, the following details are not often shown in P&ID:

- Instrument root valves
- Control relays
- Manual switches
- Primary instrument tubing and valves
- Pressure, temperature and flow data
- Elbow, tees and similar standard fitting
- Extensive explanatory notes

2.5.8 STANDARDS AND RULES FOR PREPARATION OF P&IDs

P&IDs are prepared according to a set of general rules and standards for its universal acceptance with some minor differences between regions. The basic purpose of a P&ID is to provide maximum information of a process consisting of static/rotating equipment and interconnecting piping along with the instrumentation and control system. Therefore, each equipment, piping line, instrument, etc. shall be labelled using specific conventions of nomenclature, which may look odd to novices in the field, but once learned, reading is quite easy, simple and straightforward.

The common standard followed for preparation of P&ID for process industry is ISO 15519.

The ISO 15519 series consists of standards for specification of various types of diagrams used in the process industry, intended for standardizing information contained in each of the diagrams and to make reading of the same unambiguous among users spread across the world.

The specification describes preparation of different types of diagrams and use of graphical symbols, letter codes and reference designation in diagrams. The standard addresses all process industry fields, for example, chemical, petrochemical, power, pharmaceutical, foodstuff, pulp and paper and covers PFD, process and instrument diagram (PID), process control diagram (PCD) and typical diagram (TYD).

Apart from the above, process industry practice (PIP), a consortium of process industry owners and engineering construction contractors who serve the industry, have developed standards for preparation of P&IDs. PIP document, PIC 001-Piping and Instrumentation Diagram, details what a P&ID should contain.

The three most important items identified and described in a P&ID are the equipment (static and rotating), process lines and instrumentation connections.

2.5.9 EQUIPMENT DESIGNATION

The very purpose of tagging or labelling an equipment using a unique number is to identify the equipment quickly and is usually done using a letter followed by 3- to 5-digit number. The letter designates the type of equipment such as V = Vessels, E = Heat Exchanger, H = Heater, P = Pump, T = Tank, etc. Out of the following 5-digit numbers, the first two digits are used to represent the process and the following three digits give the sequential identification number from 001 to 999. In relatively small process plants, the equipment designation is limited to letter + three digits. For example, V 102 designates a Pressure Vessel identified as the 102nd equipment in the process/utility section.

2.5.10 LINE DESIGNATION

A nomenclature used in line designation is also similar to the one used in designating equipment, to specify interconnecting piping (in both process and utility piping) as well. The most widely used designation is something like "00-XX-00000-X00X-X0".

- The first two digits indicate the line size in inches (e.g., 12 for line size of 12 inches).

- The two letters (XX) that follow designate the process system in which the piping falls (e.g. PG = Process Gas, PL = Process Liquid, IA = Instrument Air, VA = Atmospheric Vent).
- Out of the following five digits (00000), the first two digits (00) represent the process system like 10, 20, 30, etc. to represent systems like process gas, process liquid, gas dehydration and so on.
- The next three digits (000) are a sequential identifier 001–999 to represent each of the lines.
- The next part is an alphanumeric section representing the PMS (X00X) consisting of letters and numerals used based on the standard practices and nomenclature followed by the organization. For example, Piping Class C 95 B is derived from the nomenclature, letter C representing Carbon Steel Pipe, 9 designating piping Class 900, 5 designating corrosion allowance of 3 mm, and B stands for NACE Compliance required for lines operating in sour environment envisaged in oil and gas industry.
- The consultants are quite often free to use their nomenclature and in such cases, the same shall be followed by the client who utilizes the services of consultant for design/construction of the plant. Many major clients, especially those in oil and gas, have their own PMS, included in the contract, which may need some more additions during detailed engineering phase. In such cases, consultants are required to follow the same methodology adopted in naming the additional piping classes (if any to be developed) considered essential during detailed design.
- The last 2 alphanumeric X0″ represent the type of insulation and its thickness. For example, P1″ represents personnel protection insulation of 1″ thickness. The other common insulations identified are H for Heat Conservation, T for Tracing, etc. In case of bare line (without any insulation), two characters are often excluded from line number designation.

The above explains the significance of line number and shows the host of information carried by the number such as its size and schedule, design boundaries, material specification and type/thickness of insulation.

2.5.11 INSTRUMENT LINES

Instrument lines indicate the flow of information between various instruments and how signal is transferred between instruments. Different types of signal lines include equipment to instrument connection line such as electrical, hydraulic, pneumatic and capillary. These lines are differentiated using solid dashed lines with special qualifiers added to describe service of the lines.

2.5.12 INSTRUMENT DESIGNATION

Compared to equipment and line designations, instrument designation is more complicated. However, it becomes simple once the basic philosophy used in designating instruments is understood. Instruments are often represented as either circles or as squares inscribed with fitting circles called balloons. They contain two main information: the top indicating service of the instrument and the bottom, a unique identification (Figure 2.5).

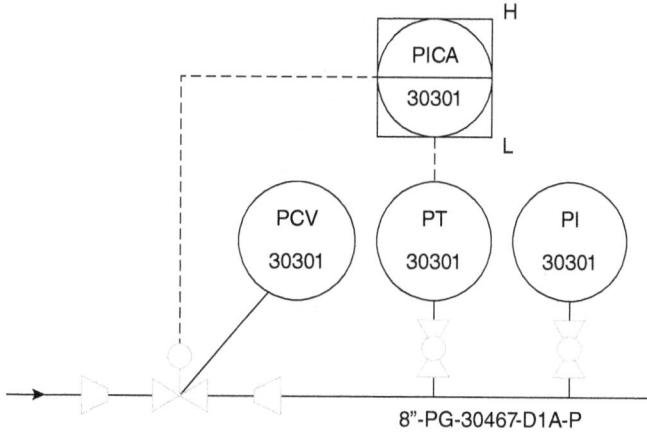

FIGURE 2.5 Instrument designation and nomenclature (Walker, 2009).

By looking at top and bottom designators, it is easy to locate the loop where instrument is placed. Common instrument designations used are shown in the table below:

Typical Instrument Functions Nomenclature			
Pressure			
PS	Switch		
PSL	Switch	Low	
PSH	Switch	High	
PSLL	Switch	Low	Low
PSHH	Switch	High	High
PSXL	Switch	Extra	Low
PSXH	Switch	Extra	High
PAL	Alarm	Low	
PAH	Alarm	High	
PC	Controller		
PI	Indicator		
PIC	Indicator	Controller	
PICA	Indicator	Controller	Alarm
Temperature			
TS	Switch		
TSL	Switch	Low	
TSH	Switch	High	
TSLL	Switch	Low	
TSHH	Switch	High	
TSXL	Switch	Extra	
TSXH	Switch	Extra	
TAL	Alarm	Low	
TAH	Alarm	High	

(*Continued*)

Typical Instrument Functions Nomenclature

Pressure

TC	Controller		
TI	Indicator		
TIC	Indicator	Controller	
		Flow	
FAL	Alarm	Low	
FAH	Alarm	High	
FI	Indicator		
FIC	Indicator		
FISL	Indicator	Switch	Low
FSL	Switch	Low	
FSH	Switch	High	

2.5.13 STANDARD SYMBOLS

It is very important to use standard symbols while preparing P&ID for understating the same universally, with respect to equipment, valves and other inline piping components included in P&ID. Though there are differences in nomenclature used by any two companies, the differences can be picked up easily by an experienced engineer (Figures 2.6–2.11).

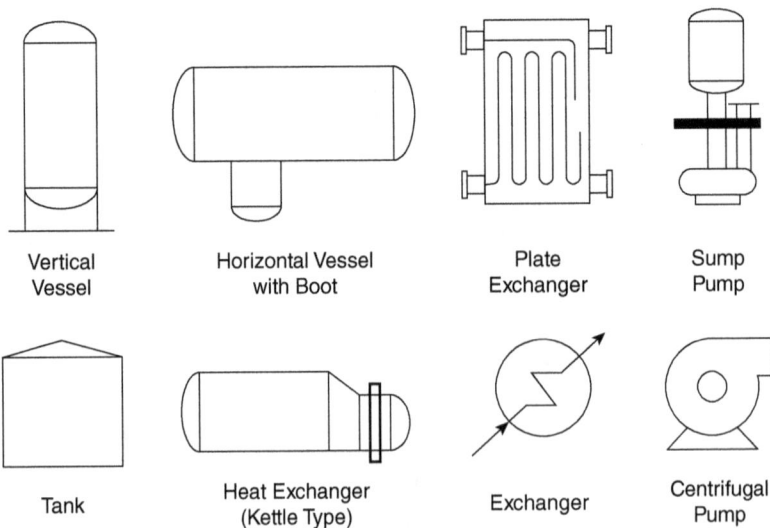

| Vertical Vessel | Horizontal Vessel with Boot | Plate Exchanger | Sump Pump |

| Tank | Heat Exchanger (Kettle Type) | Exchanger | Centrifugal Pump |

FIGURE 2.6 Common static/rotating equipment representation (Walker, 2009).

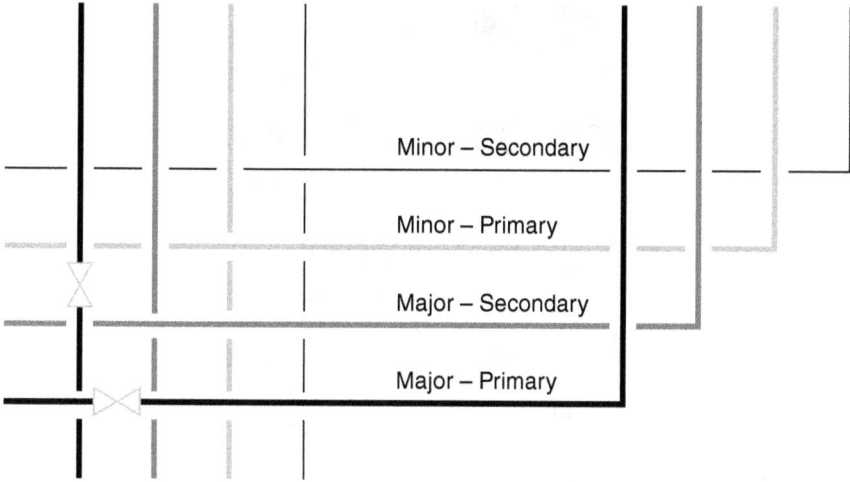

FIGURE 2.7 Line designation using colours and layers (ACAD) (Walker, 2009).

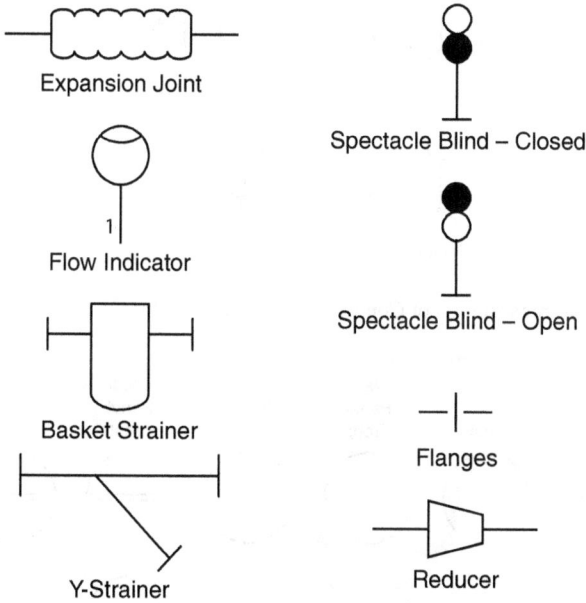

FIGURE 2.8 Representation of piping elements (Walker, 2009).

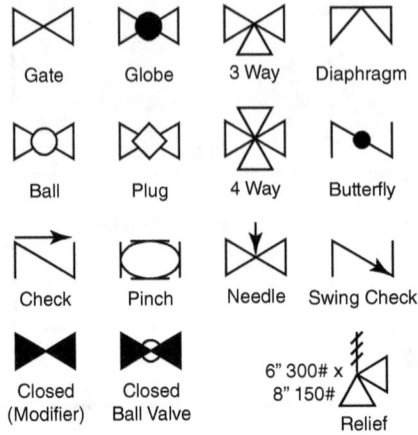

FIGURE 2.9 Representation of valves (Walker, 2009).

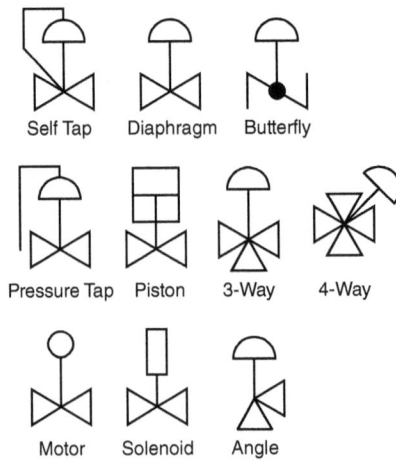

FIGURE 2.10 Representation of Control Valves (Walker, 2009).

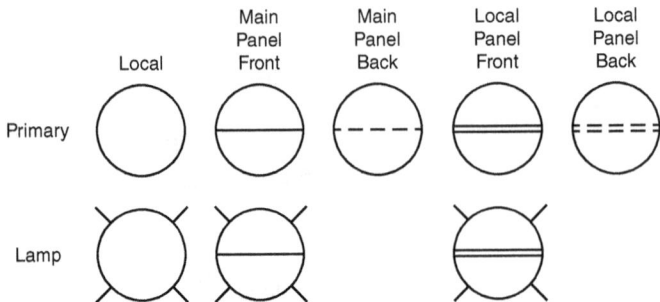

FIGURE 2.11 Instrument types (Walker, 2009).

2.6 PIPING ENGINEERING DELIVERABLES

	Specifications	Data Sheets	Drawings	Lists	Calculations	Data Files
	Piping Line Class Material	Specialty Item Data Sheets	Specialty Item (Vendor Catalogue or certified drawings)	Piping Line List (Originate and Maintain)	Maximum Allowable Flange ratings	Relational Database (3D geometry for electronic modelling)
	Piping Specialty Items and Datasheets			Piping Specialty Items List (Originate and Maintain)	Pipe Wall Calculations	
	Piping Pressure Testing			Piping Commodity Catalogue (Originate and Maintain)		
	Piping System Internal Cleaning (Standard)					
	Piping System Internal Cleaning (High Purity)					
Piping Material Engineering	Positive Material Identification, Traceability and Certification					
	Piping Painting					
	Piping Commodity and Safety Marking					
	Galvanizing					
	Piping Insulation (Hot, Cold and Acoustical)					
	Coating and Wrapping of Underground Piping					
	Welding – General Systems					
	Welding – Hygienic System					
	Piping System NDE (Non-Destructive Examination)					

(Continued)

	Specifications	Data Sheets	Drawings	*Lists	Calculations	Data Files
Piping Design	Plant Layout and Piping Design		Piping Standards	Interference Report	Dead Weight Loads for Structure	Electronic Drawings
	Electronic Design Model (3D-PDS)		Plot Plans	Piping Isometric Index		Final Electronic Databas
	Field Survey/Close Range Photogrammetry		Pipeline Transportation	Tie-In List		
	Piping Tie-Ins		Piping Layout Studies and Vessel Orientations			
	Shop Fabrications and Handlings		Plant Equipment Location Control Plan (Equipment General Arrangement Plant)			
			3D Plant Design Model Index			
	Field Fabrication and Installation		Piping Plan Drawings			
	Heat Tracing (Piping, Equipment and Instruments)		Piping Isometrics			
			Revamp Demolition Drawings			
			Tie-In Location Index			
			Tie-In Isometrics			
			Heat Tracing Manifold Index			
			Heat Tracing Plans			

	Specifications	Data Sheets	Drawings	Lists	Calculations	Data Files
Piping Design			Heat Tracing Details Heat Tracing Isometrics Piping Stress Sketches			
Piping Material Control	Commodity Purchase Specifications (Pipe, Fittings, Flanges, etc.) Field Receiving Geographic Colour Code			Piping Material Summary (Preliminary, Intermediate and Final) Shortage and Overage Reports	Piping Material Cost Estimations	Material History (Item and Source)
Piping Stress Engineering	Piping Flexibility Design Criteria Pre-engineered and Engineered Piping Support Elements Metallic Expansion Joints Elastomer Expansion Joints Pipe Restraint and Shock Arrestors		Pre-engineered Pipe Support Elements	Piping Flexibility Analysis Log Calculation Data Files Piping Flexibility Analysis	Special Calculations	Engineered Spring Support Engineered Expansion Joints

2.7 PIPING KEY PLAN

Piping layout is dimensional drawing, showing plan of pipelines laying view between starting and end points. This incorporates size of pipe (inches); thickness in inch or schedule (such as STD, XS and XXS); fitting detail such as bend, elbow, tees, reducers and valves; elevation of piping points; and all details to lay pipe between the two points in consideration. It is a top view drawing which shows the arrangement of piping along with all piping components including process equipment, ladder, pipe supports, cages, platforms, pipe rack, etc.

This shall not be confused with the Plot Plan, which shows plan view of the land with a prominent reference point so that location of project site, and that of specific equipment can be plotted at site. As mentioned this also provides a top view drawing which used to show the total arrangement of equipment, building pipe racks, etc., required in the plant and also with empty space for anticipated future expansions, which consists of plant and non-plant facilities (Figure 2.12).

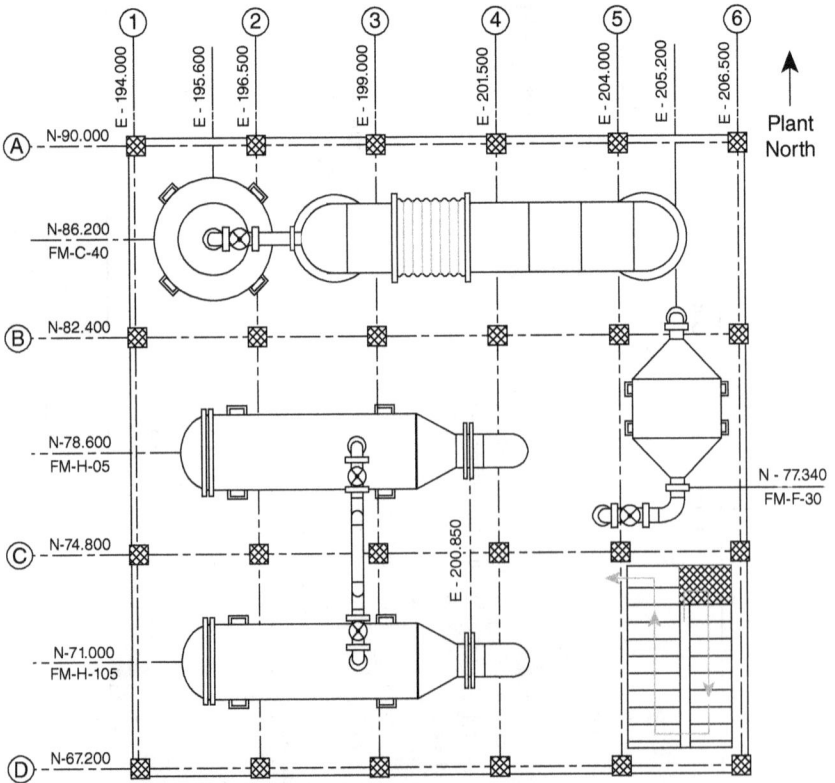

FIGURE 2.12 Piping key plan (Sölken, n.d.).

2.8 PIPING ISOMETRIC DRAWINGS (ISOMETRICS)

Piping isometric drawing is the isometric view of pipeline between two points in consideration. It is drawn to scale and gives precise bill of materials (BOM) needed for procurement of pipes and fittings for the project. Nowadays, it is a 3D drawing created in 3D modelling and produced as a 2D drawing to have clarity in direction of piping and dimensions with a proper BOM for all inline equipment falling within the piping under consideration (Figures 2.13–2.16).

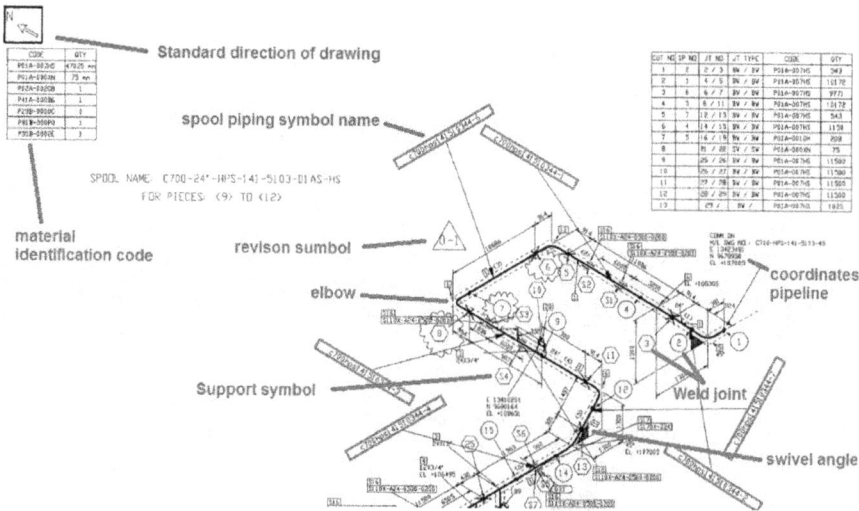

FIGURE 2.13 Piping isometric drawing (Engineering Construction, 2016).

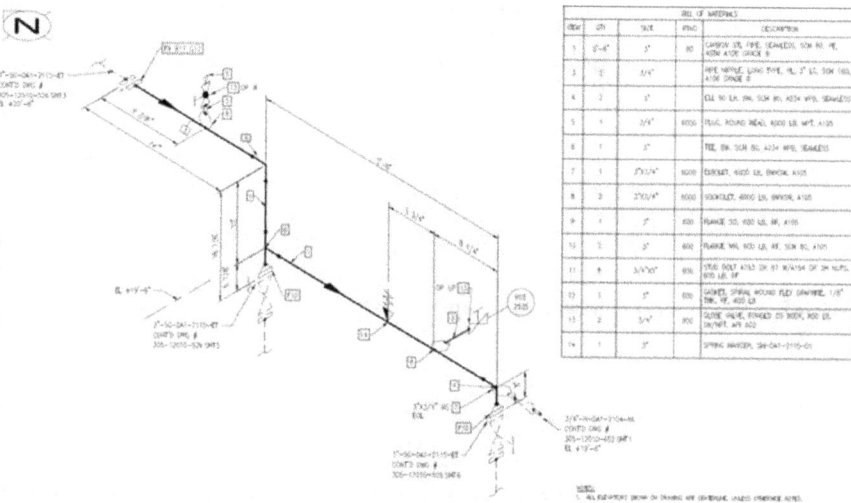

FIGURE 2.14 Piping isometric drawing with bill of materials (Unitel Technologies, 2021).

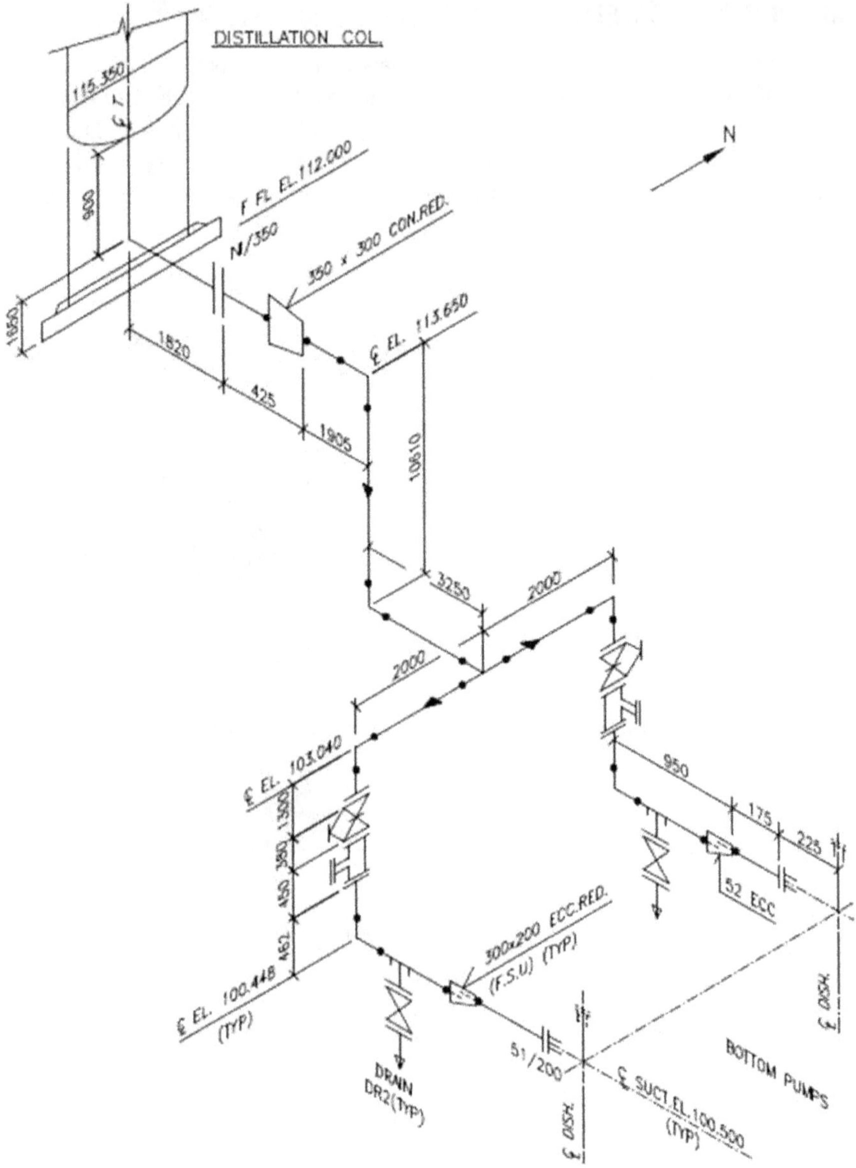

FIGURE 2.15   Isometric drawing with dimensions.

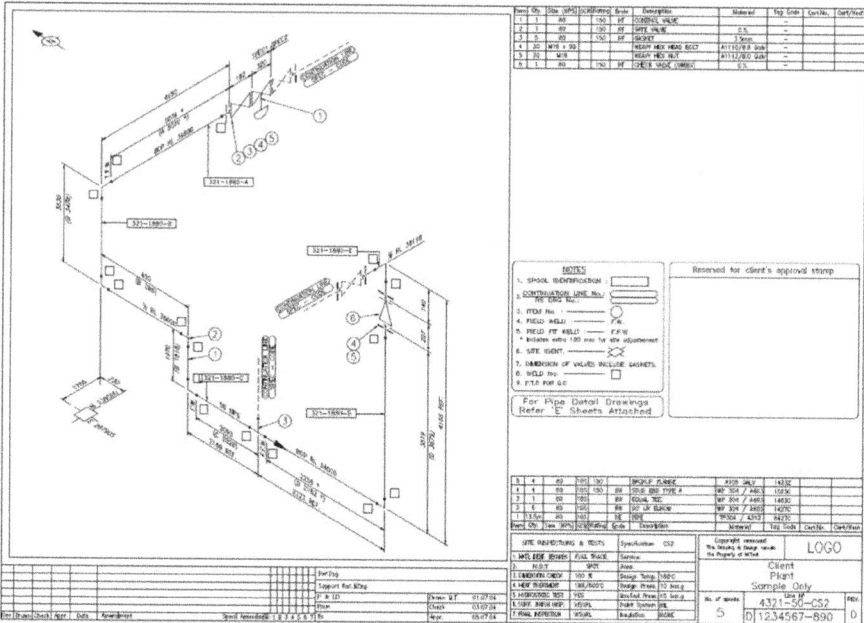

FIGURE 2.16 Typical isometric drawing used in piping.

2.9 PIPING MATERIAL SPECIFICATION

The very purpose of PMS is to reduce the large number of piping components from many different schedules. Because of the same, a piping class is often suitable for a large number of services and pressure/temperature ratings as well. There is no typical format for PMS. It varies from EPC contractor to contractor. However, an overview of the usual contents of a PMS for a project is as below. PMS shall essentially contain the following information about each component required in each piping class:

- Types of piping elements envisaged such as pipe, fittings, flanges, valves including different types, gaskets, fasteners, etc.
- Material specification for all those components and manufacturing specifications based on size and schedule ranges.
- Pressure and temperature ratings the piping can be used for.
- Branch connection table for each class.
- Corrosion allowances for each piping element.

As mentioned earlier under P&ID, the line numbers contain the specific PMS. The methodology to develop PMS and the MTO is dealt with in detail in ensuing sections.

3 Piping Isometric Drawings (Isometrics)

3.1 SIGNIFICANCE OF ISOMETRIC DRAWINGS

As mentioned earlier, plant piping drawing development starts with finalization of P&ID, layout of individual equipment and pipe racks based on process, safety, operational and maintenance considerations. Generally, in piping documents' development process for a process plant, conceptual 2D drawing is first created. In this all pipelines shall be at zero z-coordinate. This is then converted into a 3D scale model, to check interference of pipe lines (now replaced with 3D modelling, wherein a walk-through is possible). The 2D General Arrangement (GA) drawings are used for fabrication on site. A GA drawing is generally on A0- or A1-size paper and contains a lot of lines, and pipe routing from that drawing may not be easy to understand. Moreover, size of drawings (A0 or A1) makes it difficult to handle. For the pipe fabricator to understand GA drawing clearly, he may have to refer to 3D scale model as well. Unfortunately, this 3D modelling cannot be put on paper and 3D scale models lack dimensions. To circumvent these two issues and to help the pipe fabricator at site, isometric drawings are prepared.

Isometric drawing is generally prepared in A3 size and fabricator can read and understand pipe routes very easily from this drawing. Bill of materials (BOM) provided in the isometric drawing helps pipe fabricator to draw materials from store and size the pipe to required lengths with other fittings to form pipe spools as required in the isometric drawing.

3.2 ADVANTAGES OF ISOMETRIC DRAWINGS

- Simple in nature, with only one pipe line drawn in one A3-size paper.
- Orthographic view becomes complicated when pipe runs in all three coordinates (e.g. north to south, then down and then to west, etc.), whereas isometric shall be much less complicated.
- More number of drawings are required in orthographic views than in isometric drawing to represent the same piping system.
- Sectional views are required in orthographic drawings for clarity of pipe routing.
- Unlike orthographic drawings, piping isometric drawings allow pipeline to be drawn in a manner by which length, width and depth are shown in a single view.

DOI: 10.1201/9781003328124-3

- Isometrics are usually drawn from information found on a plan and sectional elevation views of piping key plan. The symbols that represent fittings, valves, flanges, etc. are redefined to depict it as lines/curves as symbology for isometric drawings.

The isometric drawing is commonly referred in piping construction as "ISO" and is oriented on the grid relative to the **north arrow** found on plan drawings. In ISO vertical lines are represented by vertical lines in the drawing, whereas the two other directions (x- and y-axes on same horizontal plane) are at 30° from the horizontal on either side of the vertical line. All directions of the pipe spool shall match the three isometric axis so defined. Since ISOs are not drawn to scale, accurate dimensioning is required to specify exact lengths of piping segments of spool indicated in the ISO. Pipe length between fittings and flanges is determined through calculations using the three coordinates. While vertical lengths of pipe are calculated using centre line elevations of pipe lines, horizontal lengths are calculated using north–south and east–west coordinates of the piping system. ISOs are used along with piping key plans and sectional elevations, but typically it is used to supplement plan drawings. Further, ISOs are used as fabrication and shop drawings for fabrication of piping spools in transportable sizes followed by erection at site.

3.3 COMPARISON OF ORTHOGRAPHIC AND ISOMETRIC DRAWINGS

See Figures 3.1 and 3.2.

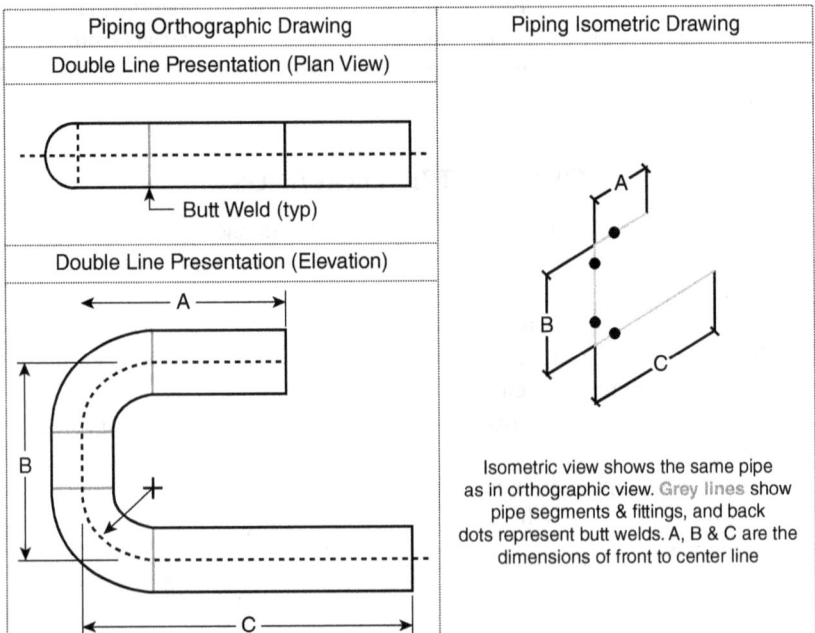

FIGURE 3.1 Comparison of orthographic and isometric drawings (Sölken, n.d.).

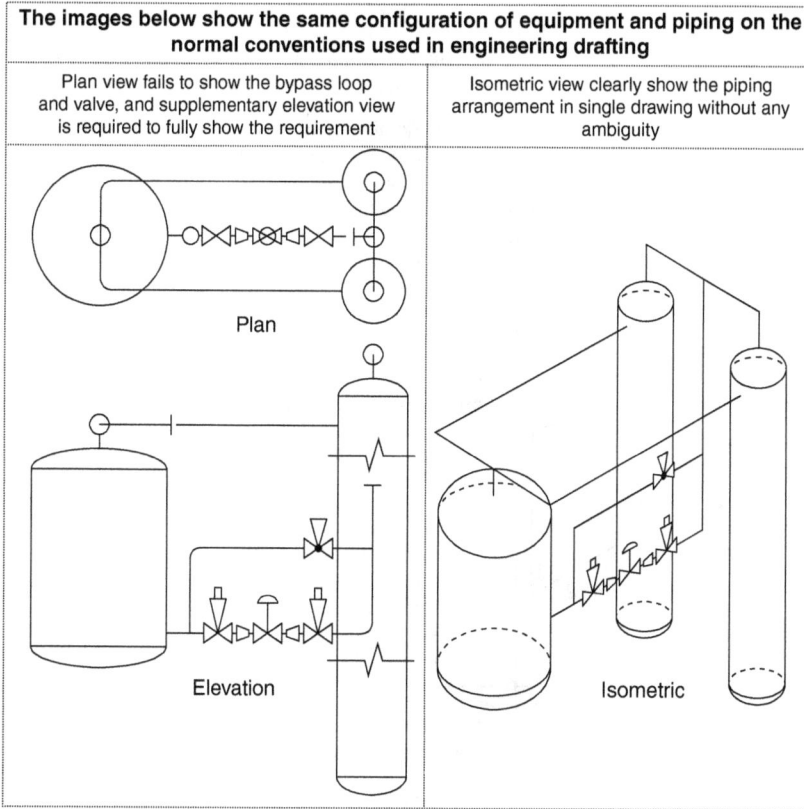

The images below show the same configuration of equipment and piping on the normal conventions used in engineering drafting

| Plan view fails to show the bypass loop and valve, and supplementary elevation view is required to fully show the requirement | Isometric view clearly show the piping arrangement in single drawing without any ambiguity |

Plan

Elevation

Isometric

FIGURE 3.2   Isometric, plan and elevation presentations of a piping system (Sölken, n.d.).

3.4   ISOMETRIC VIEWS IN MORE THAN ONE PLANE

Figures 3.3 and 3.4 show some examples of isometric drawings. Auxiliary lines in the shape of a cube, ensure better visualization of pipe routing

3.5   PIPING ARRANGEMENT SYMBOLS

Symbols used in single-line isometric drawings to represent piping components and elements such as pipes, fittings, flanges, vales and other common inline equipment are shown below. Though not exhaustive, symbols for most commonly used fittings are captured in ensuing sections.

3.5.1   PIPING ARRANGEMENT SYMBOL (PIPES)

As mentioned earlier, isometric piping drawings are not scale drawings and hence they need to be dimensioned. Pipes in ISOs are shown as single lines. In other words, the straight lines between two fittings or valves or any other piping element represent pipe section.

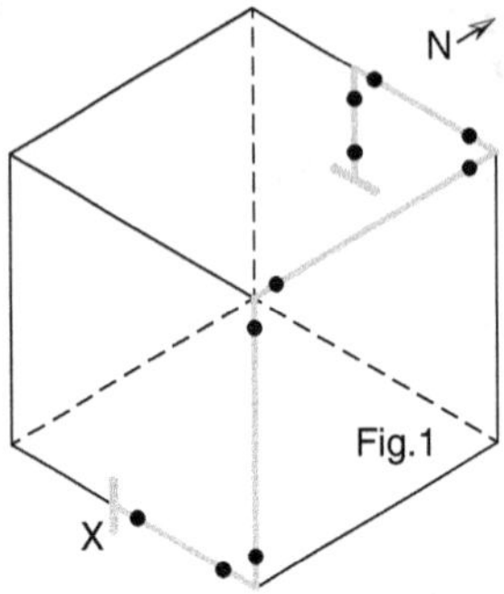

Figure 1 shows a piping, which runs through three planes. Piping begins and ends with a flange.

- Routing starting point X
- Pipe runs to east
- From there upwards
- Pipe runs to north
- Pipe runs to west
- Pipe goes down

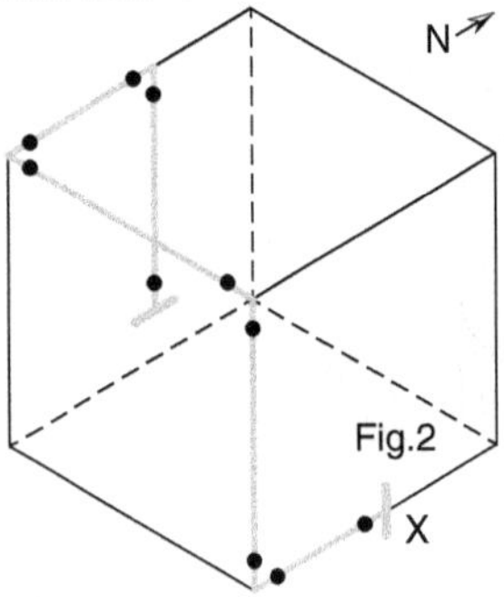

Figure 2 is almost identical to the drawing above except for the length of pipe which goes down, which is longer, shown in a different perspective. As this pipe in isometric view, runs behind the other pipe, this must be indicated by a break in the line.

- Routing starting point X
- Pipe runs to south
- Pipe runs up
- Pipe runs to west
- Pipe runs to north
- Pipe goes down

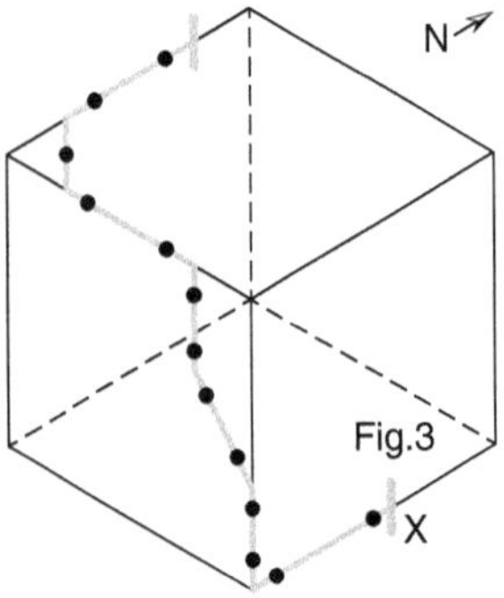

Figure 3 shows a pipe that runs through three planes and in two planes it make a bow

- Routing starting point X
- Pipe runs to south
- Pipe runs up
- Pipe runs up and to west
- Pipe runs up
- Pipe runs to west
- Pipe runs to north-west
- Pipe runs to north

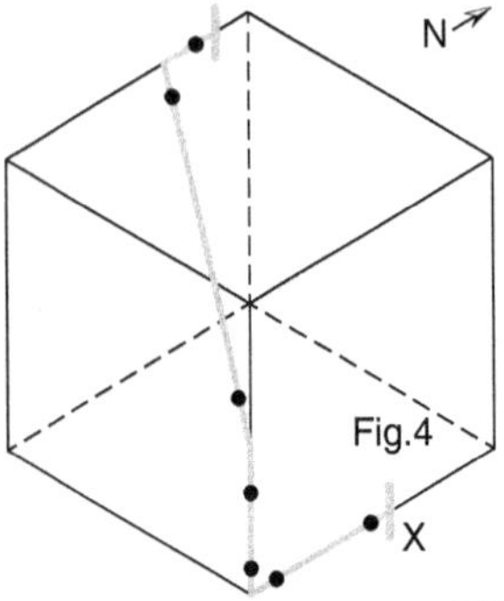

Figure 4 shows a pipe that runs through three planes, from one plane to an opposite plane.

- Routing starting point X
- Pipe runs to south
- Pipe runs up
- Pipe runs up and to north-west
- Pipe runs to north

FIGURE 3.3   Isometric views in more than one plane (Sölken, n.d.).

Figure 5 shows a pipe, where the hatch indicates that the middle leg runs to east.

- Routing starting point X
- Pipe runs up
- Pipe runs up and to east
- Pipe runs up

Figure 6 shows a pipe, where the hatch indicates that the middle leg runs to the north.

- Routing starting point X
- Pipe runs up
- Pipe runs up and to north
- Pipe runs up

The two drawings above show, that changing from only the hatch, a pipeline receives a different direction. Hatches are particularly important in isometric views.

Figure 7 shows a pipe, where the hatches indicates that the middle leg runs up and to north-west.

- Routing starting point X
- Pipe runs up
- Pipe runs up and to north-west
- Pipe runs to north

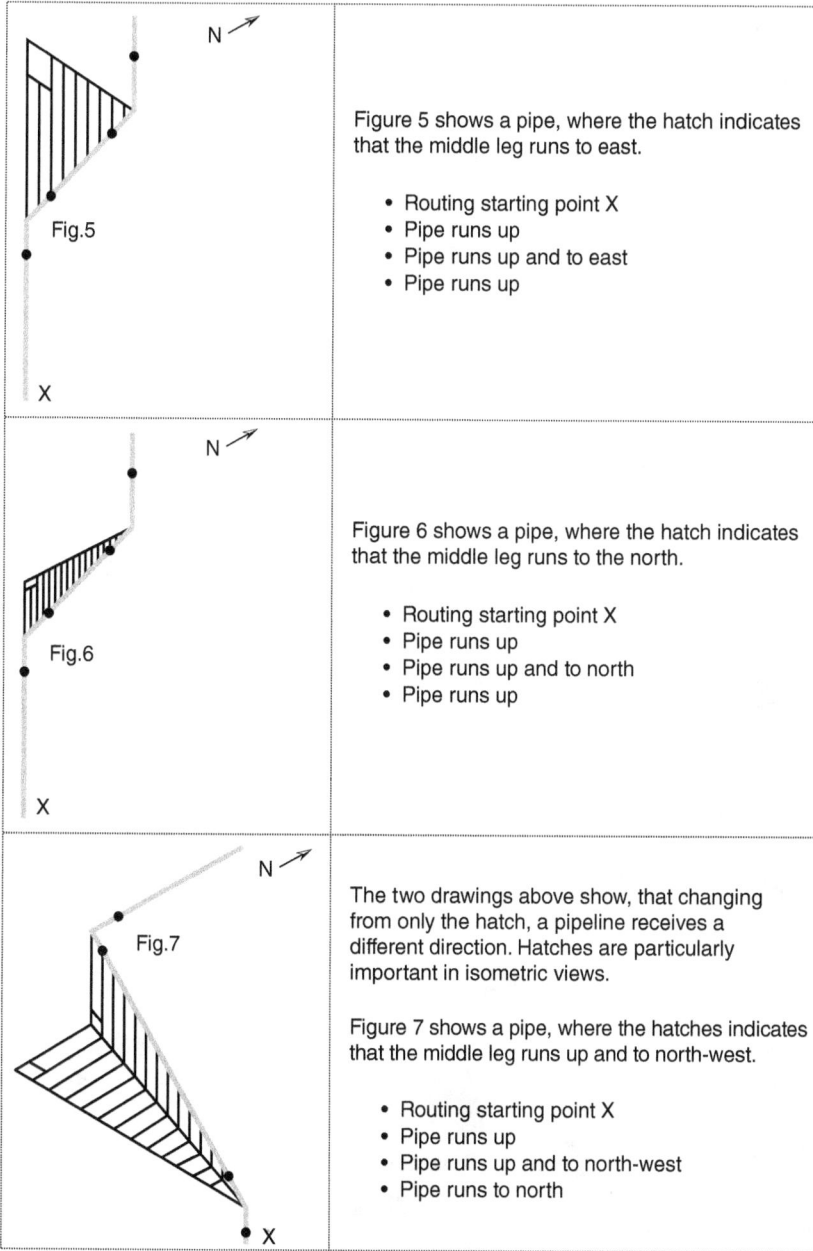

FIGURE 3.4   Hatches on an isometric drawing (Sölken, n.d.).

3.5.2 PIPING ARRANGEMENT SYMBOLS (FITTINGS)

See Figure 3.5.

3.5.3 PIPING ARRANGEMENT SYMBOLS (FLANGES)

See Figure 3.6.

3.5.4 PIPING ARRANGEMENT SYMBOLS (VALVES)

See Figure 3.7.

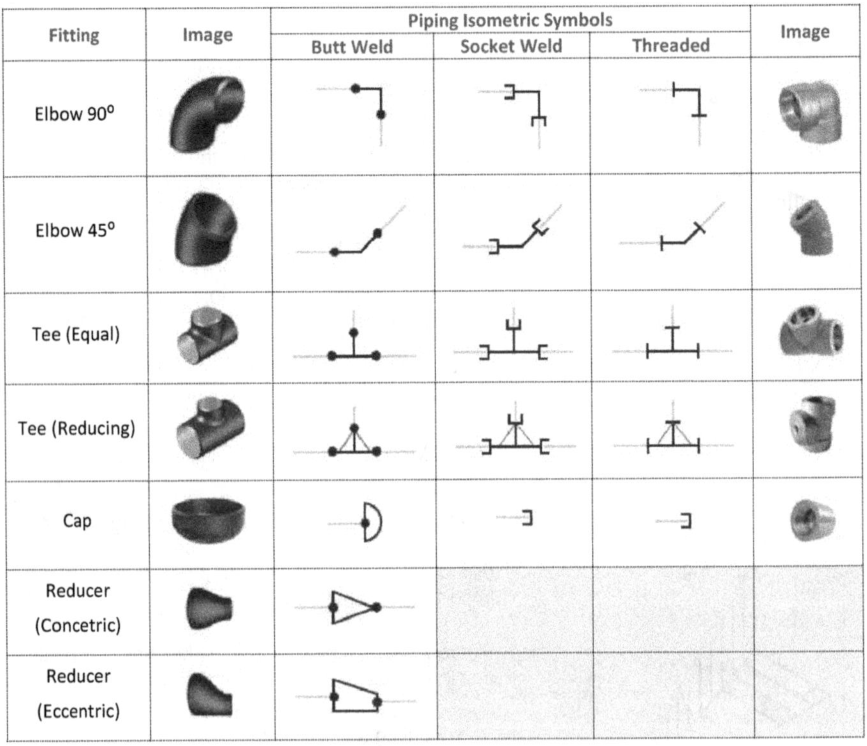

Fitting	Image	Piping Isometric Symbols			Image
		Butt Weld	Socket Weld	Threaded	
Elbow 90°					
Elbow 45°					
Tee (Equal)					
Tee (Reducing)					
Cap					
Reducer (Concetric)					
Reducer (Eccentric)					

FIGURE 3.5 Piping arrangement symbols (fittings) (Sölken, n.d.).

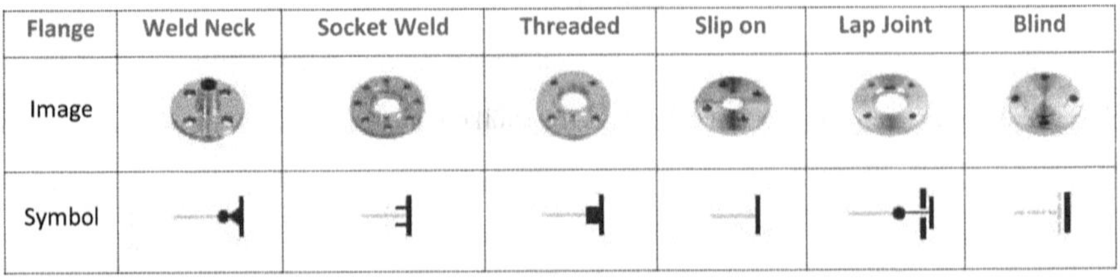

Flange	Weld Neck	Socket Weld	Threaded	Slip on	Lap Joint	Blind
Image						
Symbol						

FIGURE 3.6 Piping arrangement symbols (flanges) (Sölken, n.d.).

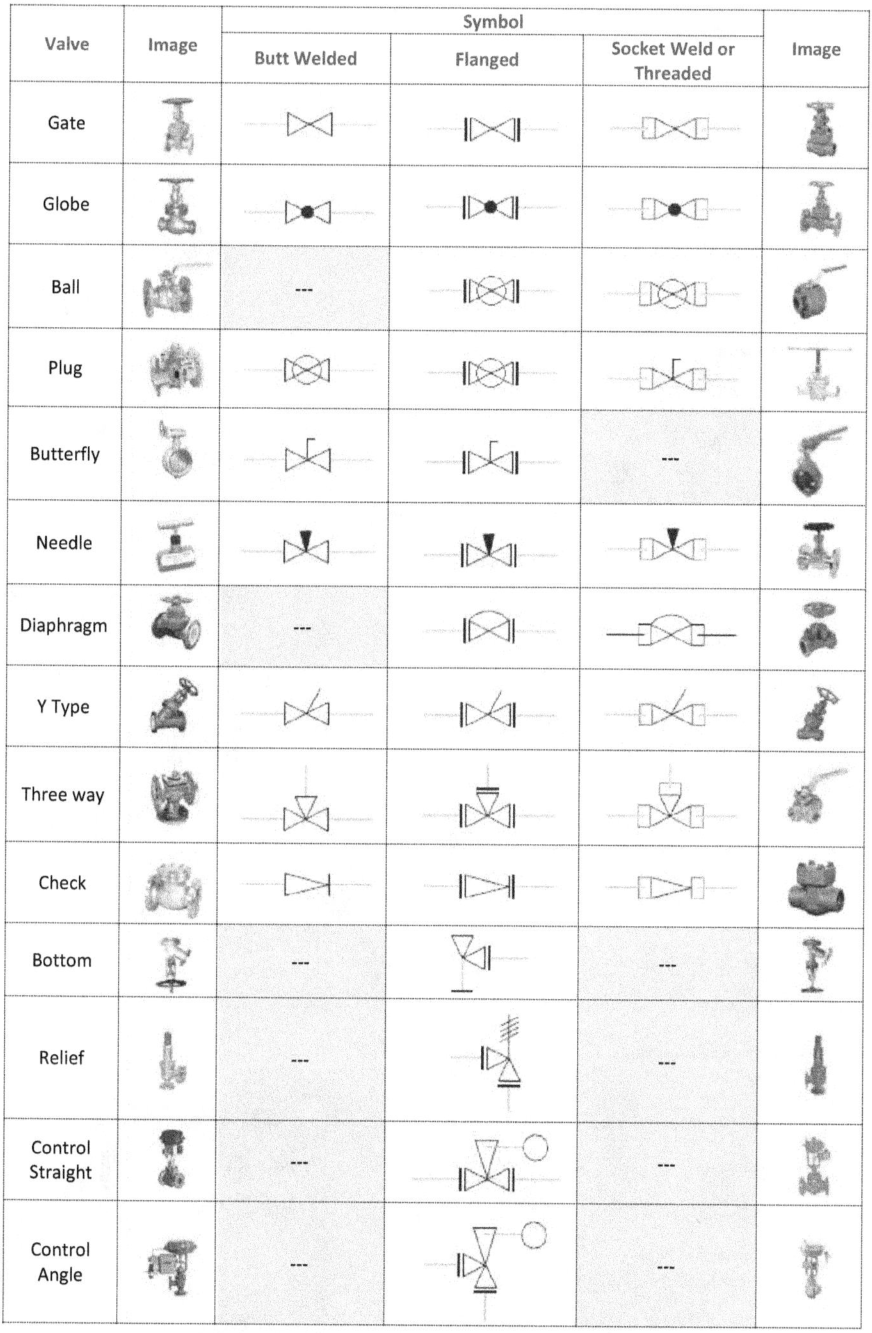

Valve	Image	Symbol			Image
		Butt Welded	Flanged	Socket Weld or Threaded	
Gate		▷◁	‖▷◁‖	⊏▷◁⊐	
Globe		▷●◁	‖▷●◁‖	⊏▷●◁⊐	
Ball		---	‖▷⊗◁‖	⊏▷⊗◁⊐	
Plug		▷⊗◁	‖▷⊗◁‖	⊏▷⊗◁⊐	
Butterfly		▷◁	‖▷◁‖	---	
Needle		▷▼◁	‖▷▼◁‖	⊏▷▼◁⊐	
Diaphragm		---	‖▷◁‖	⊏▷◁⊐	
Y Type		▷◁	‖▷◁‖	⊏▷◁⊐	
Three way		▷◁	‖▷◁‖	▷◁	
Check		▷▷	‖▷◁‖	⊏▷⊐	
Bottom		---	▷◁‖	---	
Relief		---	‖▷◁	---	
Control Straight		---	▷◁ ○	---	
Control Angle		---	▷◁ ○	---	

FIGURE 3.7 Piping arrangement symbols (valves) (Sölken, n.d.).

3.5.5 PIPING ARRANGEMENT SYMBOLS (MISCELLANEOUS FITTINGS)

See Figure 3.8.

Miscellaneous	Image	Symbol
Branch Outlet Weldolet ®		
Branch Outlet Nippolet ®		
Flanged Branch Outlet Flangolet		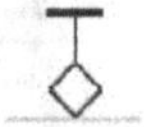
Spade		 SP
Spectacle Blind		 SB
Hammer Blind		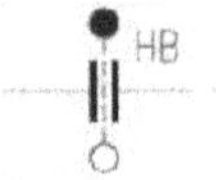 HB
Spacer		
Restriction Orifice		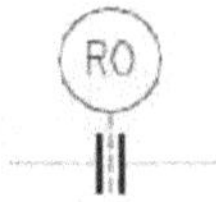 RO
Field Weld		 FW
Butt Weld		
Pipe to Pipe Connection		
Pipe bend with Special Radius		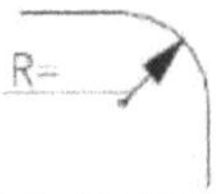 R=
Sight Glass		 SGI
Direction of Hand Wheel Wrench		
Y Type Strainer		

FIGURE 3.8 Piping arrangement symbols (miscellaneous fittings) (Sölken, n.d.).

(Continued)

Conical Strainer		
Conical Strainer (Built in)		
Orifice Assembly (typical) showing position of taps		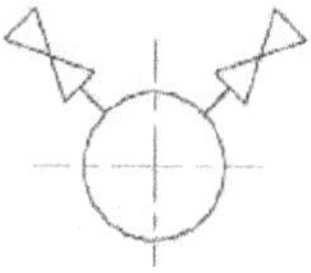
Meter run Orifice Assembly (typical) Flanged / Butt Welded)		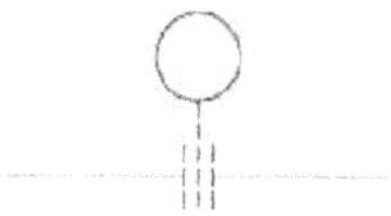

FIGURE 3.8 (*Continued*) Piping arrangement symbols (miscellaneous fittings) (Sölken, n.d.).

3.6 PREPARATION OF PIPING ISOMETRIC DRAWINGS OR ISOMETRICS

Piping layout for a process plant is often developed in both plan and elevation views. Sectional details are also included whenever further clarity is required for easy understanding and construction of the piping. Such drawings are called the GA drawings of piping. As explained in Section 3.3, in piping within buildings or in units with complex configurations, orthographic views seldom illustrate details and give rise to doubts during erection of piping. In all such cases, isometric drawings are drawn which will be very easy to understand and also to carry out fabrication.

3.7 SIZE OF PIPING DRAWINGS

The paper sizes in use in technical documentation as per ISO 216 standards are as shown in Figure 3.9. (All dimensions in cm.)

Of the above sizes, A1 size is commonly used for drawing piping key plans (rarely A0 size is also used, but the size is unfriendly to handle). For isometric drawings, A3 size is predominantly used for initial issue of drawing. For subsequent markups, red lining, etc., in connection with preparation of as-built isometrics and test pack attachments, isometrics are often produced in A4, considering the ease in handling.

3.8 TYPICAL DRAWINGS – STYLE/PRESENTATION AND CONTENTS

The general style and contents of each of the piping drawing are shown in sample documents below.

3.8.1 PIPING KEY PLAN – AREA MATCH LINES

See Figure 3.10.

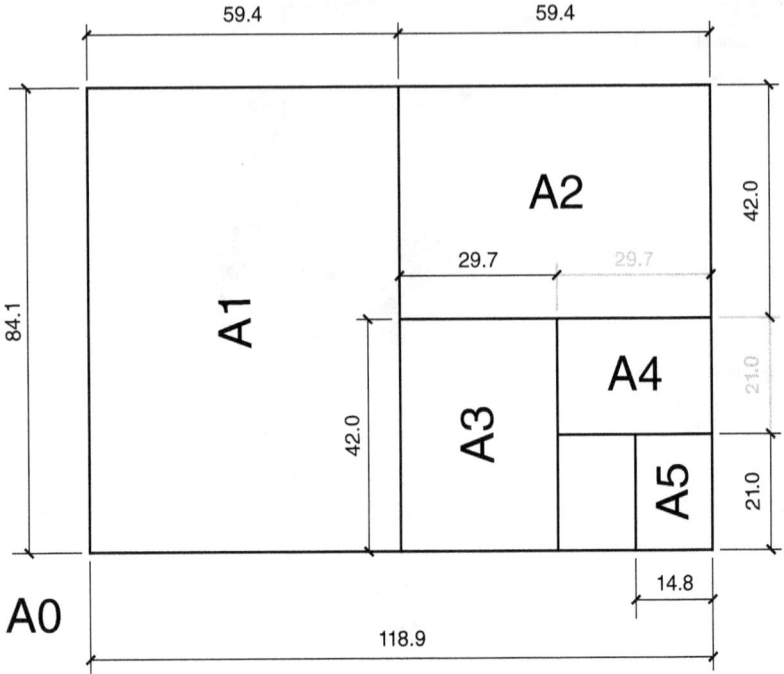

FIGURE 3.9 Sizes of piping drawings (CAD Standard, n.d.).

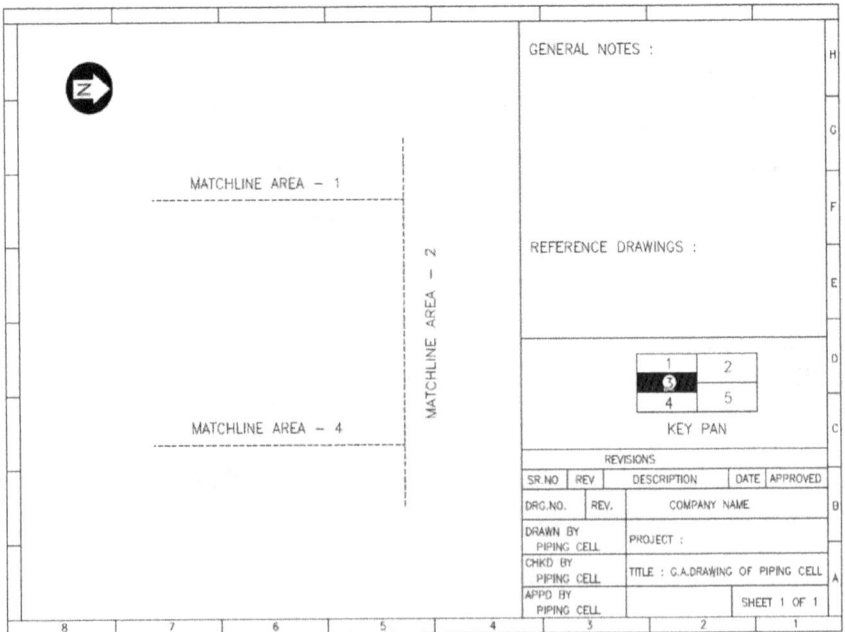

FIGURE 3.10 Piping key plan with match lines.

3.8.2 PIPING KEY PLAN

See Figure 3.11.

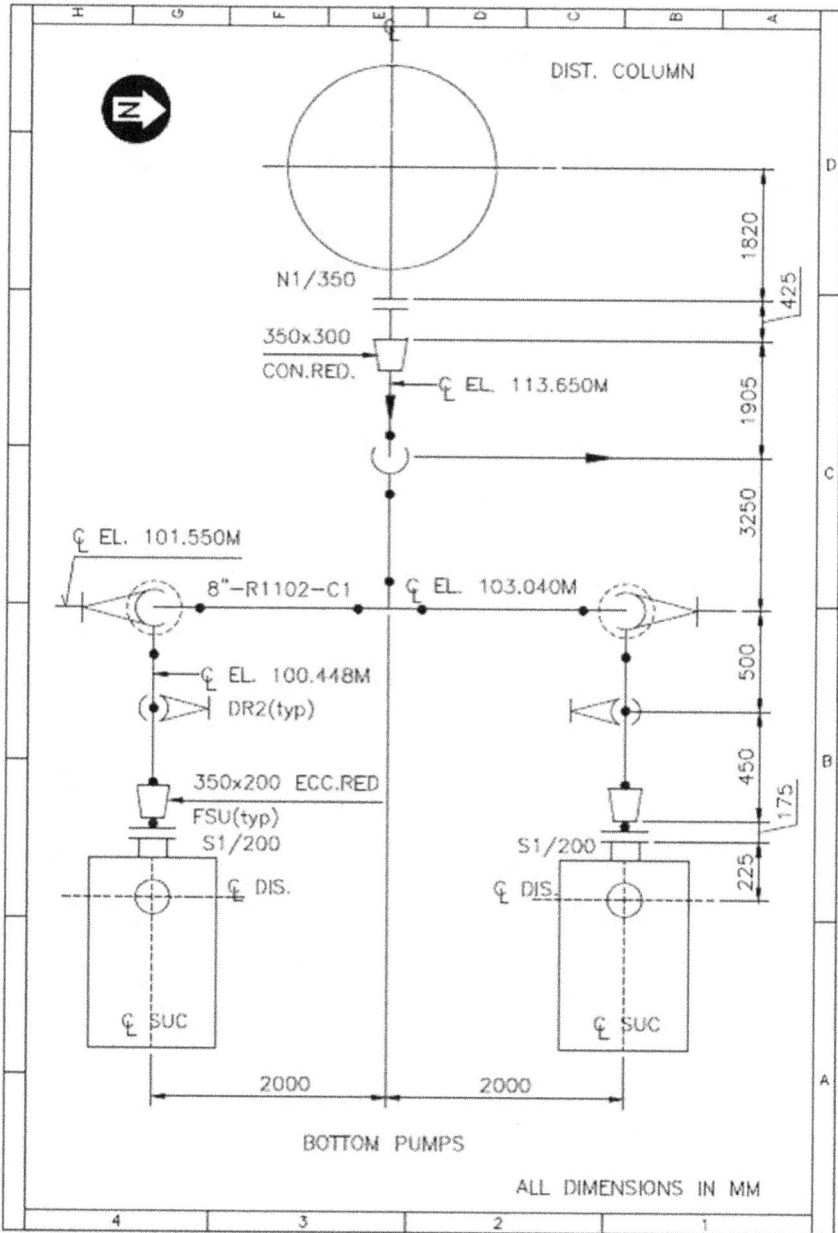

FIGURE 3.11 Piping key plan.

3.8.3 PIPING ISOMETRIC DRAWING

See Figure 3.12.

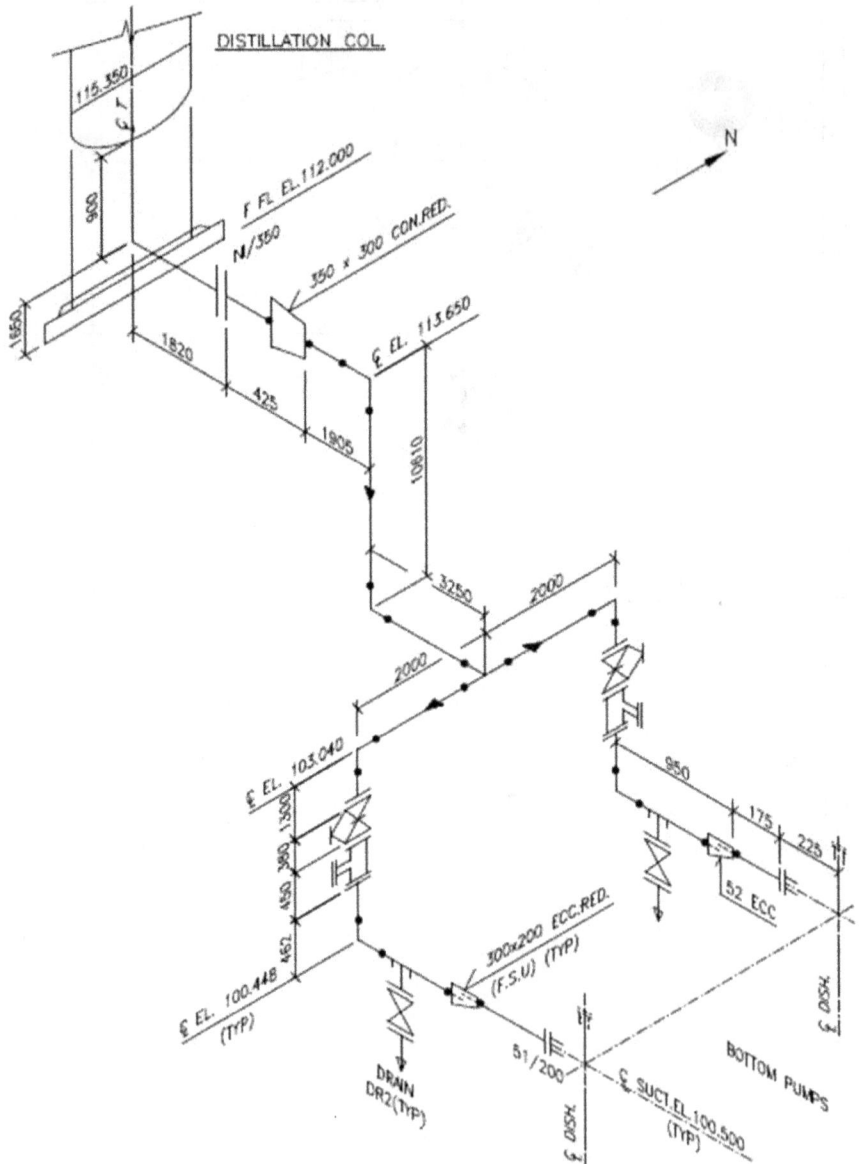

FIGURE 3.12 Piping isometric drawing with prominence for dimensions.

3.8.4 PIPING ISOMETRIC DRAWING WITH BOM

See Figure 3.13.

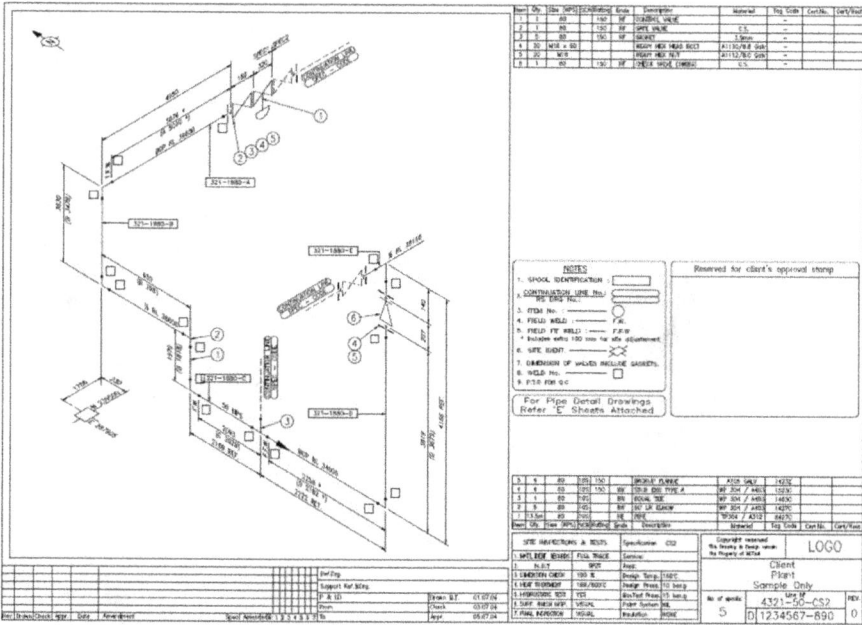

FIGURE 3.13 Piping isometric drawing with bill of materials.

4 Piping Material Specification (PMS) and Material Take-Off (MTO)

Process plant piping involves thousands of piping components with different specifications, configuration thicknesses and quantities, which makes piping material management very complicated. The easy escape route is to order extra quantity in each commodity. The wide variety of material specifications, pressure temperature ratings, configurations and process requirements call for thousands of entirely different piping components, which are not often interchangeable. So ordering even a very small percentage extra quantity under each item tag may attract a huge financial burden on the Engineering, Procurement and Construction (EPC) contractor, for whom the extra materials ordered may not be of use if they do not have a similar project in hand utilizing the same components.

On the other hand, if the piping installation process is held up due to shortage of a single piping component, that again may cause delay and money.

Therefore, it is absolutely essential to strike a balance between these two aspects and arrive at a fair way forward with reduced risk of shortage and too much excess quantity of procured piping components. This makes the whole piping material management risky and complicated and the task is achieved through the systematic development of documents mentioned in ensuing sections.

4.1 PIPING MATERIAL SPECIFICATION

Piping material specification (PMS) is a document prepared during the design phase of any piping project, which provides appropriate selection, specification and material grade of pipe and piping components for a given service. This document shall be valid throughout the life of the process plant and shall be one of the key reference documents for all subsequent maintenance, repair or replacement of any part of the piping.

In any engineering project, piping engineer is responsible for development of PMS. PMS is developed based on applicable piping codes (including codes and standards referred therein), engineering specification and other contractual requirements spelled out in EPC contract. A brief about sequence to arrive at PMS is given below.

The awarded contract often shall have the following documents as part of the signed contract:

- Project design basis (PDB)
- Block flow diagram (BFD)
- Process flow diagram (PFD)

DOI: 10.1201/9781003328124-4

- Material selection diagram (MSD)
- Process and instrumentation diagram (P&ID)
- Piping material specification used by the owner (PMS)

In spite of the above, in EPC contracts, the contractor is expected to carry out the entire detailed engineering once again with firm input data, meeting product and other statutory norms. Renowned consultants and established EPC contractors often have their own PMS. When a PMS (by the client) is included in the contract, EPC contractor is bound to expand the PMS as required by engineering according to methodology and nomenclature followed by the client in their PMS.

In order to develop PMS for a project, the following documents are absolutely essential:

- PMS included as part of contract.
- Other piping requirements included in the contract by way of attached specifications, technical requirements, etc.
- International standards and specifications referred in the contract as applicable and those standards referred in applicable piping code (namely ASME B 31.3).

PMS shall essentially contain the following information about each component required in each piping class. Separate piping classes are developed for specific fluid service consisting essentially of material specification, pressure and temperature range and wall thickness for each size. In brief PMS shall essentially contain the following information:

- Types of piping elements envisaged such as pipe, fittings, flanges, valves including different types, gaskets, fasteners (bolts and nuts), etc. for each piping class
- Material specification for all piping components and manufacturing specifications based on size and schedule ranges
- Pressure and temperature ratings the piping can be used for
- Branch connection table for each class
- Corrosion allowances for each of the piping elements

The very purpose of piping material specification is to reduce the large number of piping components from many different schedules. Because of the same, a PMS is often suitable for a wide range of services, especially with regard to pressure/ temperature ratings.

There is no typical format for PMS. It varies from EPC contractor to contractor. However, an overview of the usual contents of a PMS for a project is shown below.

4.1.1 CONTENTS OF PMS

- Project Introduction
  - Project Overview
  - Scope of the Document

- Applicability and Coverage
  - Reference FEED Documents (*Lists out the FEED documents based on which the PMS was developed*)
- Specific Exclusions (*Exclusions such as pipelines and piping items such as control valves, motor-operated valves and safety valves, which are procured based on a detailed separate data sheets including the valve portion and the control system*)
- Abbreviations used (*Lists out all types of abbreviations used in the document to designate products and services*)
- Reference Codes, Standards and Specifications
  - Project Specifications
  - Client Standards
  - International Standards
  - Order of Preference
  - Units of Measurements
- Environmental Conditions
- Health, Safety and Environment
- Technical Requirements
- Sour Service Requirements (if any)
  - Requirements for Carbon Steels
  - Requirements for Stainless Steels
  - Requirements for Other Exotic Materials
- Welding
- Positive Material Identification
- General Requirements
  - Pipes
  - Fittings
  - Flanges
  - Valves
  - Gaskets and Valve Stem Packing
  - Fasteners
- NDE requirements
- Piping Class Service Index
- Piping Class (*detailed description of each and every piping element required in the class such as pipes, fittings, flanges, valves, gaskets, fasteners (bolting) including branch table for each class*)
- Notes Table

4.2 PIPING MATERIAL MANAGEMENT

Apart from the development of PMS, major steps involved in piping material management are the preparation of:

- Material take-off (MTO)
- Purchase specifications/datasheet for piping specials
- Material requisition (MR)

With the above three documents, enquiries can be floated for piping components and quotations are obtained and the steps below follow:

- Technical bid evaluation (TBE)
- Clarifications if any required
- Purchase order (PO)
- Vendor document review and approval (post-award of PO)

4.2.1 MATERIAL TAKE-OFF

4.2.1.1 Bill of Materials

As mentioned in Section 2.6 and shown in Section 3.8.3, each of the isometric drawings generated shall have a bill of material (BOM) at the top right hand corner. This essentially consists of a list of all materials shown on an individual isometric drawing to complete the stretch of piping in that isometric. It is not essential that each of the isometric shall contain only one pipe spool. Depending upon other constraints like transportation feasibility, easiness of erection, lifting limitations, etc., an isometric may contain one or more spools and each spool shall be uniquely numbered.

4.2.1.2 Material Take-Off

MTO is a consolidated list of materials with the quantities required for any unit of the project or for the whole piping in a project, prepared by adding up the quantities of items prescribed in the individual isometrics for the piping work. In order to facilitate purchase of right material, often the EPC contractor shall be developing a commodity specification for each of the items (by consolidating all technical requirements specified for the project). The MTO with commodity specification is often sent to the vendors for obtaining competitive quotations for probable procurement, subsequent to technical and commercial reviews.

As mentioned above, MTO of piping items is a detailed listing of piping components required for a given project and it includes the following:

- Commodity code (developed by EPC contractor),
- Size,
- Quantity and
- Commodity specification or purchase description for all the items (which includes additional requirements above those specified in applicable codes/standards).

The MTO is often prepared at many stages and is used for the following purposes:

- Making proposal,
- Material estimation and
- Preparation of purchase requisitions.

For a large project, EPC contractors are compelled to generate a few MTOs after award of contract at various stages of engineering as below.

4.2.1.3 Initial MTO

Initial MTO also called first MTO is generated manually at the very beginning of the project itself, at which point of time isometrics and 3D model would not be ready. Hence, MTO is generated based on P&ID and other applicable project specifications forming part of the contract. This is required to accelerate procurement actions to a reasonable extent, by which time drawings and 3D models would have reached a satisfactory firm level to take more accurate MTO.

Upon finalization of P&IDs, locations of equipment and pipe rack route, preparation of isometrics and 3D modelling can progress. Thereafter, detailed MTO is generated electronically using 3D modelling software (such as PDS/PDMS or other similar software). Further, the PMS mentioned earlier in electronic format is also linked to the 3D modelling software to improve the accuracy in quantity and correctness of specifications in MTO.

The key documents/inputs required to prepare initial (first) MTO:

- Piping and instrumentation diagram
- Line list
- Standard/project PMS
- Project specification and standard drawings
- Equipment drawings/data sheet
- Instrument hook-up sketches

4.2.1.4 Manual MTO Preparation Overview

See Figure 4.1.

4.2.1.5 Steps for Generating Manual MTO

- Make a folder of all P&ID as a master copy for the first bulk MTO.
- All pickable items are entered in MTO format in Excel file.
- All entries must be taken against specific line number. In case a line number is not marked on P&ID, it has to be assumed by adding suffix A, B, etc. with main header line.

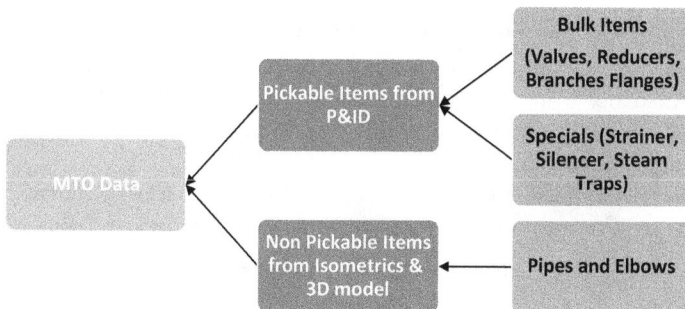

FIGURE 4.1 MTO preparation sequence (Matveev, 2022).

- Queries shall be raised for components which are not available in PMS. All such queries shall be documented and MTO shall be updated as and when queries are resolved.
- All assumptions made are documented and such items shall be marked as "Hold". All assumed data is maintained in a separate file and tracked. MTO shall be updated whenever confirmed data is received.
- Basis of MTO is properly issued version of P&ID or documented communication which serves the same purpose.

4.2.1.6   MTO for Special Items

- Prepare specialty item index in proper format.
- All special items (strainer, steam trap, sample cooler, flame arrestor, silencer, sight glass, special valves, etc.) shall be tagged in P&ID.
- Items like safety shower, hoses, coupling, etc. shall be given common tag number distinctly.
- Prepare datasheet having all piping/material data of special items.
- Send datasheet to process and related department for review and for incorporating missing data as well as for comments. All the missing data/information required to be filled by other disciplines shall be highlighted by material team.
- Prepare "Approved for Construction" (AFC) datasheet, when commented and updated datasheet is received.

4.2.1.7   MTO for Non-Pickable Items

- Pipe length and elbows quantity is generated using piping isometrics and 3D modelling.

4.2.1.8   Generation of Total MTO

- Combine pickable bulk items and non-pickable items (except piping special items).
- Combined data is processed and validated against PMS.
- Check and rectify errors.
- Final BOM is received from successful consolidation of MTO file.
- Piping specials are processed separately.

4.2.1.9   Stages of MTO during Engineering

For any process plant piping project, a few MTOs are often required to complete the project, so also some local purchase at the fag end of the project. This may be required if site modifications or changes made due to unforeseen obstructions are found during construction at site.

In almost all projects, there are normally three MTO sessions during the engineering phase of the project, termed as preliminary, secondary and final MTOs.

4.2.1.9.1   Preliminary MTO

A preliminary MTO is taken at early stages of engineering of the project, by which time only a limited amount of information is known (or firm) and very little detail has been developed. However, a preliminary MTO is prepared for three reasons:

- To assist in early quantification of piping commodities (within ±10%) to compare with estimates
- To obtain early offers (with better price) and delivery schedules
- To allocate procurement cash flow based on quantities and delivery schedules

To facilitate preparation of preliminary MTO, essentially plot plan, equipment layout plan and pipe rack plan need to be approved by all concerned. Moreover, basic principles adopted in the development of P&ID should also be agreed between all, say to around 95% confidence level. MTO is often developed by experienced piping designers.

Preliminary MTO is usually taken in a format similar to the one shown at the end of this section, which presents the number of items required for each line under each size. As software are being used now to do piping engineering, the production of MTOs has become very easy and quick compared to olden days where all these were taken manually. MTOs so prepared along with commodity specifications shall be handed over to the procurement team to proceed with the request for quotation (RFQ) by the purchasing team.

4.2.1.9.2   Secondary MTO

The primary purpose of secondary MTO is to update quantities for the issue of firm POs for piping materials such as pipes, fittings, flanges, valves, gaskets and fasteners. Apart from the above, secondary MTO provides a firm estimate so that necessary budgetary controls can be imposed.

Secondary MTO is taken when significant progress is made on the 3D design model (or other electronic design method). However, it must be done early enough to ensure that procurement (purchase and delivery) of piping material to site shall fit the overall project schedule and planning.

4.2.1.9.3   Final MTO

The principal purpose of final MTO is to identify whether any items were added lately in the project or any items were missed out in the preliminary or secondary MTOs. Yet another reason for final MTO is to firm up the final cost of piping materials for the project. The final MTO is made when last isometric has been released for construction.

In case changes in quantities are observed in the final MTO, POs are amended accordingly based on final MTO to tackle the discrepancies observed or as local purchases as feasible.

4.3   PREPARATION OF MR

MR for purchase includes:

- BOM having commodity codes, sizes, material specification and quantity
- Additional technical requirements, inspection and test guidelines, documentation requirements (vendor data requirement (VDR)) and guidelines for bid submission

4.4 PURCHASE ENQUIRY

EPC contractors shall have their own formats for enquiry. Enquiry in prescribed format along with MR is sent to reputed and reliable manufacturers for piping components with due date for closing. It is presumed that EPC contractors have an approved list of manufacturers based on either prequalification or past supply experience.

4.5 TECHNICAL BID EVALUATION

Bids received are tabulated both technically and commercially. Satisfactory bids in TBE are considered against commercial evaluations. EPC contractor places the PO with the most economical and technically acceptable bidder. During the process from receipt of enquiry until finalization of TBE, many rounds of communications between contractor and vendor may be required to ascertain the technical and commercial acceptability. Many clients require the TBE as a deliverable to be submitted for review.

4.6 PURCHASE ORDER

PO is the final contractual document between the vendor and EPC contractor to supply materials as per the technical and commercial requirements enclosed with the PO. In short, PO shall take into account all pervious communications, clarifications and confirmations obtained on submitted bid and shall be final.

4.7 VENDOR DOCUMENT REVIEW/APPROVAL

- After the PO is placed on a particular vendor, the vendor submits all the documents required as per VDR included in MR.
- These documents are reviewed and commented to ensure compliance to all specifications.
- After approval of essential vendor data such as drawing, procedures, etc., actual manufacturing of material begins.
- Inspection and testing of materials are carried out at vendor shop according to approved Inspection and Test Plan/Quality Assurance Plan.
- Material is dispatched to site after successful inspection and quality checks.
- Normally every EPC organization has its own standard format for each of the below-mentioned MTO categories:
  - Piping and Piping Component MTO
  - Special Item MTO
  - Insulation MTO
  - Paint MTO
  - Total MTO

4.7.1   TYPICAL **BOM** IN ISOMETRIC DRAWING

Each isometric drawing shall have a BOM with the following details as a minimum, reflecting all the piping elements in that drawing:

Mark	Qty (Nos)/Length (mm)	Size (NPS)	Description	Specification
1	5000 mm	20	Pipe Sch 40	SA 312TY 316
2	2	20	90° Elbow Sch 40	SA 312TY 316
3	1	20	Equal Tee	SA 312TY 316
4	4	20	SO Flange	SA 182TY 316
5	2	20	Gasket	CNAF
6	8	½″ × 55mm long	Bolt/Nut	SA 193/SA 194

4.7.2   TYPICAL **MTO**

MTO needs to be prepared for each line, which may consist of different pipe sizes with almost all piping elements reflecting in it. The sample table below is prepared for (say) 20 NB Sch 40 SS piping. The total MTO shall reflect all materials required for all lines of the above size and schedule as shown in Table 4.1 below.

In order to make a bulk material procurement requisition, each item such as pipe, fittings, flanges, valves, etc. are separated out and sent to manufacturers or supplier dealing with respective items. As mentioned earlier, PMS often codifies materials and each code used shall have a commodity specification assigned to it, which provides all details required for procurement of the component. Therefore, the purchase requisition shall essentially contain both commodity specification and the quantity of each of the items enabling the vendor to quote according to the intricacies of specification, quantity and delivery time provided.

Similarly, the commodity specification shall essentially contain material specification, diameter, wall thickness, pressure, temperature rating, etc. as applicable to individual components.

TABLE 4.1
Typical MTO for 20 NB Sch 40 SS Piping

Line No.	Pipe in metres	Elbows 45° (Nos)	Elbows 90°	Elbows 180°	Tee	Unequal Tee 20×15	Reducer Concentric 20×15	Reducer Eccentric 20×15	Flange	Gasket	Gate Valves	Globe Valve	Ball Valves	Check Valve	Plug	Pipe Cap	Bolts/Nuts 1/2"×55 Long
20-MET-05-SS	300	10	1	0	2	0	0	0	5	4	1	0	0	0	0	0	16
20-MET-06-SS	290	2	0	0	0	0	0	0	3	3	0	1	0	0	0	1	12
20-MET-07-SS	200	0	2	0	3	0	0	0	4	2	0	0	0	0	0	0	8
20-MET-08-SS	350	2	1	0	6	0	0	0	6	5	0	0	1	0	0	0	20
20-MET-09-SS	400	0	4	0	4	0	0	0	4	3	0	0	0	1	0	0	12
20-MET-10-SS	230	0	6	0	5	0	0	0	5	3	0	0	0	0	2	0	12
20-MET-11-SS	225	0	2	0	5	0	0	0	7	4	0	0	0	0	0	0	16
20-MET-12-SS	500	0	5	0	4	0	0	0	6	3	0	0	0	0	0	0	12
Total	2,495	14	21	0	29	0	0	0	40	27	1	1	1	1	2	1	108

5 Pipes

5.1 HISTORY OF PIPE

Many old civilizations were using pipes for carrying water to desired locations from nearby rivers and streams, to tide over hassles of transporting water to their dwelling places. Probably, the first pipes used might be from split bamboo tied together with a sealant and other similar wood like coconut tree stem, etc. Use of clay pipes is seen in the ancient monuments in Egypt and Jordan. It is said that the first metallic pipe was produced by the Greeks and Romans, made from lead due to its high malleability or formability. In fact, the terminology "plumbing" used in water piping evolved out of the Latin word for lead, *Plumbum*. However, lead pipes are prone for sagging between supports due to creep. Probably this might have paved the way for development of copper and brass pipes later as alternate materials. Iron as pipe material is said to have been evolved subsequent to the invention of gunpowder. Though gunpowder is in no way connected to manufacturing of iron, the explosive properties of gunpowder were said to be instrumental in developing stronger material for production of gun barrels, resulting in production of iron pipes. Production of iron underwent lot of improvisations over time and resulted in the development of a wide range of exotic alloy materials as available now, and this evolution will go on in the future as well.

5.2 PIPE AND TUBE

Pipe or tube in general parlance is the circular hollow section (with an outside diameter (OD), inside diameter (ID), with a corresponding wall thickness (WT)) used for transportation of fluids and gases from one location to another, using an external force (usually pressure generated by pump or static head).

Pipes are generally manufactured in accordance with long established industrial standards. Though there are similar standards for tubes, they are often made to custom sizes with a wider range of diameters, WTs and tolerance. Further, the terminology "tube" is applicable for non-cylindrical sections like square and rectangular sections as well.

5.3 DIFFERENCE BETWEEN PIPE AND TUBE

The obvious difference for all practical purposes lies in the way pipes and tubes are specified. Pipes are specified by "nominal pipe size" (NPS) and schedule (WT). NPS is a size standard established by the American National Standards Institute (ANSI), and shall not be confused with the various thread standards such as NPT.

NPSs from 6 to 300 mm (1/8″–12″) have hypothetical diameter between OD and ID (neither equal to OD or ID), whereas NPS 350 mm (14″) and above have

DOI: 10.1201/9781003328124-5

measured ODs that correspond to the NPS in inches. Pipes of a particular NPS are available in different WTs specified as schedules such as Sch 40, Sch 60, STD, XS and XXS.

Because of the above, many mistakenly believe NPS refers to ID for smaller pipes (in the range of 6–300 mm (1/8″–12″)), which is not true. This anomaly (as seen now) is because of the way in which standard was originally defined as described below.

The standard OD was originally defined as ID of pipe in inches plus twice WT of a common thickness used extensively at that time. At that point of time, sizes up to 12″ only were envisaged and hence this methodology was followed for sizes up to 12″. For example, 80 mm (3″) Schedule 40 NPS has an ID of 77.92 mm (3.068″) (close to the required 76.2 mm equal to 3″) with a WT of 5.49 mm, resulting in an OD of 88.9 mm. Later, OD of 80 mm NPS pipe was fixed at 88.9 mm and pipes had different WTs (schedules were produced as required resulting in various IDs more or less than 77.92 mm as shown in the Table 5.1).

As mentioned earlier, the above methodology was followed only in smaller diameter pipes in NPS range of 6–300 mm (1/8″–12″). Probably larger pipes came in production later and by which time, it was decided to equate OD of pipe to match the nominal size when expressed in inches, with varying WTs. This resulted in pipe OD equal to NPS in inches with varying IDs for all sizes of pipes from NPS 14″ to 48″. In other words, OD of pipes shall not change regardless of WT for any pipe in the range of NPS 6–1,200 mm (1/8″–48″) specified in standard tables.

Tubes are customarily specified by its OD and WT, expressed either in Birmingham Wire Gauge (BWG) or in thousandths of an inch (Thou). Moreover, measured OD and stated OD are generally within very close tolerances of each other and hence tubes turn out to be more expensive than pipes due to tighter manufacturing tolerances.

The principal uses for tube are in heat exchangers, instrument lines and small interconnections on equipment such as compressors and boilers.

The comparison of the differences in dimensions of pipe and tube is presented below with specifications for a 15 mm (1/2″) pipe and tube.

Pipe of NPS ½″ Sch 40 has an OD of 21.3 mm with a WT of 2.77 mm, whereas tube of 1/2″ × 1.5 shall have the OD of 12.7 mm (equal to ½″ itself) with a WT of 1.5 mm.

TABLE 5.1
Pipe Wall Thickness Range of NPS 80 (3.0″) Pipe

Nominal Pipe Size	Outside Diameter	Schedule Designation			Thickness		Inside Diameter	
(mm) (Inch)	(mm) (Inch)		ASME		(mm)	(Inch)	(mm)	(Inch)
80	**88.9**	5		5S	2.11	0.083	84.68	3.334
3	**3.5**	10		10S	3.05	0.120	82.80	3.260
		STD	40	40S	5.49	0.216	77.92	3.068
		XS	80	80S	7.62	0.300	73.66	2.900
		160			11.13	0.438	66.64	2.624
		XXS			15.24	0.600	58.42	2.300

5.4 MATERIAL OF CONSTRUCTION FOR PIPES

The materials used for pipes vary widely depending on type of service, life expectation and obviously direct and indirect costs in construction/maintenance. The material of construction includes concrete, glass, lead, brass, copper, plastic, aluminium, cast iron, carbon steel, steel alloys and composites. With such a gamut of materials available, selection of appropriate material for pipes and other piping components has become complicated, requiring thorough knowledge of corrosive properties of the fluid conveyed so also pressure, temperature and other salient flow characteristics, that too after considering the cost. This becomes extremely relevant as every material has its own limitations with regard to intended services. The ensuing sections of the book are written considering carbon steel piping as the model, and other exotic materials like stainless steels and other alloy steels are considered only in some sections of the book.

5.5 MANUFACTURING PROCESSES

Introduction of rolling mill technology and its development at the beginning of the 19th century spearheaded large-scale manufacture of pipes for industrial purposes. At the beginning, strips of rolled steel sheets were formed into a circular cross section by funnel arrangements or rolls, and then butt or lap welded in the same heat through forge welding process. With drastically increasing production needs over a short span of time, towards the end of the same century, various processes to produce seamless pipes came in. This affected the welded pipe production adversely, in spite of development of other welding processes in the manufacture of welded pipes. Further, developments and refining of seamless pipe production methods almost flushed out the welded pipes from market and seamless pipes dominated the market after World War II.

Extensive research and developments in welding technology thereafter reversed the situation and welded pipes regained its lost glory. As it stands, welded pipes dominate the market, approximately to the tune of two-thirds of the market share. Of the above, one-quarter contribution is from large-diameter line pipes which falls outside normal manufacturing range of seamless pipes.

5.5.1 SEAMLESS PIPES

Major changes in seamless pipe production processes occurred towards the end of the 19th century, followed by many more developments in manufacturing process over the years. As it stands, the following modern high-performance processes dominate the market.

- Continuous mandrel rolling process and push bench process in size range from 21 to 178 mm OD approximately.
- Multi-stand plug mill (MPM) with controlled (constrained) floating mandrel bar and plug mill process in size range from 140 to 406 mm OD approximately.

- Cross roll piercing and pilger rolling process in size range from 250 to 660 mm OD approximately.
- **Mandrel Mill Process**
  In Mandrel Mill Process, a solid round (billet) is used as raw material, which is then heated in a rotary hearth heating furnace and then pierced by a piercer. The pierced billet or hollow shell is then rolled by a mandrel mill to reduce OD and WT, which forms a multiple length mother tube. The mother tube is reheated and further reduced to specified dimensions by stretch reducer. Tube so produced is then cooled, cut, straightened and subjected to finishing and inspection processes before shipment as shown in the schematic in Section 5.7 (Figure 5.2).
- **Mannesmann Plug Mill Process**
  In plug mill process also, a solid round (billet) is used as raw material, which is uniformly heated in rotary hearth heating furnace and then pierced by a Mannesmann piercer. The pierced billet or hollow shell is roll reduced in OD and WT. The rolled pipe simultaneously burnished inside and outside by a reeling machine, followed by sizing in sizing mill to specified dimensions. Thereafter, pipe goes through straightener and with that, hot working of pipe is complete. Pipe (referred to as a mother pipe) after finishing and inspection becomes a finished product.

5.5.2 WELDED PIPES

Ever since start of production of strips and plate steels, attempts were made to bend the strip or plate and to connect edges to form pipe, leading to one of the oldest form of welding, forge welding and dates back to about a century and a half.

In early 1800, James Whitehouse (a British iron merchant) was granted a patent for manufacture of welded pipes. The process consisted of forging individual metal plates over a mandrel to produce an open-seam pipe, and then heating the mating edges of open seam and welding them by pressing them together mechanically in a draw bench.

Since then, welding techniques evolved significantly over the years and various gas-shielded welding processes were developed, predominantly for production of stainless steel pipes. Consequent to the far-reaching developments in energy sector over the last three to four decades, requiring large capacity and long-distance pipelines, submerged arc welding (SAW) process has gained predominance for welding of line pipes of diameters above 500 mm approximately.

- **Electric Welded Pipe**
  Steel strip in coil form (slit into required width from wide plate coil) is shaped by a series of forming rolls into a multiple length shell. The longitudinal edges so formed are then continuously joined by high-frequency resistance/induction welding. The weld of multiple length shell (pipe) is then heat treated electrically, sized and cut to specified lengths by a flying cut-off machine. The cut pipes are then straightened, squared or bevelled as required at both ends and is followed by NDT, inspection and hydrostatic testing. Pipes manufactured as above are known as long seamwelded pipes.

FIGURE 5.1 Types of pipes.

- **Spiral Welded Pipe**

  Least common (formerly) of the three methods is spiral welded pipe. Spiral welded pipe is formed by twisting strips of metal into a spiral shape, similar to a barber's pole, then welding where the edges join one another to form a spiral seam. This type of pipe is restricted to piping systems using low pressures due to its thin walls. The spiral welded steel pipe is manufactured using hot rolled steel strip which is spiralled at room temperature to be welded. Automatic SAW is the most commonly used method in the spiral tube production, and the end products come with welding seam both inside and outside (Figure 5.1).

5.6 STRENGTHS AND WEAKNESSES OF MANUFACTURING PROCESSES

Each of the three methods depicted above used in the production of pipes has its own advantages and disadvantages with regard to their properties, cost and availability, a summary of which is provided in the table below.

Manufacturing Process	Advantages	Disadvantages
Seamless	The obvious advantage of seamless pipes is that they don't have a weld seam. Traditionally, weld seam in pipes has been viewed as a weak spot, vulnerable to failure and corrosion.	Thickness variations possible.
	Since weld is not present, pipe is more uniform in strength.	Internal surface finish is a major issue and possibility of more surface defects.
	Pipes with a wide range of wall thicknesses are produced commercially.	Longer lead time required often culminating in significant price fluctuation during the lead time.
	Seamless pipes provide peace of mind. Although there should be no issues with the seams of welded pipes supplied by reputable manufacturers, seamless pipes prevent any possibility of a weak seam.	
	Seamless pipes have better dimensional control such as in ovality, or roundness, than welded pipes.	

(Continued)

Manufacturing Process	Advantages	Disadvantages
Butt Welded Pipes	Butt welded pipes formed from rolled plates have more control over thickness compared to seamless pipes and hence are generally more consistent in thickness than that of seamless pipes.	Probability of defects in weld, in spite of numerous quality control checks imposed during manufacture to eliminate it.
	For many years, fear of defects in weld persisted. However, improvements introduced in manufacturing process (especially in welding) boosted strength and performance of weld seam to levels indistinguishable from that of parent metal of the pipe.	
	Possible visual inspection of inside surface (before forming) ensures a defect free on the inside surface to a considerable extent.	
	Welded pipes are typically more cost-effective than their seamless equivalents.	
	Welded pipes are usually more readily available than seamless.	

5.7 SCHEMATIC PRESENTATIONS OF VARIOUS MANUFACTURING PROCESSES

See Figure 5.2.

5.8 SIZE OF PIPE

Pipes are designated through NPS and a schedule reflecting applicable WT.

5.8.1 NOMINAL PIPE SIZE

NPS is a North American set of standard sizes for pipes used for all kinds of pressures and temperatures and is based on earlier "Iron Pipe Size" (IPS) system. As mentioned earlier, in IPS system, the size represented the approximate ID of pipe in inches. For example, IPS 6″ pipe is one whose ID is approximately 6″. With two WTs added, OD shall be more than 6″. The end users since then started designating pipes as 2″, 4″, 6″, 8″ and so on. At the beginning, each size of pipe was produced in one thickness, which later was termed as standard (Std.) or standard weight (Std. Wt.) and based on that the OD of pipe was standardized in inches. Initially, production range of pipes sizes was limited and the above standardization of OD of pipes was made applicable in the range from 6 to 350 mm (1/8″–12″), which was the usual production range at that time.

Later as industrial needs increased, manufactures were encouraged to produce pipes with thicker walls to accommodate higher pressures and temperatures, which resulted in

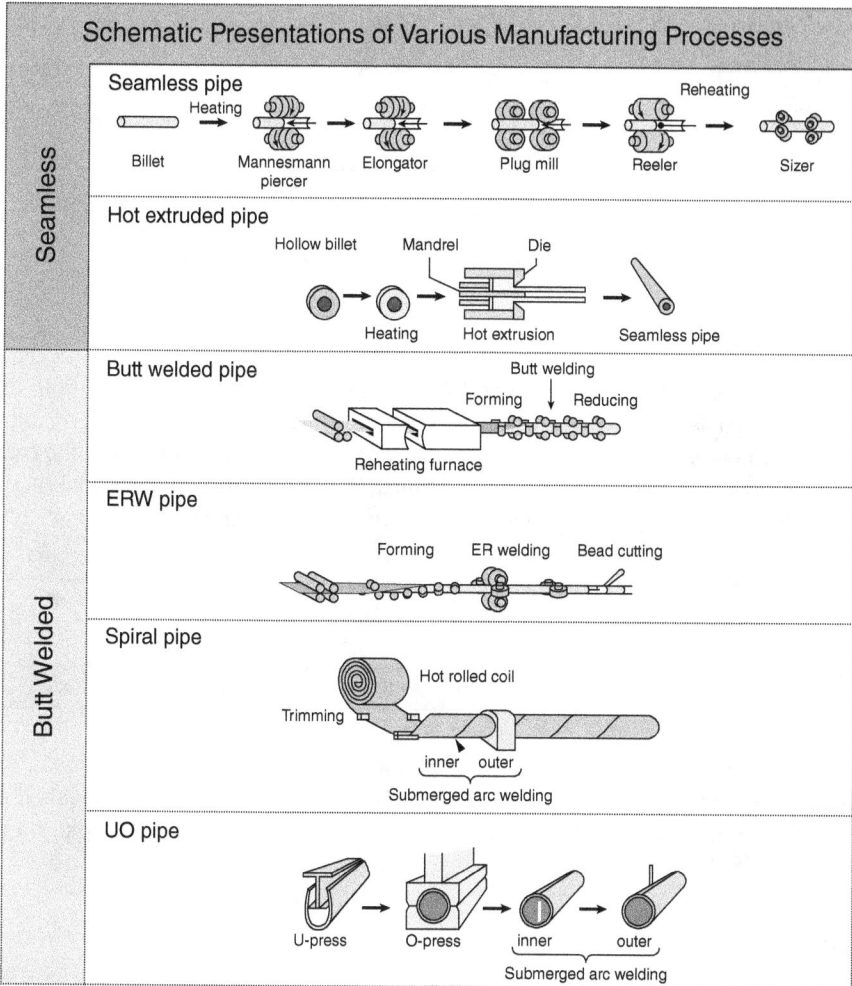

FIGURE 5.2 Schematics of pipe manufacturing processes (JFE 21st Century Foundation, 1991).

pipes known as extra strong (XS) or extra heavy (XH) and double extra strong (XXS) or double extra heavy (XXH) walls. The important aspect during this developmental phase was that the "OD which was standardized based on STD wall thickness" was fixed as the standard in the pipe diameter range from 6 to 350 mm (1/8″ to 12″) range.

It appears that larger diameter pipes were added later in production line. This may be the reason for maintaining the OD of pipe to specified dimension (in.) for NPS sizes from 400 mm (14″) and above. This implies that the OD of a NPS 350 mm (14″) pipe is 355.6 mm (14″).

Now coming to the NPS specification in metric system, approximate conversion figures to give a round figure was used as shown in the table below.

NPS	Imperial	1/8	¼	3/8	½	¾	1	1¼	1½	2	2½	3	3½	4
	Metric	6	8	10	15	20	25	32	40	50	65	80	90	100

NPS	Imperial	4½	5	6	7	8	9	10	11	12	14	16	18
	Metric	115	125	150	175	200	225	250	275	300	350	400	450

NPS	Imperial	20	22	24	26	28	30	32	34	36	42	48
	Metric	500	550	600	650	700	750	800	850	900	1,050	1,200

5.8.2 Pipe Schedule

In IPS system, there were only three WTs for each size of pipe. In early 1900, the American Standards Association created a system with a wide range of WT in smaller steps based on demand from industry. Along with change from IPS to NPS, the term schedule (Sch.) was introduced to specify nominal WT of pipe. By adding schedule numbers to the IPS standards, today pipes are produced with a wide range of WTs, namely:

Schedule	5	5S	10	10S	20	30	40	40S	60

Schedule	80	80S	100	120	140	160	STD	XS	XXS

ASME standards provide WT (along with working pressures and many more information) of each schedule of pipes under each size. In this regard, ASME/ANSI B 36.10 is the applicable standard for welded and seamless wrought steel pipes and ASME/ANSI B 36.19, that for stainless steel pipes.

5.8.3 Length of Pipe

Seamless pipe is produced in single and double random lengths. Single random lengths vary from 16'-0" to 20'-0" long. Pipe 2" and below is found in double random lengths measuring 35'-0" to 40'-0" long. Of late, it is seen that pipes are made up to length of 48' in seamless and up to 80' in electric resistance welded pipes. The limiting factors for diameter, WT and maximum length of pipes produced at a mill is principally based on the capacity of equipment at mill and the transportation feasibility, especially on length.

5.9 METHODS OF JOINING PIPES

The industry mainly uses three types of joining extensively, especially in steel piping consisting of carbon, stainless and alloy steels. They are Butt Welded (BW), Screwed (SCRD) and Socket Welded (SW) connections (Figure 5.3).

Methods for Joining Pipes		
Butt Welded Joint	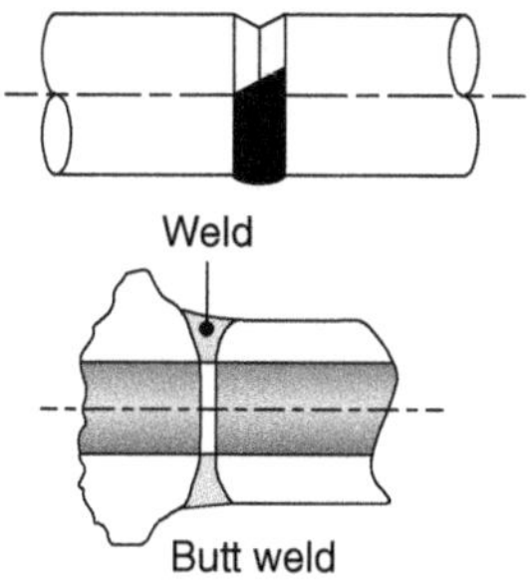 Weld Butt weld	In butt welded joints, the Beveled Ends (BE) in pipe to pipe or pipe to fittings joint are welded using appropriate welding process and technique. The details regarding welding processes, techniques, bevel preparations etc., are described in detail in ensuing sections.
Socket Welded Joint	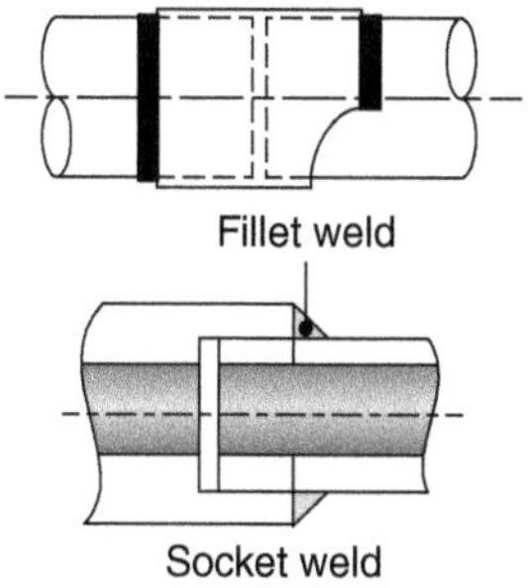 Fillet weld Socket weld	The second method of joining carbon steel pipe is by Socket Welding (SW). In this type of joints, square cut pipe is inserted inside socket weld fitting and a socket weld (fillet weld) between pipe and fitting is applied. While assembling pipe inside socket of the fitting, collar provided inside the fitting shall be limiting the inserted length of pipe inside socket. The amount of length of pipe lost in inserting it inside the socket welded fitting is smaller compared to pipe length lost in making threaded joints.
Screwed or Threaded Joint	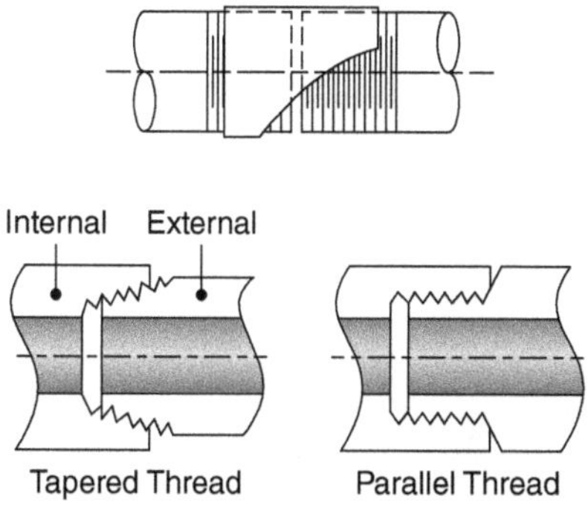 Internal External Tapered Thread Parallel Thread	Yet another common method of joining pipe is by using Threaded End (TE) connections. This type of connection is typically used in pipes of 3" and smaller sizes. Threaded connections are also referred to as Screwed connections. With the tapering threads made on ends of a run pipe, it is very easy to assemble internally threaded fittings to form the piping link. Screwed pipe and its mating fittings shall have threads that are either male or female with same screw pitch.

FIGURE 5.3   Pipe joining methods (Envestis SA, 2018).

6 Pipe Fittings

Fittings are fabricated piping elements used to change the direction of fluid flow in required direction. Examples are fittings used to change direction such as elbows and bends, branch from a main pipe (tee) or to make a reduction in line size (reducer), etc. Since the entire piping system between two equipment or stations is subject to more or less the same kind of pressure and temperature applied by the fluid under transit, each element of the piping system in that loop shall be capable of providing designed service life and hence to withstand chemical and physical characteristics of fluid. Because of this reason, pipe fittings are also available in many pressure and temperature ratings (as in the case of pipes) and hence selection of the same shall be made as per piping specifications developed for the specific industry. Almost all the oil and gas and petrochemical companies have their own piping specifications developed in consultation with engineering consultants for the range of piping materials used in their industry, in line with the international codes and standards applicable to their process domain.

As mentioned earlier, a pipe fitting is defined as a part used in piping system, for changing direction, branching or for change of pipe diameter, and which is mechanically joined to the system. The range of pipe fittings available is very wide, that too matching the pipe wall thicknesses. However, three broad classifications possible is based on the type of end connections described in Chapter 5.

- Butt weld (BW) fittings whose dimensions, dimensional tolerances, etc., are defined in the ASME B 16.9 standards. Lightweight corrosion-resistant fittings are made to MSS SP43.
- Socket weld (SW) fittings Classes 3000, 6000 and 9000 are defined in the ASME B 16.11 standards.
- Threaded (THD), screwed fittings Classes 2000, 3000 and 6000 are defined in the ASME B 16.11 standards.

6.1 BW PIPE FITTINGS

Piping systems designed with BW fittings have many inherent advantages over other types of connections like SW and THD.

- The welding between a pipe and a fitting can be considered as a permanently leak-proof joint.
- The continuous metal structure formed between pipe and fitting through weld adds strength to the system.

DOI: 10.1201/9781003328124-6

- Smooth inner surface at the weld possible through the appropriate selection of welding process reduces pressure loss and turbulence and minimizes effects of corrosion and erosion at joint.
- Above all, welded system utilizes a minimum of space.

6.1.1 BEVELLED ENDS

In order to facilitate welding between pipe and fitting, ends of butt-welded fittings need to be bevelled on both ends to be joined. The configuration of the bevel depends upon actual wall thickness. Single (plain) bevel and double (compound) bevel as shown in Figure 6.1 are the most common bevel preparations used in piping.

Bevel joint design is basically guided by piping design code, namely ASME B 31.3 and applicable engineering specifications developed for the specific project. Therefore, fittings need to be ordered with required bevel preparation, which normally will be reflected in the commodity description developed as a part of the Piping Material Specification prepared for individual projects.

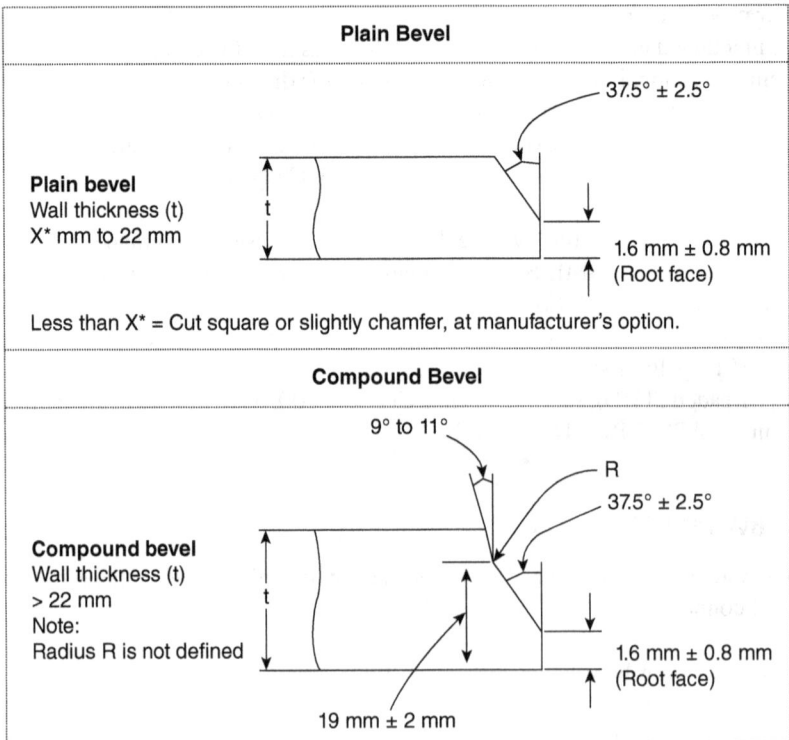

Plain Bevel

37.5° ± 2.5°

Plain bevel
Wall thickness (t)
X* mm to 22 mm

1.6 mm ± 0.8 mm
(Root face)

Less than X* = Cut square or slightly chamfer, at manufacturer's option.

Compound Bevel

9° to 11°

R

37.5° ± 2.5°

Compound bevel
Wall thickness (t)
> 22 mm
Note:
Radius R is not defined

1.6 mm ± 0.8 mm
(Root face)

19 mm ± 2 mm

FIGURE 6.1   Bevelled ends (Sölken, n.d.).

6.2   MATERIAL OF CONSTRUCTION

Pipe fittings are manufactured with a wide range of materials. The most common materials are carbon steel, stainless steel, other exotic alloy steel, cast iron, aluminium, copper, glass, rubber, the various types of plastics, composites, etc. The material of a fitting is basically guided by choice of material for pipe; in most cases, a fitting is of the same material as the pipe or nearly similar in chemical and physical properties, so that the quality of weld between them can be properly ensured.

Figures 6.2, 6.3, 6.6 and 6.7 that follows provide a very brief description of most commonly used pipe fittings as a quick reference guide to readers.

6.3   TYPES OF BW FITTINGS

See Figure 6.2.

6.4   O'LET TYPE WELDED BRANCH CONNECTIONS

Branch connection fittings (also known as O'lets) are fittings which are welded directly to main running pipe to provide a branch of either same size or smaller. The main pipe onto which the branch connection is welded is called the Run or Header and the other which provides a diversion in flow is the branch. The O'let size is designated by size of the branch and O'let branch connections are available in all sizes, types, bores and classes, in a wide range of materials commercially.

Bonney Forge was one of the pioneers in the supply of branch fittings (O'lets) for many years. Apart from the built-in compensation for opening made on run pipe, Bonney Forge branch connections help in avoiding fillet welds and sharp corner reinforcement tapering at sides, thus preventing abrupt change in thickness, where fitting joins run pipe.

Bonney Forge fittings are made to meet 100% reinforcement requirement of applicable piping codes like ASME B 31.1, B 31.3, B 31.4 and B 31.8. They also meet 2001 edition of MSS-SP-97 Standard – "Integrally Reinforced Forged Branch Outlet Fittings".

6.4.1   Types of O'lets

See Figure 6.3.

6.4.2   Welding Requirements of O'let Fittings

The quantum of welding required between O'let and run pipe was always a matter for debate, with respect to its adequacy to impart required strength to the joint, making it on par with standard pipe fittings.

ASME B 31 Series of Codes and ASME Boiler and Pressure Vessel Code provide guidelines with regard to requirements of reinforcement of branch connections and

Type	Description	Picture	Remarks if any
Elbows			
Long Radius 90°	Purpose of Elbow is to change flow direction of fluid in piping system 90° and 180° elbows available in LR & SR versions, whereas 45° elbow is available in LR version only		Elbows are divided into two groups based on the distance over which they change direction; (the center line of one end to the other face). This is known as the "center to face" distance and is equivalent to the radius through which the elbow is bent. Center to face distance for a Long Radius (LR) elbow is always 1½ x Nominal Pipe Size (NPS) = (1½D)", while the center to face distance for a Short Radius (SR) elbow is 1x NPS.
Long Radius 180°		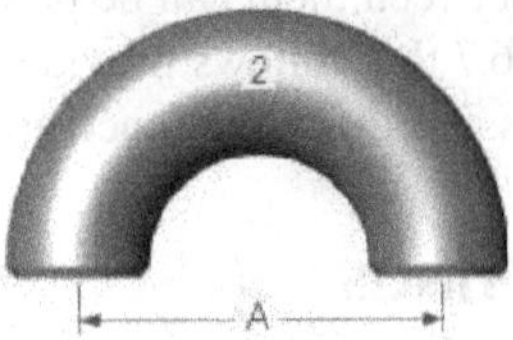	
Short Radius 90°		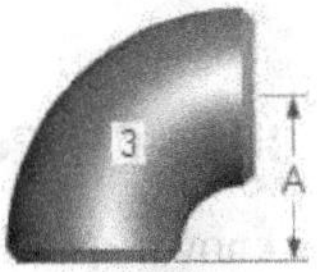	
Short Radius 180°		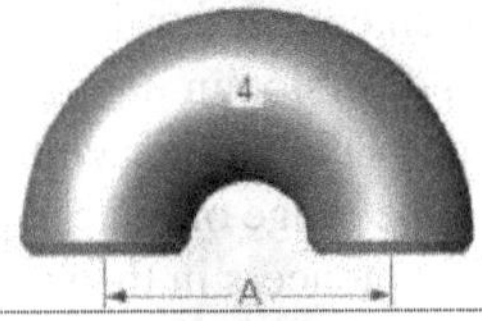	
Long Radius 45°			Radius of a 45° elbow is the same as the radius of 90° LR Elbow (1½D).
Tee			
Tee Equal	Purpose of a Tee is to make a 90° branch from the main run of pipe.		Equal tee (or straight tee) when the branch has the same diameter as run-pipe.
Tee Reducing			Reducing tee is used when branching pipe has a smaller diameter compared to run-pipe.

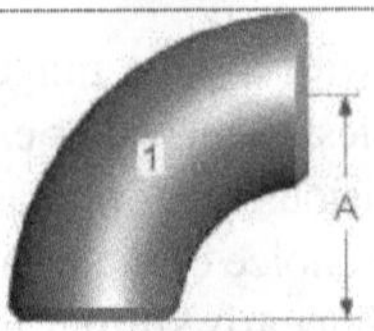

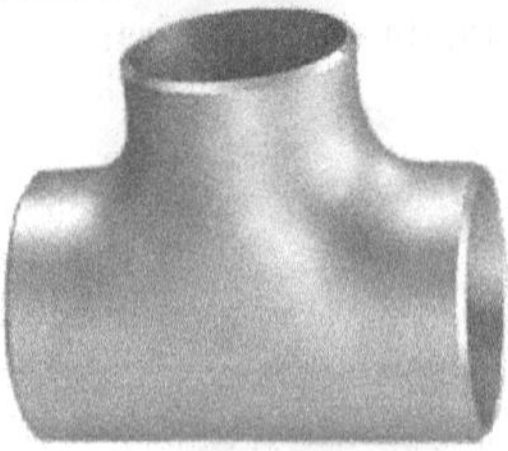

FIGURE 6.2 Types of butt-welded fittings.

(Continued)

Cross	In addition to defined tees, there are straight and reducing crosses.			Straight crosses are generally stock items, reducing crosses are often difficult to obtain. Not generally used in oil & gas / petrochemical industry.
Reducer	Concentric	Reducers are used to reduce pipe diameter without affecting direction of flow.	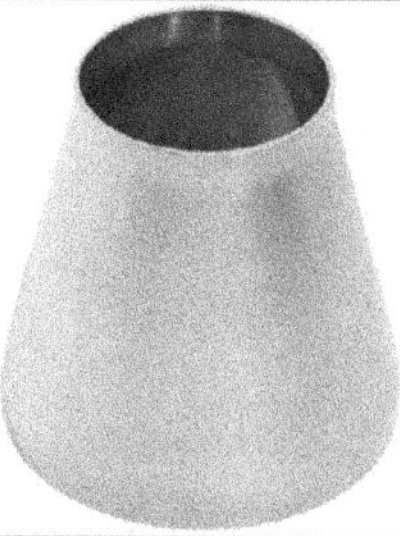	Concentric reducer, is usually used in vertical piping.
	Eccentric			Eccentric reducer is used in horizontal piping.
Cap	Used to terminate piping.			Cap, as it is shown in image on left, is available for all pipe dimensions, and is used many times for other purposes.
Stub End	A Stub Ends are always used with a Lap Joint Flange, as a backing flange; both are shown on image on the right.		 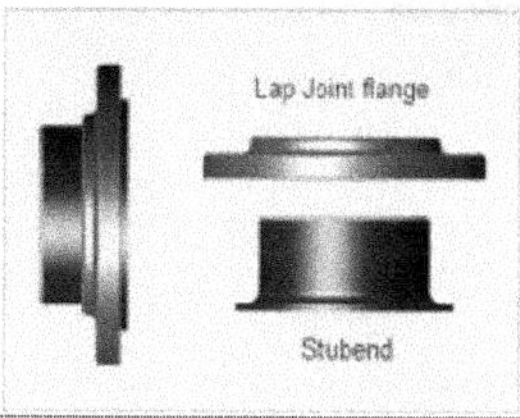	This type of flange connection is used in low-pressure and non-critical applications, and is a cheap method of flanging. For example, in a stainless steel or other exotic alloy pipe system, a carbon steel back up flange can be used, because they do not come in contact with the product in pipe, which drastically reduces cost.

FIGURE 6.2 (*Continued*)   Types of butt-welded fittings.

Weldolet

Weldolet is the most common of all branch connections, and is welded onto run pipe to obtain a branch. The ends are beveled to facilitate this process, and therefore weldolet is considered a butt-weld fitting. Weldolet's are designed to minimize stress concentrations and provide integral reinforcement.

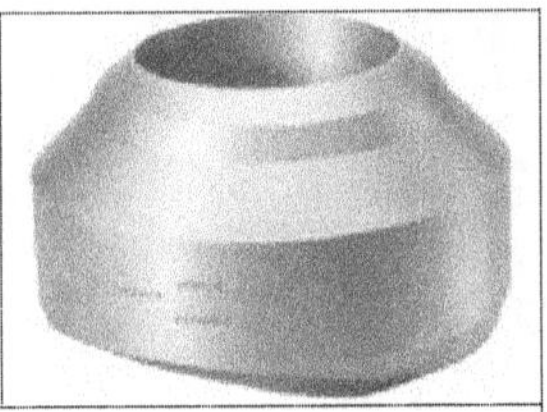

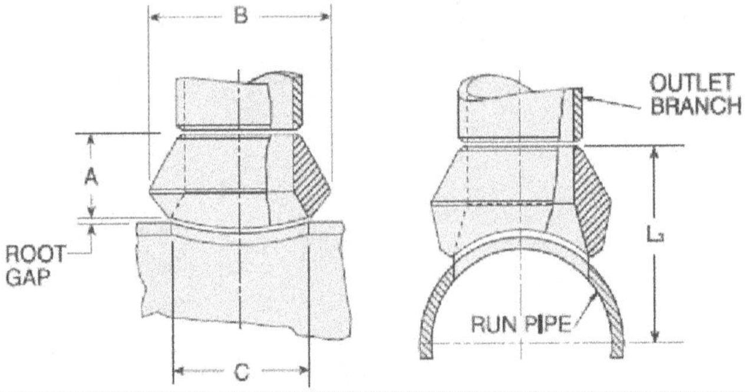

Sockolet

Sockolet is similar to basic Weldolet. The difference is in the attachment of branching pipe. Brach pipe is attached Sockolet using socket weld (instead butt weld as in weldolet) with the socket provided in the Olet.

Sockolet is considered a socket weld fitting, and manufactured in 3000#, 6000# and 9000# classes.

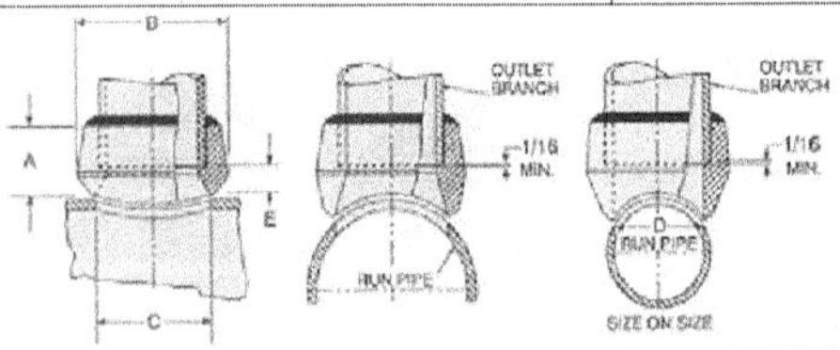

Thredolet

Thredolet is also similar to basic Weldolet, with a threaded branch. Here, instead of socket for welding the branch pipe, internal threading is provided for attaching branching pipe.

Thredolet is considered a threaded fitting, and manufactured in 3000# and 6000# classes.

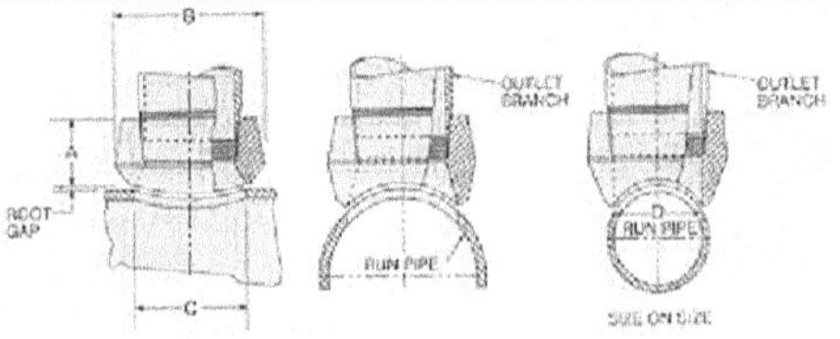

FIGURE 6.3   Types of O'lets.

(*Continued*)

Latrolet

Latrolet is used for 45° lateral connections on running pipe through butt welding, meeting reinforcement requirements as per code

Branch end connection available for Butt weld, Socket Weld or Threaded applications and in 3000# or 6000# classes.

BUTT WELD   THREADED   SOCKET-WELD   FULL PENETRATION WELD

Elbolet

Elbolet® is used on 90° Long Radius Elbows (can be manufactured for Short Radius Elbows) for thermowell and instrumentation connections. Available for Butt-Weld, Socket Weld and Threaded applications to meet specific reinforcement requirements, and in 3000# and 6000# classes.

BUTT-WELD   THREADED   SOCKET-WELD   FULL PENETRATION WELD

Nipolet

Nipolet® is a one piece fitting for valve take-offs, drains and vents. Manufactured for Extra Strong and Double Extra Strong applications in 3.1/2in to 6.1/2in lengths. Available with male-socket weld or male threaded outlets.

Sweepolet

Sweepolet® is a contoured, integrally reinforced, butt-weld branch connection with a low stress intensification factor for low stresses and long fatigue life. The attachment weld is easily examined by radiography, ultrasound and other standard non-destructive techniques. Manufactured to meet specific reinforcement requirements by users.

FIGURE 6.3 (*Continued*) Types of O'lets.

(Continued)

Sweepolet

Insert Weldolet

Insert Weldolet® is another contoured butt-weld branch connection used in less critical applications. Like the Sweepolet®, the attachment welds are easily examined by radiography, ultrasound and other standard non-destructive techniques. Manufactured to meet your specific reinforcement requirements.

Brazolet

Brazolet® is designed for use with KLM and IPS brass or copper tubing. Available with socket or threaded connections.

Coupolet

Coupolet® fittings are designed for use in fire protection sprinkler systems and other low pressure piping applications. Manufactured with NPT female threads usually for 300# service.

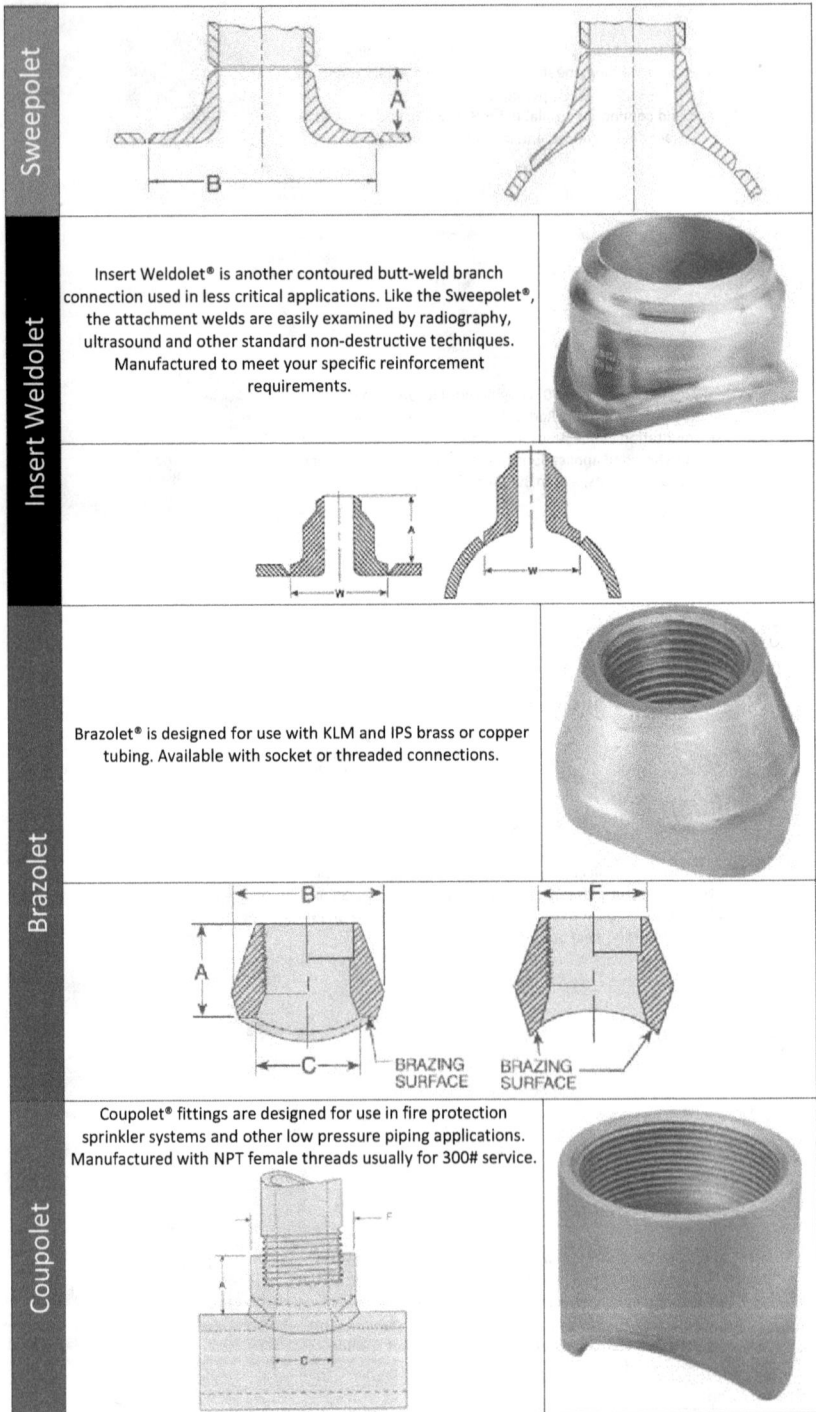

FIGURE 6.3 (Continued)   Types of O'lets.

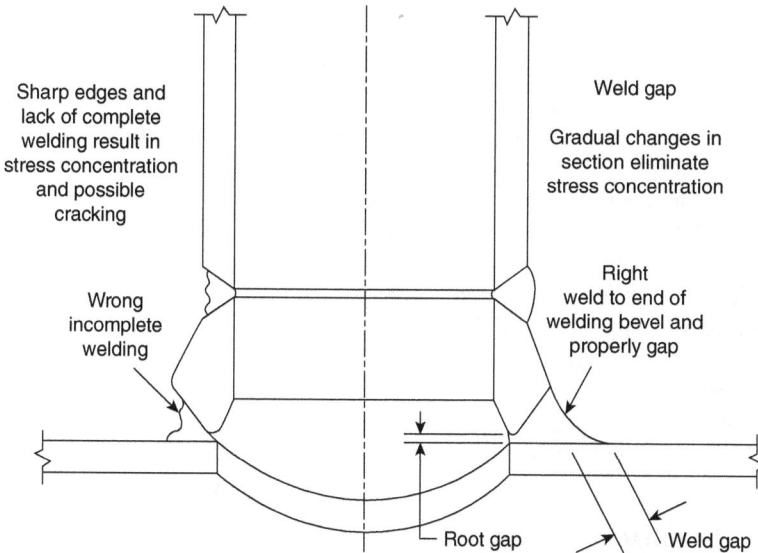

FIGURE 6.4 Welding requirements for O'let fittings (Sölken, n.d.).

nozzle pipes. Those codes require a full penetration groove weld up to the outer edge of the manufacturer's weld bevel and that the weld shall be properly capped.

Noncompliance to the above may lead to initiation of cracks at the top of the weld of branch connection. Moreover, unfilled welds or abrupt contours in weld between run pipe and O'let also may serve as stress risers, causing initiation and propagation of cracks, especially in piping designed for severe fatigue services.

Figure 6.4 describes the requirements for successful welding of an integrally reinforced branch fitting. The right weld profile shall have the following characteristics:

- The fitting is welded fully, up to outer edge of the weld bevel.
- The groove weld has a smooth transition weld cap that provides a smooth transition to main (run) pipe wall.
- The use of excessive weld metal shall be avoided as this is similar to having insufficient weld at the cross section.

6.5 SW FITTINGS

As seen above, in SW pipe fittings, connecting pipe is inserted into a recessed area of the fitting and a fillet weld is provided between fitting and pipe. In contrast to BW fittings, SW fittings are mainly used for small bore piping, usually below NPS 2 or smaller. SW joint is a good choice for high leakage integrity piping systems and the great structural strength of such joint is an important design consideration. However, the fatigue resistance is lower for these joints compared to BW joints, due to the use of fillet welds. The abrupt change in weld geometry also contributes to lower fatigue strength of such joints. In spite of all these, SW joints are better than most of the other mechanical joining methods, except BW joints.

6.5.1 TECHNICAL DETAILS

- SW fittings are a family of high-pressure pipe fittings used in a multitude of industrial piping systems in a variety of industries.
- These fittings are used for piping systems conveying a variety of flammable, toxic, corrosive or expensive fluids, wherein no leakage is permitted, that too under a wide range of pressures and temperatures.
- Pipe fittings of this category are available in the same range of pipe sizes and wall thicknesses as specified in ASME Codes.
- SW fittings are in permanent piping systems in process plants which are designed to provide good fluid flow characteristics.
- SW fittings are produced to several ASTM standards and applicable ASME standard is ASME B 16.11.
- SW fittings are available in three pressure rating classes, namely, Classes 3000, 6000 and 9000.

6.5.2 ADVANTAGES/DISADVANTAGES OF SOCKET WELD FITTINGS

Advantages	Disadvantages
Pipe need not be bevelled for weld preparation.	The fitter or welder (in case welded without a tack weld) shall ensure a gap of 1.6 mm (1/16″) between pipe and shoulder of the socket for expansion.
Temporary tack welding is not necessary for alignment, because in principle, the fitting ensures proper alignment between pipe and fitting.	The expansion gap and internal crevice left unfilled in socket-welded joint promotes corrosion, especially in low-velocity and stagnant lines (used intermittently) in corrosive or radioactive applications.
Weld metal can never penetrate into the bore of pipe, causing burn through as in butt-welded joints.	Socket welding is generally unacceptable in many services, as gap and crevice left back serves as spots for accumulation of sediments, often difficult to remove through washing from hygiene point of view.
Quality of fillet weld between pipe and fitting can be ensured only through examination techniques like ultrasonic testing (UT), magnetic particle (MP) or liquid penetrant (LP) testing methods.	Radiography of the joint is often used to ascertain the availability of the expansion gap and not as a means for ensuring quality of fillet weld.
Socket weld joints can be used in place of threaded fittings so as to reduce the risk of leakage.	
Construction costs are lower compared to butt-welded joints due to non-necessity of exact fit-up requirements and elimination of special machining as required in butt weld end preparation.	

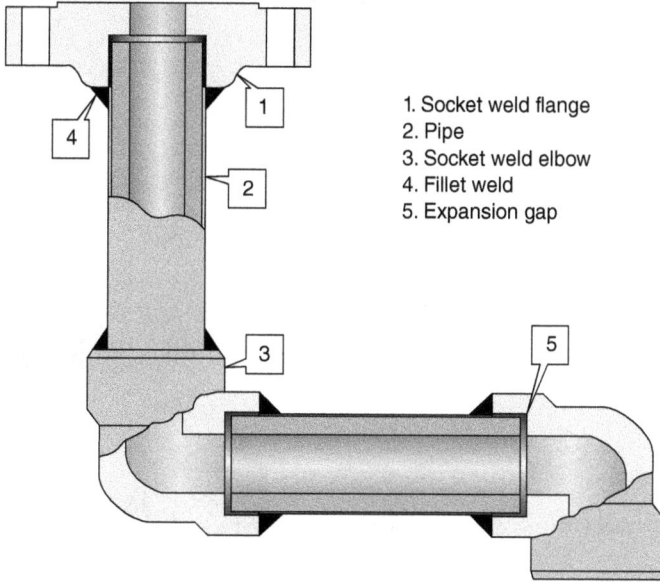

1. Socket weld flange
2. Pipe
3. Socket weld elbow
4. Fillet weld
5. Expansion gap

FIGURE 6.5 Typical socket weld joint (Sölken, n.d.).

6.5.3 TYPICAL SW JOINT

A typical SW joint is shown in Figure 6.5 with parts identified for easy reference. The figure shows various SW configurations possible in piping.

6.5.4 TYPES OF SW FITTINGS

See Figure 6.6.

6.6 THD FITTINGS

THD connections are probably one of the oldest of joining methods used in piping systems. THD fittings are also used for smaller pipe sizes (NPS 2 and below) as in the case of SW fittings.

In order for it to be commercially useful, the threading is unified according to ASME B 1.20.1, which provides all required dimensions, including number of threads per inch, pitch diameter and normal engagement lengths for all pipe diameters:

- THD piping is commonly used in low-cost, noncritical applications such as domestic water, fire protection and industrial cooling water systems.
- THD fittings are also produced commercially in a wide range of materials, ranging from carbon steel, alloy steel and a host of non-ferrous materials as well.
- THD fittings are available in three pressure ratings: Classes 2000, 3000 and 6000.

Full Coupling	Used to join pipe to pipe or to a nipple etc.	
Half Coupling	Directly welded to the run pipe, to make a branch connection	
Reducing Coupling	Used to join two different size pipes	
Reducer Insert	Enable quick and economic combinations of pipe reductions to be made using standard Socket Weld fittings. Manufactured to MSS SP-79.	
Union	It is a screwed joint design consisting of three interconnected pieces, with two of them having threads including the centerpiece that draws the ends together when rotated. (MSS SP-83) Unions should be screwed tight before the ends are welded to minimize warping of the seats.	
Elbow 90°	To make 90° change in direction in the run of pipe	

FIGURE 6.6   Types of socket weld fittings.

(*Continued*)

Elbow 45°	To make 45° changes in direction in the run of pipe	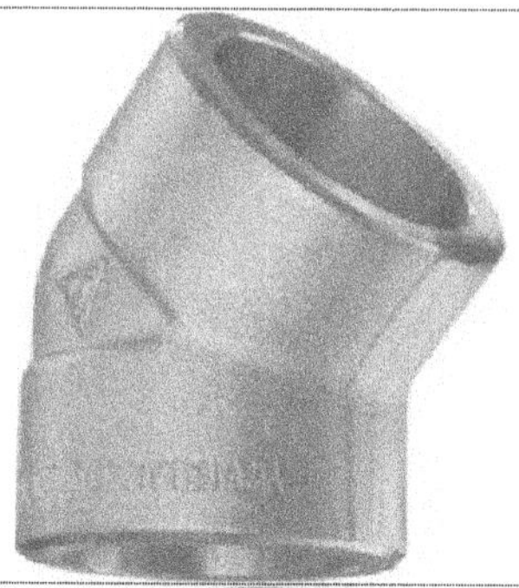
Tee Straight	To make 90° branch from the main run of pipe	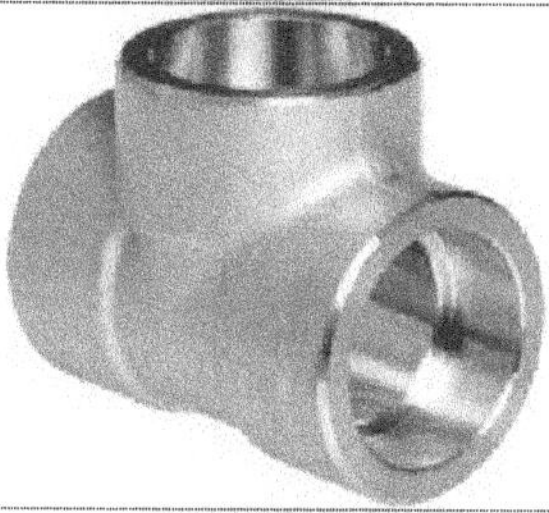
Cross	To make two diametrically opposite 90° branch from the main run of pipe.	
Cap	To cap or seal a pipe	

FIGURE 6.6 (*Continued*)   Types of socket weld fittings.

6.6.1 NPT Thread

The American National Pipe Thread (Tapered) is the most widely used threading in THD connections. As mentioned earlier, dimensions of threading are standardized as per ASME B 1.20. NPT has tapered male and female threads on mating components, which provides both a strong mechanical connection and hydraulic seal when used with PTFE (Teflon) tapes or any other joining compound.

6.6.2 Types of THD Fittings

See Figure 6.7.

Elbow 90°	Used to make 90° changes of direction in the run of pipe.	
Tee	Used to make 90° branch from the main run of pipe.	
Cross	Used to make two 90° branch from the main run of pipe in diametrically opposite direction in same plane	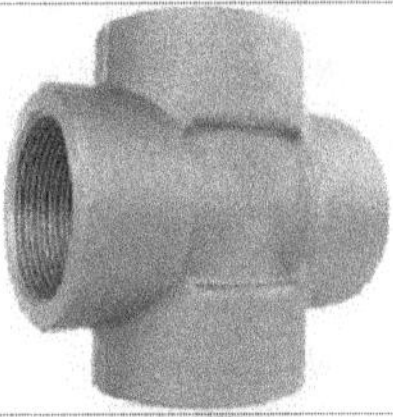
Elbow 45°	Used to make 45° change of direction in the run of pipe	
Full Coupling	Used to connect either pipe to pipe or pipe to nipple or similar.	
Cap (End Cap)	Used to seal the threaded end of pipe.	
Half Coupling	This fitting can be directly welded to the run pipe, to make a threaded branch connection.	

FIGURE 6.7 Types of threaded fittings.

(*Continued*)

Square Head Plug	Used to seal the threaded (internally) end of a fitting.	
Hexagonal Head Plug	Used to seal the threaded (internal) end of a fitting.	
Round Head Plug	Used to seal the threaded (internal) end of a fitting.	
Hexagonal Head Bushing	Used to reduce the size of a threaded fitting.	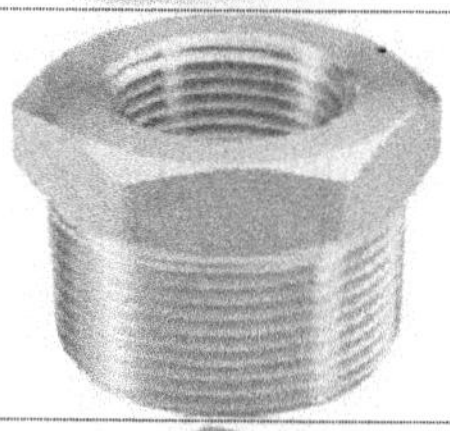
Union	Used for easy installation & maintenance purposes. Consists of three interconnected pieces.	

FIGURE 6.7 (*Continued*) Types of threaded fittings.

6.7 MITERED ELBOWS

Under the fittings category, it will not be complete if mitered elbows are missed out. Mitered elbow is not really a fitting, but manufactured using pipe segments cut at various angles and joined together (by welding) to effect angular shift in piping. These elbows can be manufactured to any angle and using any number of pipe segments, but the usual ones used in 90° elbow are shown in Figure 6.8.

The above configurations are classified as one, two and four weld miters. The number of welds used depends on smoothness of flow required through turn and also weld proximity requirements specified in project documents. It may be noted that turbulence created in flow through a miter with two bends compared to that with three or four weld mitre, one with four welds, shall be creating the least turbulence.

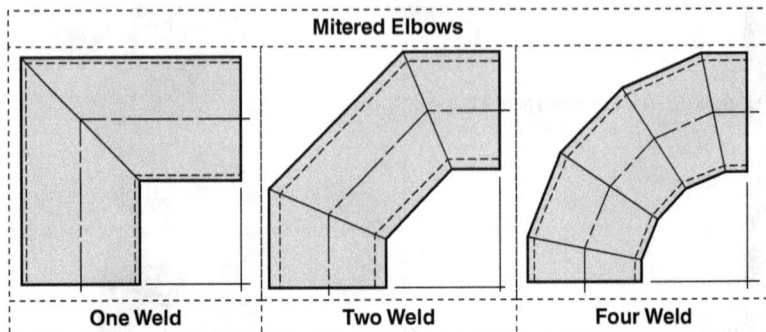

FIGURE 6.8   Mitered elbow.

6.8   HEADER AND BRANCH CONNECTIONS

Header and branch connections in piping can be manufactured in two different methods. The predominantly used method is by the use of standard "Tee" pipe fitting, equal or reducing. When equal or straight tee is used, all three outlets shall be of the same pipe size, implying that header and branch connection shall be of the same size, whereas when unequal tee is used to fabricate a header and branch, size of the branch shall be smaller than the header.

In the case of header and branches using unequal tees, a callout as shown in Figure 6.9 is required on reducing tee to identify header and branch sizes, with size of header indicated first.

6.8.1   STUB-IN CONNECTIONS

Another method of making a branch connection is by using the **stub-ins**. The stub-in is most commonly used as an alternative to the reducing tee. As in the case of mitered elbows, the stub-in is also not an actual fitting but rather a description of how the

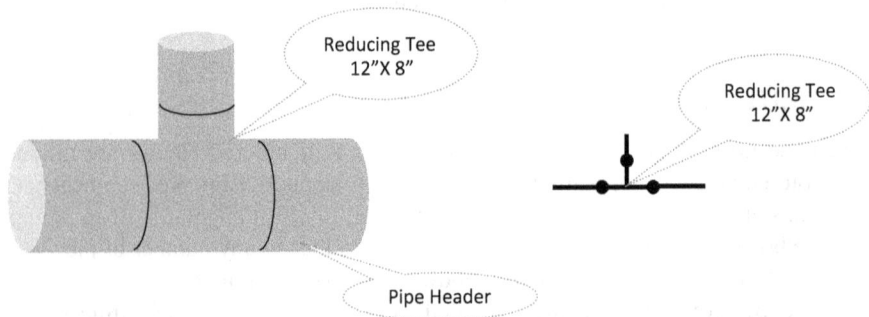

FIGURE 6.9   Header and branch connections.

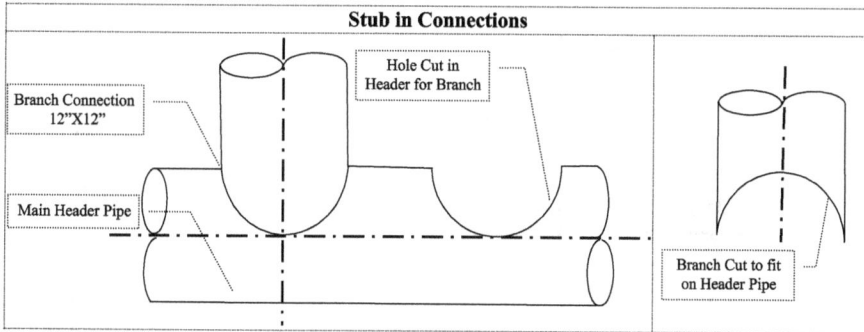

FIGURE 6.10 Stub-in connections.

branch connection is manufactured. A hole is bored into the header pipe, either the size of the outer diameter (OD) or inner diameter (ID) of the branch, and the branch is then stubbed into it. The two pipes are fitted together and then welded. Although the branch connection can be the same pipe size or smaller as header, it cannot be larger. Figure 6.10 depicts the attachment of a stub-in.

Yet another important aspect considered in making headers with stub ends is the closeness between them. A general rule is to allow a minimum of 75 mm (3″) between welds. This means a minimum of 75 mm (3″) should be allowed between the outsides of branches made from a common header, and a header should be attached no closer than 75 mm (3″) to fitting. Figure 6.11 provides minimum measurements recommended between branches and fittings on a 450 mm (18″) header.

6.8.2 STUB-IN REINFORCEMENTS

Even though the use of stub-in is limited by the pressure, temperature and commodity within run pipe, its use is becoming increasingly more popular. Its chief advantage over standard tee is its low cost. Not only can the cost of purchasing a fitting be avoided, but the stub-in requires only one weld, whereas tee requires three. When internal conditions such as pressure or temperature of the commodity or external forces such as vibrations or pulsations are imposed on a stub-in, special reinforcement may be necessary to prevent branch from separating from the header. Three reinforcing alternatives often used are listed below.

- **Reinforcing Pad**
  Fabricated from a ring cut steel plate, curved to match OD of the header pipe, with a hole at the centre to accommodate branch pipe. After completion of welding of branch pipe, reinforcement plate is slipped onto the branch and seated properly with the header and welded both to header and branch to impart required strength.

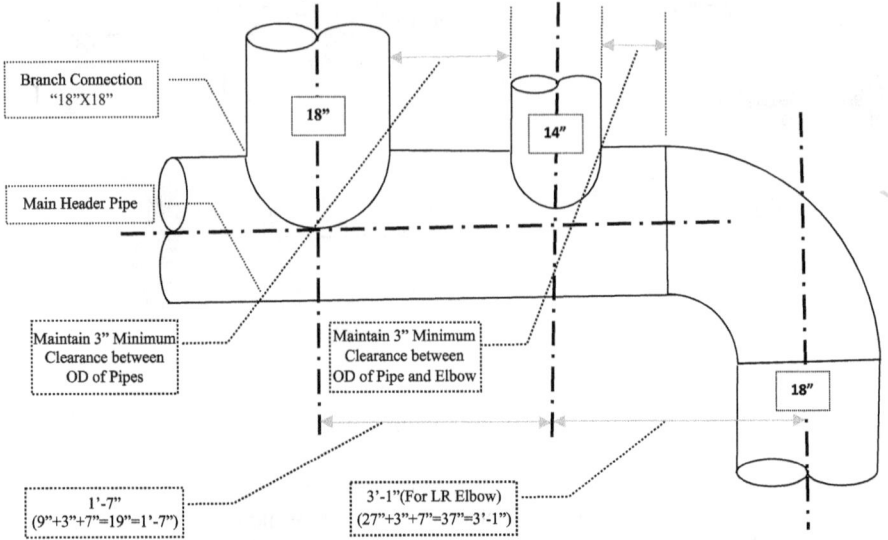

Branch Connection "18"X18"

18"

14"

Main Header Pipe

Maintain 3" Minimum
Clearance between
OD of Pipes

Maintain 3" Minimum
Clearance between
OD of Pipe and Elbow

18"

1'-7"
(9"+3"+7"=19"=1'-7")

3'-1"(For LR Elbow)
(27"+3"+7"=37"=3'-1")

FIGURE 6.11   Spacing between stub-in joints.

- **Welding Saddle**

  This is a purchased reinforcing pad with a short neck designed to give additional support to the branch. Figure 6.12 shows a typical welding saddle.

- **O'lets**

  This again is a purchased fitting and was touched upon earlier in this chapter. They are available as weldolets, sockolets, threadolets, laterolets and elbowlets depending on their configuration. Commonly used O'lets were described and shown in Fig 6.3.

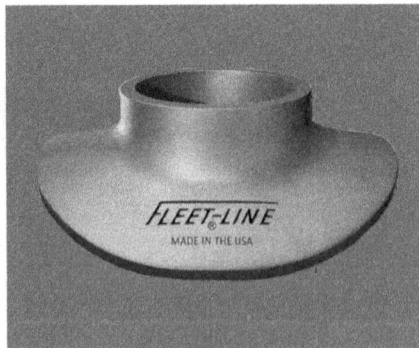

FIGURE 6.12   Typical welding saddle. (Picture Courtesy: Steel Forgings Inc.)

7 Flanges, Gaskets and Fasteners

7.1 FLANGES

Flange is a ring-shaped component designed to be used as an alternative to welded or threaded connection in piping system in process industries. Flanged connections are used as an alternative to welding, because they can be easily dismantled for shipping, routine inspection, maintenance or replacement. Flanged connections are preferred over threaded connections because threading large bore pipe is not an economical or reliable operation. Therefore, flange is considered as a vital component of any piping system.

As mentioned, flanges are primarily used where a connecting or dismantling joint is needed. Flanged joints may be provided at any place as required, either at the joint between pipe to fittings, valves, equipment or any other integral component within the piping system. Every mechanical equipment like pressure vessels (static equipment) or pump (rotating equipment) shall have a minimum of two nozzles as the inlet and outlet. The vessel or pump jurisdiction ends with the first flanged connection. Therefore, the interconnecting piping between equipment also starts and ends with mating flanges at the beginning and end of piping connecting the two equipment (Figure 7.1).

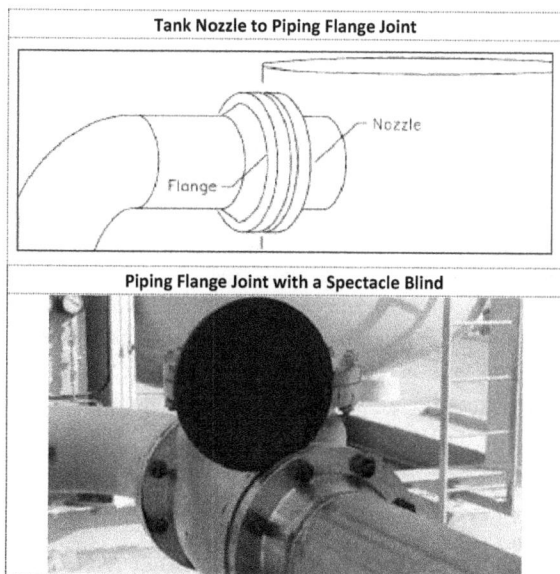

FIGURE 7.1 Flange connections (Rits Int'l Korea LLC, n.d.; Excellence Engineering Corporation, 2015; and Sölken, n.d.).

DOI: 10.1201/9781003328124-7

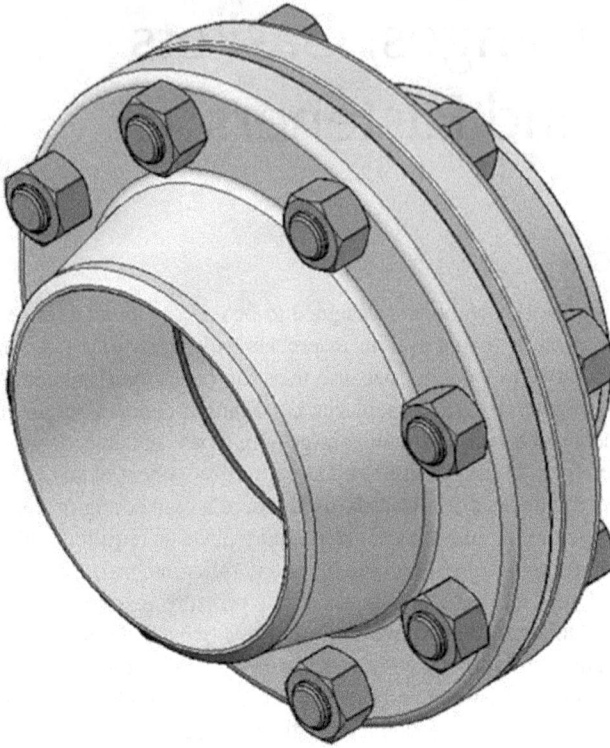

FIGURE 7.2 Typical flange joint (Sölken, n.d.).

7.1.1 BOLTED FLANGE CONNECTIONS

An acceptable, satisfactory and leak-proof bolted flange connection comprises of many components, characteristics and processes. The components involved are flanges, gaskets and bolts/nuts and characteristics concerned are pressure, temperature and fluid medium handled by piping. Apart from proper selection of flanges, gaskets and bolts based on the above aspects, the process of assembling flange connection also plays a vital role in obtaining a satisfactory, leak-tight flange connection. The process of making a flange connection includes proper placement of the designated gasket, tightening of bolts in pre-specified sequence, and that too to the specified torque for flange/gasket combination. The typical flange connection is shown in Figure 7.2.

7.1.2 FLANGE RATING

Flange rating indicates the maximum pressure permitted by pressure piping code for flange under consideration for the specific temperature range for which it is designed, and also the entire piping in that loop. Flanges are sized according to pressure ratings established by the American National Standards Institute (ANSI). Many times, these pressure ratings are called pound ratings. Flanges are generally available in seven ratings, 150#, 300#, 400#, 600#, 900#, 1500# and 2500# made from a variety of materials.

The flange selection as to the pound rating is decided based on the pressure and temperature of the piping system with safety margins as specified in design specification applicable to the piping segment in question.

7.1.3 FLANGE FACING

Face of flange is the mating surface of the flange with yet another flange of another piping spool, a valve, a vessel or a pump as the case may be. Face of the flange is machined smooth (with varying degrees of roughness) to make it leak-proof when assembled with gasket in between at recommended bolting torque according to pre-decided sequence.

Many types of flange faces are commercially produced and three main types predominantly in use are considered here:

- Flat face (FF)
- Raised face (RF)
- Ring-type joint (RTJ)

Because of differences in configurations and dimensions, RTJ and RF flanges vary considerably, whereas differences between RF and FF types are only marginal.

- **Flat Face**
  As evident from the name, face of flange is flat and level in FF flanges and is commonly available in 150# and 300# ratings. Apart from normal use, this type of flange is extensively used as a connection to cast iron flanges with pound ratings 125# and 250# found on some valves and mechanical equipment like pumps from yesteryears. As brittleness is one of the major drawbacks associated with cast iron, the full surface contact provided by FF flange reduces probability of cracking of cast iron flange of mating equipment.
- **Raised Face**
  RF type of flanges is the most extensively used type and hence is commercially available in all the seven-pound ratings mentioned earlier. As evident from the name, flange has a distinct RF for mating with the connecting flange. RF thickness varies with flange rating. For pound ratings 150# and 300#, the RF thickness is 1.6 mm (1/16″) and that for 600# and 900# pound rating is 6 mm (1/4″). Serrations with specified roughness are often provided on the RF to ensure positive grip of the gasket and leak-proofing upon tightening of mating flanges against gasket with specified torque and sequence.
- **Ring-Type Joint**
  RTJ flange in common parlance is also known as ring joint flange. Compared to flat gaskets used as a sealing component, RTJ flanges use a round metallic ring with either oval or octagonal cross section.

Ring-type gaskets are available in a few other standard configurations, developed for specific applications.

While tightening bolts, metal ring gets compressed inside groove provided on flange face, creating a continuous sealing line of surface. This is the costliest type of flange connection but very effective for high-pressure piping systems. In this design, apart from bolting load, internal pressure of fluid inside also acts favourably in sealing the joint. Though this type of flanges is available for all pound ratings, flanges with RTJ faces are normally used in piping systems rated 400# and higher.

7.1.4 FLANGE TYPES

The innumerable types of uses paved way for the development of a wide variety of flanges, each with its own advantages and disadvantages. The wide range of availability makes selection between these types tricky from an economic point of view and hence to be done judiciously to control cost of project.

For the sake of completeness, a brief overview of following most common types of flanges is provided below (Figure 7.3):

- Slip On
- Lap Joint
- Weld Neck
- Blind
- Threaded
- Socket Weld
- Orifice

7.2 FASTENERS FOR FLANGED CONNECTIONS

Bolts hold mating flanges (in piping, vessels or any other rotating equipment) together and provide the force required to seal the joint. Pressure rating of a flange shall determine the dimensions, spacing and number of bolts required. As nominal pipe size and pressure ratings change, bolt circle diameter, spacing and number of bolts change. ANSI standards require all flanges straddle in either horizontal, vertical or north–south centrelines of pipe and equipment in erected piping, unless otherwise indicated in drawing. To ensure that bolt holes on flanges, nozzles or valves align properly, holes are equally spaced along specified bolt circle diameter, with number of bolts in multiples of 4, starting with a minimum of 4.

7.2.1 TYPES OF BOLTS

Two types of bolts are used extensively in industry for flanged connections: stud bolts and hexagonal (Hex) bolts. Stud bolt is a threaded rod with two heavy hexagonal nuts, whereas hexagonal bolt has a hexagonal integral head at one end and tightened using nut provided at the other end.

Quantity of bolts required for a flange connection shall be determined by the number of bolt holes in designated flange. Diameter and length of bolts shall be dependent on the flange type and pressure (pound) rating of flange.

Length of bolts is defined in ASME B 16.5 and length in inches is equal to effective thread length measured parallel to the axis of the bolt.

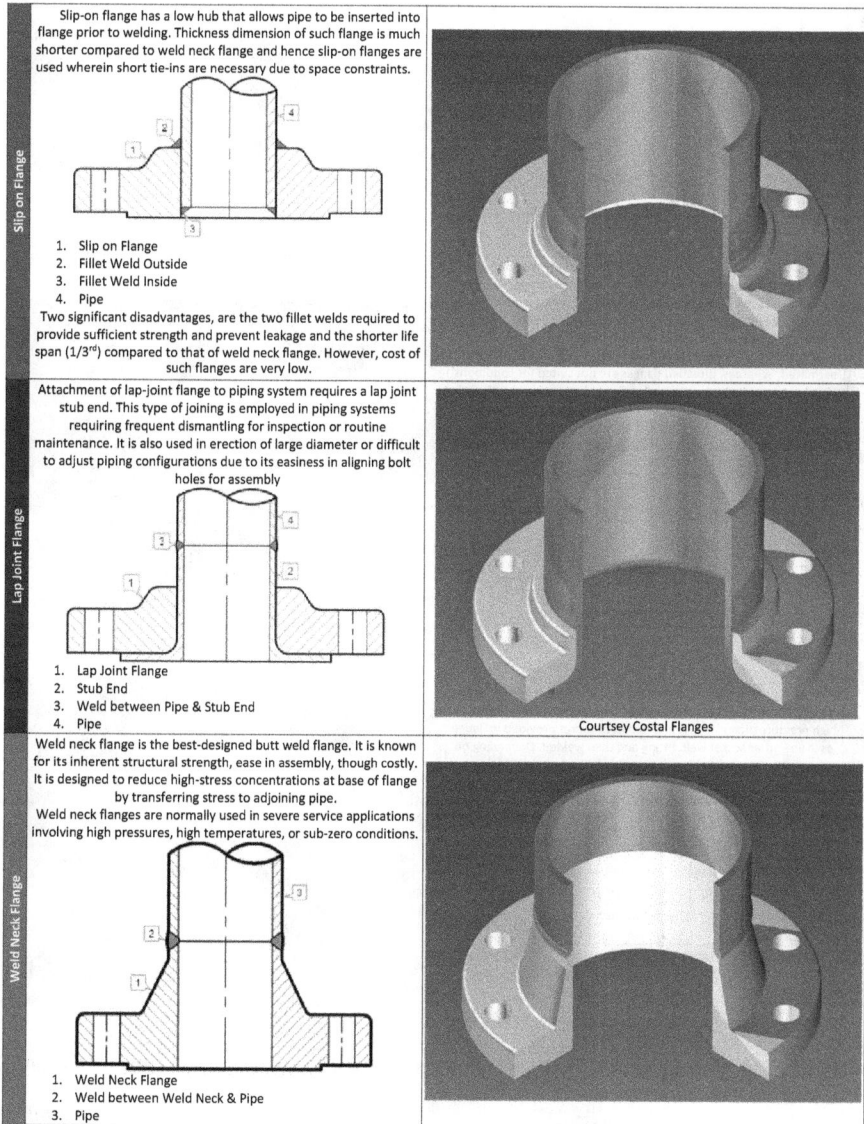

Slip on Flange — Slip-on flange has a low hub that allows pipe to be inserted into flange prior to welding. Thickness dimension of such flange is much shorter compared to weld neck flange and hence slip-on flanges are used wherein short tie-ins are necessary due to space constraints. 1. Slip on Flange 2. Fillet Weld Outside 3. Fillet Weld Inside 4. Pipe Two significant disadvantages, are the two fillet welds required to provide sufficient strength and prevent leakage and the shorter life span (1/3rd) compared to that of weld neck flange. However, cost of such flanges are very low.	
Lap Joint Flange — Attachment of lap-joint flange to piping system requires a lap joint stub end. This type of joining is employed in piping systems requiring frequent dismantling for inspection or routine maintenance. It is also used in erection of large diameter or difficult to adjust piping configurations due to its easiness in aligning bolt holes for assembly 1. Lap Joint Flange 2. Stub End 3. Weld between Pipe & Stub End 4. Pipe	Courtsey Costal Flanges
Weld Neck Flange — Weld neck flange is the best-designed butt weld flange. It is known for its inherent structural strength, ease in assembly, though costly. It is designed to reduce high-stress concentrations at base of flange by transferring stress to adjoining pipe. Weld neck flanges are normally used in severe service applications involving high pressures, high temperatures, or sub-zero conditions. 1. Weld Neck Flange 2. Weld between Weld Neck & Pipe 3. Pipe	

FIGURE 7.3 Types of flanges (Sketches: Sölken, n.d.; Images: Coastal Flange, 2022).

(Continued)

Blind Flange		
Blind flange serves the purpose of a plug or cap. It is used to terminate the end of a piping system. Blind flanges have the face thickness of the mating flange, with a matching type of face surface and with same number of bolting. Blind flanges are also used to close a nozzle opening on static equipment. Being a bolted joint, the blind flange provides easy access for inspection and maintenance (especially when applied on vessels) unlike pipe caps which are often welded. 1. Blind Flange 2. Stud Bolt 3. Gasket 4. Mating Flange (Weld Neck)		

Threaded Flange		
This type of flange is similar to slip-on flange, with bore of flange threaded instead of plain bore to insert externally threaded pipe. Advantage of this type of joint is that it can be assembled without welding. Because of the same, threaded flanges are well-suited for extreme pressure regimes at atmospheric temperatures and in areas categorized as highly explosive, wherein welding is not usually permitted. Whereas, threaded flanges are not suited for conditions involving temperatures, especially where severe cyclic conditions exist, which may cause the joint to leak through threads. To circumvent this, seal welding is applied after tightening of connection pipe. However, feasibility to weld and its effectiveness is still questionable and depended on service conditions. 1. Threaded Flange 2. Threads 3. Pipe or Fitting		

Socket Weld Flange		
Socket weld flange is also similar to the slip-on flange and was originally developed for use on small diameter below 100 NPS (4") in high-pressure piping. Pipe is inserted in to socket provided in flange as in any other socket weld fitting and then welded. Depending on feasibility, internal weld is also provided, which makes the joint stronger and cause less corrosion due to low possibility for stagnancy at the socket. 1. Socket Weld Flange 2. Fillet Weld Outside 3. Pipe X Expansion Gap		

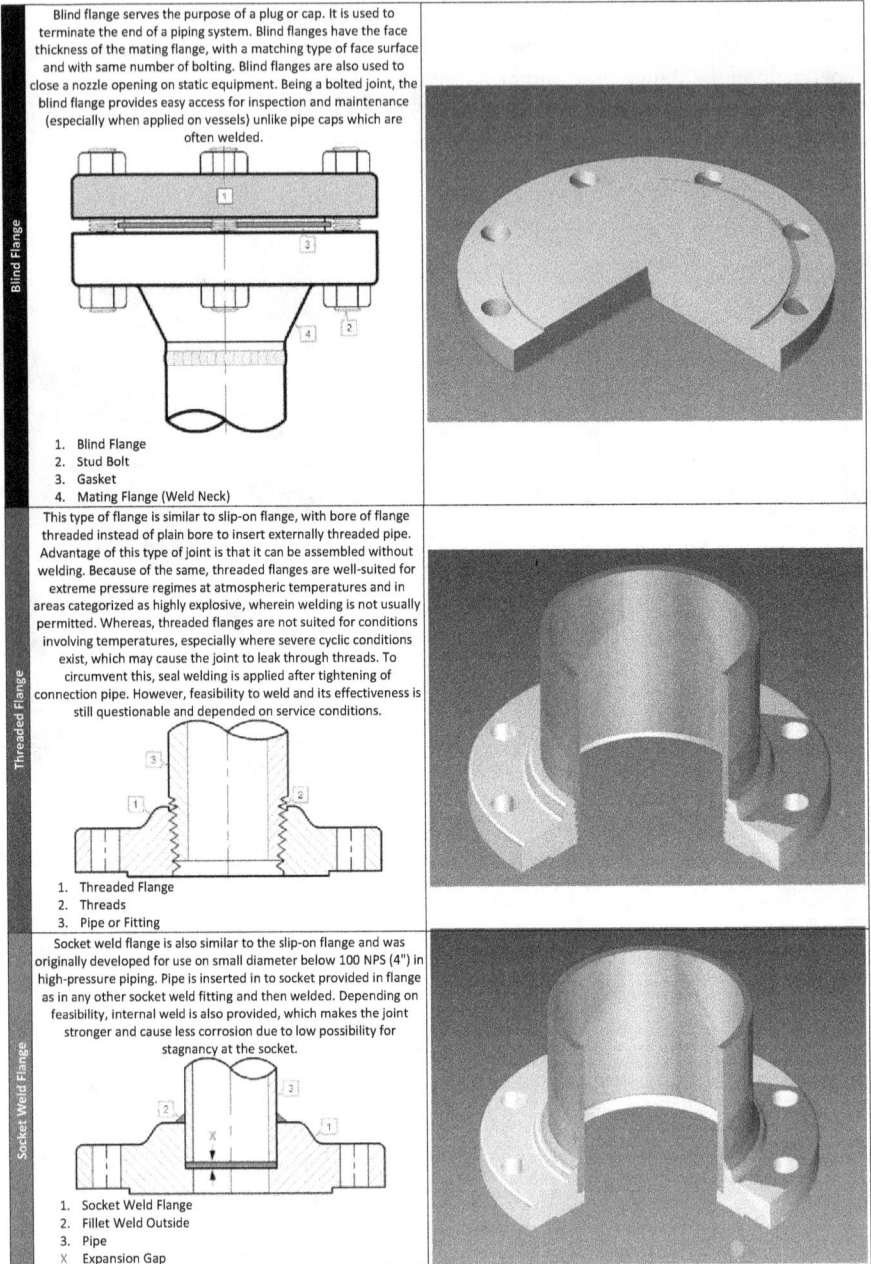

FIGURE 7.3 (*Continued*) Types of flanges (Sketches: Sölken, n.d.; Images: Coastal Flange, 2022).

(*Continued*)

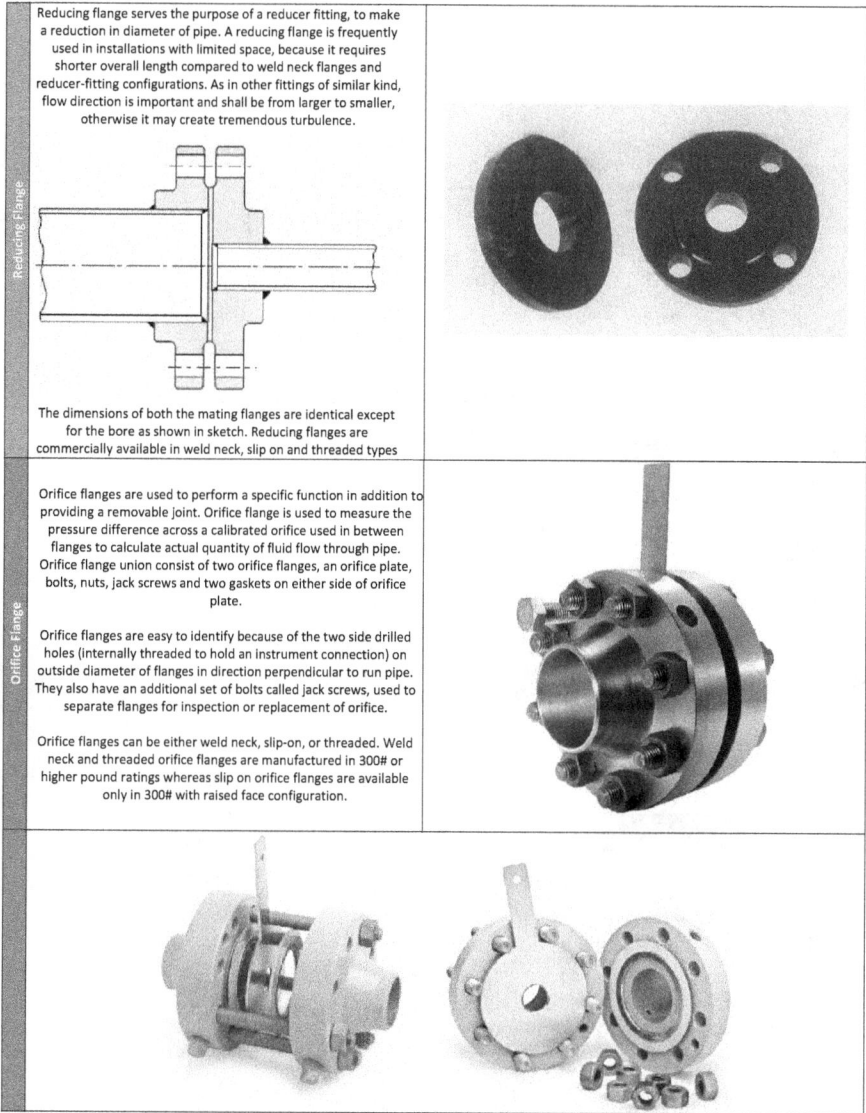

Reducing flange serves the purpose of a reducer fitting, to make a reduction in diameter of pipe. A reducing flange is frequently used in installations with limited space, because it requires shorter overall length compared to weld neck flanges and reducer-fitting configurations. As in other fittings of similar kind, flow direction is important and shall be from larger to smaller, otherwise it may create tremendous turbulence.

The dimensions of both the mating flanges are identical except for the bore as shown in sketch. Reducing flanges are commercially available in weld neck, slip on and threaded types

Orifice flanges are used to perform a specific function in addition to providing a removable joint. Orifice flange is used to measure the pressure difference across a calibrated orifice used in between flanges to calculate actual quantity of fluid flow through pipe. Orifice flange union consist of two orifice flanges, an orifice plate, bolts, nuts, jack screws and two gaskets on either side of orifice plate.

Orifice flanges are easy to identify because of the two side drilled holes (internally threaded to hold an instrument connection) on outside diameter of flanges in direction perpendicular to run pipe. They also have an additional set of bolts called jack screws, used to separate flanges for inspection or replacement of orifice.

Orifice flanges can be either weld neck, slip-on, or threaded. Weld neck and threaded orifice flanges are manufactured in 300# or higher pound ratings whereas slip on orifice flanges are available only in 300# with raised face configuration.

FIGURE 7.3 (*Continued*) Types of flanges (Sketches: Sölken, n.d.; Images: Coastal Flange, 2022).

7.2.2 HEXAGONAL (HEX) NUTS

Dimensional data for nuts are defined in ASME B 18.2.2, whereas threading is as per ASME B 1.1 (Figure 7.4).

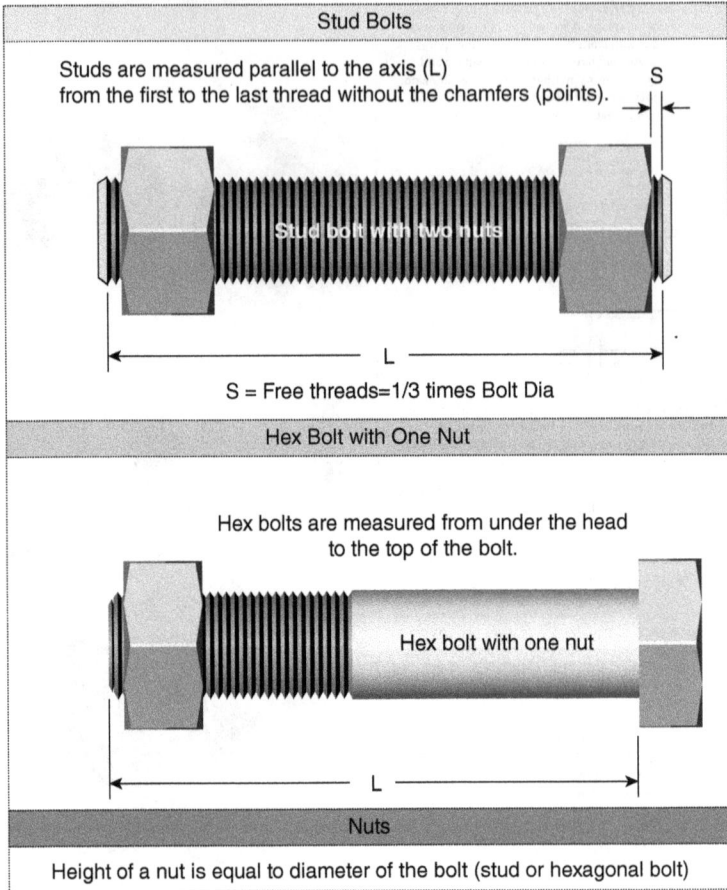

FIGURE 7.4 Stud/bolt with nuts (Sölken, n.d.).

7.2.3 THREADS OF STUD BOLTS

Bolt threading is defined in ASME B 1.1 Unified Inch Screw Threads (UN and UNR Thread Form). The most commonly used thread profile is the symmetrical "V", with an included angle of 60°. This form is widely used in unified thread (UN, UNC, UNF, UNRC, UNRF) form as ISO/metric threads. Symmetrical threads are easy to produce and inspect and are used extensively in all general-purpose fasteners.

Thread series cover designations of diameter/pitch combinations that are measured by the number of threads per inch (TPI) applied to a single diameter.

7.2.4 STANDARD THREAD PITCHES

- Coarse thread series (UNC/UNRC) are the most widely used thread system in bolts and nuts. Coarse threads are used for threads in low-strength materials such as iron, mild steel, copper and softer alloys, aluminium, etc.

- Fine thread series (UNF/UNRF) are commonly used in precision applications which use higher strength material for bolting.
- 8-Thread series (8UN) is the specified thread forming method for several ASTM standards including A193 B7, A193 B8/B8M and A320. This series is mostly used for diameters of 1″ and above.

7.2.5 Coatings on Fasteners

There are many finishes or coating applied to fasteners: some are corrosion protective, some are decorative or there may be no added coating at all. Coatings on fasteners are covered in many international standards and specifications.

- **Plain Finishes (Black – Self-Colour)**
  This essentially is "as produced" finish on carbon steel products having an oil residue applied which provides some shelf life but no real corrosion protection when in use. Nowadays, less than 20% of carbon steel fasteners would be purchased in plain finish. Stainless steel, brass and other non-ferrous materials protect themselves through a reaction of the surface to oxygen, creating a protective oxide film.
- **Corrosion Protective Coatings (Zinc Plating)**
  Most economic and common fastener finish, comprising a thin coating of zinc applied either by electroplating or mechanically. A shiny silver-grey appearance, it will normally be enhanced by a chemical chromate passivation conversion which applies a harder surface film. This can be clear (bluish tinge) or iridescent yellow which is thicker and gives marginally better protection.
  Clear is referred to as zinc, zinc clear, and blue zinc. Yellow is referred to as zinc plate gold (ZPG), zinc yellow chromate (ZYC), zinc dichromate, zinc yellow pass, etc.
- **Cadmium Plating**
  Formerly a popular electroplated or mechanically applied finish, looking like but giving slightly better protection than zinc and providing increased lubricity. Very seldom used today due to its toxicity and environmental non-acceptability. If specified, it is usually through habit, error or ignorance and possible confusion with zinc.
- **Galvanizing**
  A very heavy coating of zinc applied by hot dipping in a bath of molten zinc, then centrifuge spinning for even distribution and removal of excess, or mechanically cold welding a zinc powder in a barrel rumbling process. The hot dip finish is rougher and duller than electroplated finishes but because of the thickness achieved, gives considerably enhanced protection. Often it is wax coated to provide assembly lubrication.
- **Phosphating**
  A thin, dull grey phosphate coating obtained by insertion in a solution containing phosphoric acid. Gives a lower level of protection than zinc in mild environments, but gives an excellent base for painting or organic lubrication. Often used in automotive industry.

7.2.6 COATING THICKNESS

With sacrificial protective coatings, thicker the deposit, the longer the protection; however, there are practical and economic limitations to the thickness applied.

- **Zinc Electroplating** can provide thicknesses from a negligible flash of colour, for appearance, through normal commercial coatings of 3–5 μm, to specified heavy coatings up to 12 μm (0.0005″). Electroplating does not give an even cover; thicker concentration of deposit occurs on corners, points, thread crests and thinner concentrations on thread flanks and roots. This may cause thread galling on coatings above 8 μm average and adjustment by over-tapping of the nut may be required.
- **Hot Dip Galvanizing** shall allow much heavier coatings, normal commercial coating is approximately 50 μm, which necessitates over-tapping of mating thread and is the maximum practical to avoid serious compromise of fastener's strength. Unlike electroplating, concentration of deposits is in thread roots and internal corners. For this reason, thread diameters of less than M10 are not normally galvanized unless a subsequent light re-roll of the thread is performed. Nuts supplied with galvanized bolts will have over-tapped threads to allow for galvanized build-up on bolt threads and to reduce assembly galling.
- **Mechanical Coating** will result in a more even deposit and point of over-tapping will be raised above 15 μm. Comparable thicknesses can be achieved but costs are generally much higher.

7.2.7 LIFE EXPECTANCY OF COATINGS

Service life of coatings prior to first signs of corrosion will vary considerably depending upon thickness and environment. Experience suggests the following:

	Coatings	
Environment	**Heavy Zinc and Yellow Chromate** **12 μm (Av.)**	**Hot Dip Galvanized** **50 μm Minimum**
Heavily Polluted Industrial Area	Less than 1 year	Less than 5 years
Coastal Area	Less than 2 years	Less than 30 years
Inland Rural Area	4+ years	40+ years
Dry Indoor Area	20+ years	Not normally used

7.3 GASKETS

The primary purpose of any flanged assembly is to connect piping systems in such a manner as to produce a leak-free and durable connection, safe to the environment in the long run. Hazardous and combustible materials and extreme pressure and temperature require the utmost in safety precaution. Creating a leak-proof seal between two connecting metal surfaces in an industrial setting is almost impossible and gaskets placed in between mating flanges perform this vital function.

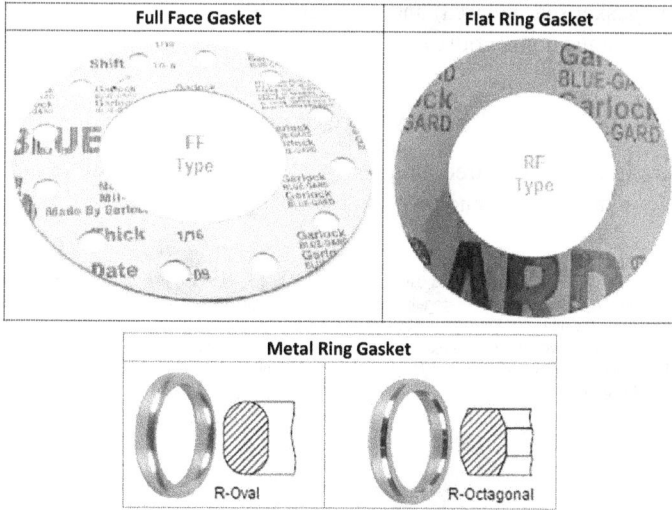

FIGURE 7.5 Classification of gaskets based on configuration (Garlock, n.d.).

Using a gasket material softer than two adjoining flanges is an excellent way to eliminate the possibility of fluid escape. Gaskets can be made of materials such as asbestos, rubber, neoprene, Teflon, lead, copper and so on. When bolts are tightened and flange faces are drawn together, gasket material will conform to any imperfections on flange faces to create a uniform seal.

There are three types of gaskets that can be found in piping systems. They are full face, flat ring and metal ring. Full face gaskets are used on FF flanges. Flat ring gaskets are used on RF flanges. Metal rings are used on RTJ flanges.

7.3.1 CLASSIFICATION BASED ON CONFIGURATION

A gasket's thickness shall be accounted for when dimensioning piping system (Figure 7.5). Thickness varies with the type and pressure ratings used, and can be seen in varying thicknesses starting from 1.5 mm (1/16″). At every occurrence of a flange bolting to a nozzle, two flanges joining one another, two valves joining one another or a flange connecting to a valve, a gasket thickness shall be added to the length of piping. Though this occupies only a very small space (a few mm), gaskets cannot be ignored. This dimension will vary depending on the size and pound rating of the flange. This is an important consideration to keep in mind when dimensioning piping runs that have RTJ connections. For each instance of a gasket or ring, gap spacing dimension must be reflected in the dimensions shown on a piping drawing. *Tick* marks are often used to indicate each location where a gasket or ring gap has been included in the dimensioning of piping configuration.

7.4 GASKETS FOR FLANGED CONNECTIONS

As mentioned, gasket is the component in a flange connection used to create a static seal between two flange faces under a wide range of operating conditions (pressure and temperature). Gaskets fill the microscopic spaces and irregularities on flange faces, and then it forms a seal to hold liquids and gases without any leak. Defect-free flange faces (gasket seating area) and a defect-free gasket are essential prerequisites for a leak-free flange connection.

7.4.1 TYPES OF GASKETS BASED ON MATERIAL OF CONSTRUCTION

Gaskets are classified into three main categories based on material of construction: non-metallic, semi-metallic and metallic types (Figure 7.6).

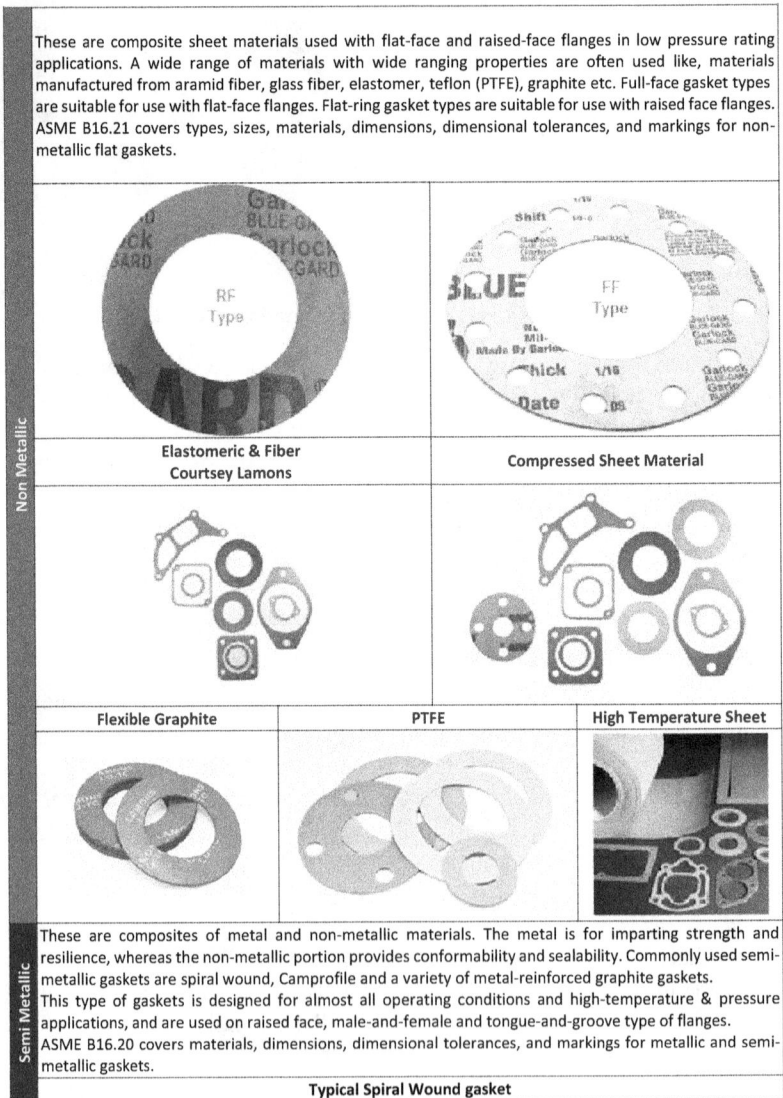

FIGURE 7.6 Types of gaskets based on material of construction (*Spiral wound gaskets*: Sölken, n.d.; *Spiral wound stacked*: Ace Alpha International FZE, 2015; *Kammprofile*: Seal & Design, 2022; *Corrugated metal*: Ferguson, 2023; *Jacketed metal*: Metal Gaskets, n.d.; *High temperature*: Seal & Design, 2022; *Defender RG Sealing*: Lamons, 2023; and *RTJ Gaskets*: AMG Sealing, n.d.).

(*Continued*)

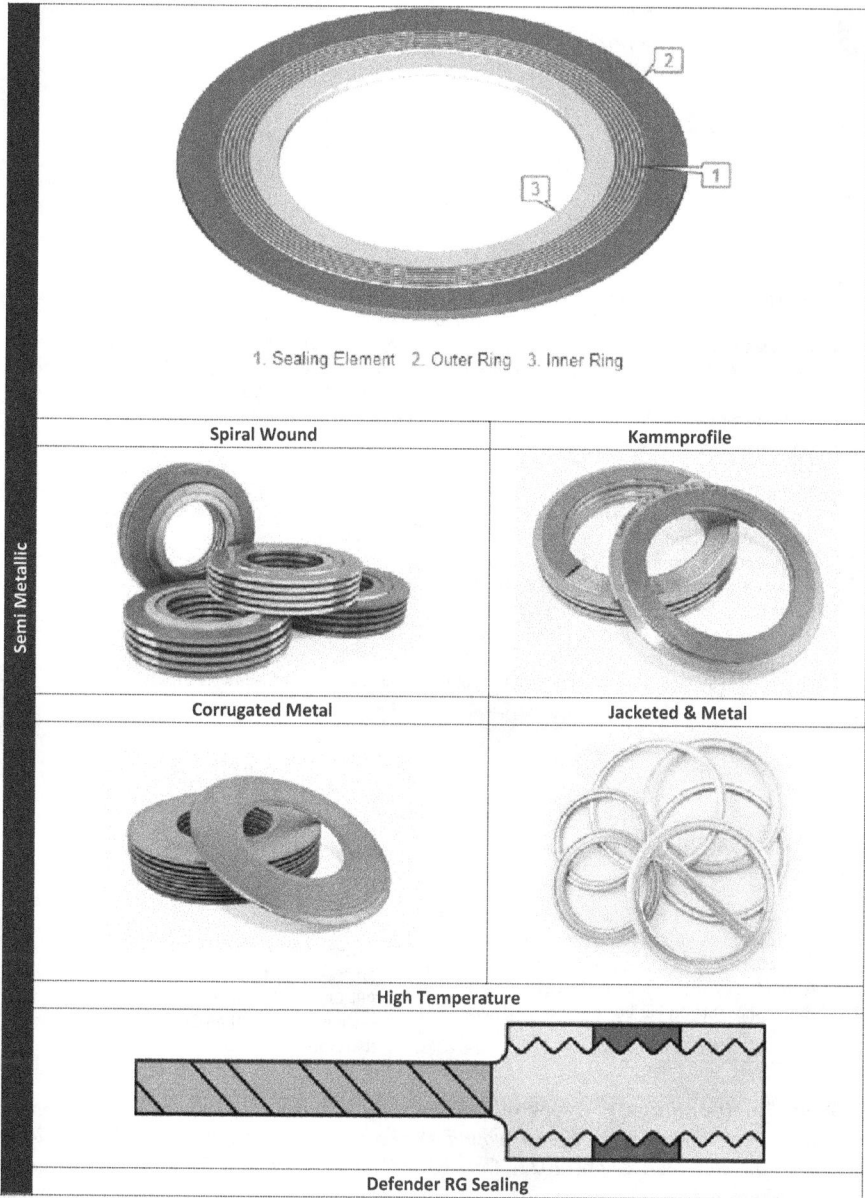

1. Sealing Element   2. Outer Ring   3. Inner Ring

Spiral Wound

Kammprofile

Corrugated Metal

Jacketed & Metal

Semi Metallic

High Temperature

Defender RG Sealing

FIGURE 7.6 (Continued) Types of gaskets based on material of construction (*Spiral wound gaskets*: Sölken, n.d.; *Spiral wound stacked*: Ace Alpha International FZE, 2015; *Kammprofile*: Seal & Design, 2022; *Corrugated metal*: Ferguson, 2023; *Jacketed metal*: Metal Gaskets, n.d.; *High temperature*: Seal & Design, 2022; *Defender RG Sealing*: Lamons, 2023; and *RTJ Gaskets*: AMG Sealing, n.d.).

(*Continued*)

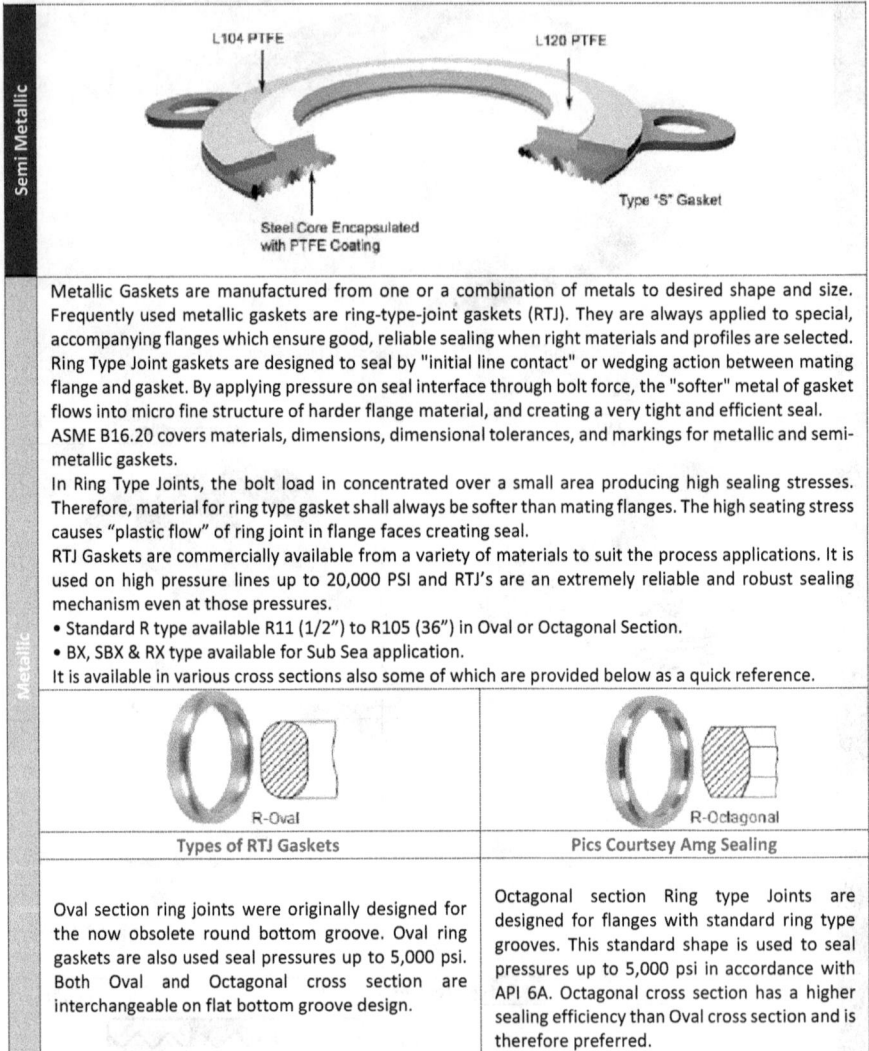

Semi Metallic

L104 PTFE   L120 PTFE

Steel Core Encapsulated with PTFE Coating   Type 'S' Gasket

Metallic

Metallic Gaskets are manufactured from one or a combination of metals to desired shape and size. Frequently used metallic gaskets are ring-type-joint gaskets (RTJ). They are always applied to special, accompanying flanges which ensure good, reliable sealing when right materials and profiles are selected. Ring Type Joint gaskets are designed to seal by "initial line contact" or wedging action between mating flange and gasket. By applying pressure on seal interface through bolt force, the "softer" metal of gasket flows into micro fine structure of harder flange material, and creating a very tight and efficient seal.

ASME B16.20 covers materials, dimensions, dimensional tolerances, and markings for metallic and semi-metallic gaskets.

In Ring Type Joints, the bolt load in concentrated over a small area producing high sealing stresses. Therefore, material for ring type gasket shall always be softer than mating flanges. The high seating stress causes "plastic flow" of ring joint in flange faces creating seal.

RTJ Gaskets are commercially available from a variety of materials to suit the process applications. It is used on high pressure lines up to 20,000 PSI and RTJ's are an extremely reliable and robust sealing mechanism even at those pressures.

• Standard R type available R11 (1/2") to R105 (36") in Oval or Octagonal Section.
• BX, SBX & RX type available for Sub Sea application.

It is available in various cross sections also some of which are provided below as a quick reference.

R-Oval	R-Octagonal
Types of RTJ Gaskets	Pics Courtsey Amg Sealing
Oval section ring joints were originally designed for the now obsolete round bottom groove. Oval ring gaskets are also used seal pressures up to 5,000 psi. Both Oval and Octagonal cross section are interchangeable on flat bottom groove design.	Octagonal section Ring type Joints are designed for flanges with standard ring type grooves. This standard shape is used to seal pressures up to 5,000 psi in accordance with API 6A. Octagonal cross section has a higher sealing efficiency than Oval cross section and is therefore preferred.

FIGURE 7.6 (Continued) Types of gaskets based on material of construction (*Spiral wound gaskets*: Sölken, n.d.; *Spiral wound stacked*: Ace Alpha International FZE, 2015; *Kammprofile*: Seal & Design, 2022; *Corrugated metal*: Ferguson, 2023; *Jacketed metal*: Metal Gaskets, n.d.; *High temperature*: Seal & Design, 2022; *Defender RG Sealing*: Lamons, 2023; and *RTJ Gaskets*: AMG Sealing, n.d.).

(Continued)

риの

BX Ring Type Joints are designed for pressures up to 20,000 psi, suitable only for use with API type BX flanges and grooves. Gasket has a square cross section with beveled corners. Average diameter of ring joint is slightly greater than that of flange groove. This way, when the ring joint is seated, it stays pre-compressed by outside diameter, thus creating high seating stress.

RX Ring Type Joints are designed for pressures up to 5,000 psi, they are pressure activated ring joints designed to use the fluid pressure to increase sealability. The outside sealing surface of the ring joint makes the initial contact with the flange. As the internal pressure rises the contact pressure between ring joint and flange also increases. This is sometimes referred to as a pressure activated ring joint due to the shape of the gasket. High seating pressures are created increasing the sealability. This design characteristic makes the RX more resistant to vibrations, pressure surges and shocks that occur during oil well drilling.

IX Seal rings are designed to be used in Norsok Compact Flange Connections (CFC). Available in different materials, IX rings are supplied with colour coded PTFE coating to denote base material supplied. Unlike all of the other RTJ's supplied,

FIGURE 7.6 (Continued) Types of gaskets based on material of construction (*Spiral wound gaskets*: Sölken, n.d.; *Spiral wound stacked*: Ace Alpha International FZE, 2015; *Kammprofile*: Seal & Design, 2022; *Corrugated metal*: Ferguson, 2023; *Jacketed metal*: Metal Gaskets, n.d.; *High temperature*: Seal & Design, 2022; *Defender RG Sealing*: Lamons, 2023; and *RTJ Gaskets*: AMG Sealing, n.d.).

(*Continued*)

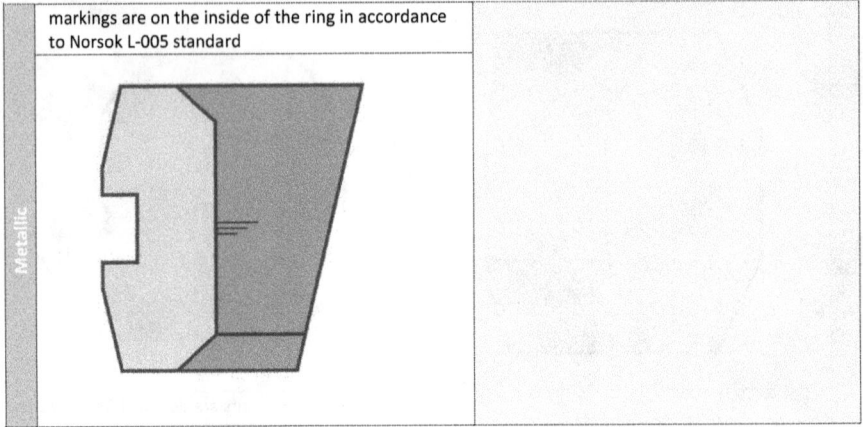

markings are on the inside of the ring in accordance to Norsok L-005 standard

Metallic

FIGURE 7.6 (*Continued*) Types of gaskets based on material of construction (*Spiral wound gaskets*: Sölken, n.d.; *Spiral wound stacked*: Ace Alpha International FZE, 2015; *Kammprofile*: Seal & Design, 2022; *Corrugated metal*: Ferguson, 2023; *Jacketed metal*: Metal Gaskets, n.d.; *High temperature*: Seal & Design, 2022; *Defender RG Sealing*: Lamons, 2023; and *RTJ Gaskets*: AMG Sealing, n.d.).

8 Valves

Though valve plays a vital role in the life of an individual, many very seldom notice it; the obvious example is the valves in the human heart. Leaving apart the human body, in day-to-day life as well, every one come across valves quite often, but still they may not notice it. Each time water is turned on and off in your washbasin or toilet, how many are really aware that this activity was the outcome of the functioning of a valve? Similarly, consider the cases of water faucet, cooking range, accelerator pedal of car, which are some of the most common applications in day-to-day life wherein valves are used.

Yet another widely seen but least recognized type of valve is the fire hydrant valves. Fire hydrants are connected to municipal water supply systems. Though the valves of fire hydrants are located below ground, they can be operated from an above-ground location when needed in emergency situations.

By classical definition, valve is the mechanical device used in piping system to control flow and pressure within the process system and hence one of the most essential components of a piping system that conveys liquids, gases, vapours, slurries, etc.

Different types of valves are available based on type of control device used like gate, globe, plug, ball, butterfly, check, diaphragm, pinch, pressure relief, control valves, etc. Each of these types has a number of models, each with different features and functional capabilities. Some valves are self-operated while others are operated either manually or with an actuator (pneumatic or hydraulic).

The following are the functions of a valve.

- Stopping and starting flow
- Reduce or increase flow
- Controlling direction of flow
- Regulating a flow or process pressure
- Relieve a pipe system of a certain pressure

Though, by definition, a valve is a device that controls the flow of a fluid, it can also control rate, volume, pressure and direction of flow of fluid in pipes. Use of valves is not limited to fluids alone; valves can also control liquids, gases, vapours, slurries or dry materials and are used to turn on or off, regulate, modulate or isolate the flow. The sizes of commercially manufactured valves (of standard configuration) range from fraction of an inch to as large as 12 feet or more. Similarly, complexity varies from simple brass valve used in water lines sold by local hardware shops to a precision-designed, highly sophisticated control valve made of exotic or noble metal alloys used in applications like highly corrosive fluids and nuclear reactors. Valves are also used to control the flow of a wide range of commodities ranging from light

DOI: 10.1201/9781003328124-8

gases to highly corrosive chemicals, superheated steam, toxic gases, abrasive slurries, radioactive materials and so on. Valves are designed to serve satisfactorily (with minimum maintenance) in all the above environments and are designed to withstand temperatures from cryogenic region to that of molten metal exceeding 1,500°F, that too within the pressure range from extreme vacuum to several thousand pounds per square inch (psi).

8.1   CLASSIFICATION OF VALVES

Valves can be classified in many ways like the movement of closure mechanism, pressure rating, material of construction and so on.

Valves are classified as follows based on the mechanical motion of the closure device:

- *Linear motion valves*: The valves in which the closure member, as in gate, globe, diaphragm, pinch and lift Check Valves, moves in a straight line to allow, stop or throttle the flow.
- *Rotary motion valves*: When the valve-closure member travels along an angular or circular path, as in butterfly, ball, plug, eccentric- and Swing Check Valves, the valves are called rotary motion valves.
- *Quarter turn valves*: Some rotary motion valves require approximately a quarter turn, 0° through 90°, motion of the stem to go to fully open position from a fully closed position or vice versa.

Mechanical movement of closure device in commonly used valve types:

Valve Types	Type of Motion		
	Linear	Rotary	Quarter Turn
Gate	√		
Globe	√		
Plug		√	√
Ball		√	√
Butterfly		√	√
Swing check		√	
Diaphragm	√		
Pinch	√		
Safety	√		
Relief	√		

8.2   CLASS RATINGS

The pressure–temperature ratings within which it is safe to operate a valve is designated by class numbers. The widely used ASME Specification B 16.34 for Valves – Flanged, Threaded and Welding End – defines the following:

Pressure Rating Designation is the number used to designate pressure–temperature rating of the valve. The standardized designations are Classes 150, 300, 600, 900, 1500, 2500, 4500, etc. Class 400 is also available but not much used in flanged-end valve designation and is regarded as an intermediate class designation.

The above pressure–temperature ratings designated by classes are further identified as Standard, Special or Limited Class with the following restrictions:

- Flanged-end valves shall be rated only as Standard Class.
- Class 4500 applies only to welding-end valves.
- A class designation greater than Class 2500 or a rating temperature greater than 538°C (1,000°F) applied to threaded-end valves is beyond the scope of B 16.34.
- Threaded and socket welding-end valves larger than NPS 2½ are beyond the scope of B 16.34.

8.3   COMMON TYPES OF VALVES

Valves are manufactured in a wide range of sizes, body configurations and pound (class) ratings to meet a wide variety of applications. As mentioned earlier, valves are also manufactured with varying types of end preparations that allow them to be connected easily with interconnecting components or piping. The valves are available with screwed, socket-weld, bevelled or flanged-end connections. Flanged valves are manufactured to have raised, flat or ring-type flange faces (Figure 8.1).

Gate Valve	Gate valve is the most frequently used valve in piping systems. It is a general service valve that is used primarily for on-off, non-throttling applications. When fully opened, gate valve creates minimal obstruction to flow. Gate valves control fluid flowing through pipe with a flat, vertical wedge, or gate, that slides up or down as the valve's hand wheel is turned. As hand wheel is rotated, wedge will slide through valve body to block or release flow. Since valve is designed to operate either fully opened or closed position, gate valve shall not be operated in a partially opened/closed position. A partially opened gate valve may accelerate erosion caused by fluid within pipe and may damage the valve seat in a short period of time. Turbulence created by flow of fluid through partially opened valve may also cause the wedge to vibrate creating a "chattering" noise.
Globe Valve	Globe valves are primarily used in situations where throttling of fluid is required. By simply rotating hand wheel, the rate at which fluid flows through the valve can be adjusted to any desired level. Having the valve seat parallel to line of flow is an important feature of globe valve. This feature makes globe valve efficient when throttling fluid flow as well as yielding minimal disc and seat erosion. This configuration, however, creates a large amount of resistance within the valve. Design of globe valve body forces the flow of fluid to change direction within the valve itself. This change in direction creates considerable pressure drop and turbulence. Globe valve is therefore not recommended when flow resistance and pressure drop are to be avoided.
Angle Valves	Like globe valve, angle valve is also used for throttling. As shown in Figure on right, flow entering and leaving the valve is at a 90° angle. If the pipe is required to take a 90° turn, angle valve is used to eliminate the need for a 90° elbow and additional fittings required to do so. Angle valves and globe valves are typically installed such that the fluid shall flow in an upward direction through valve body. This upward flow direction will maintain pressure under the disc seat, enabling easier operation and reduces the erosive action on seat and disc. For high temperature fluids like superheated steam, the flow direction is reversed. When the valve is closed, under normal flow direction, temperature on the lower side of the disc is significantly higher than that on the upper side. As the valve stem is on the upper side of the disc, it will be cooler and this temperature differential causes the valve stem to contract, thereby lifting the disc off the seat. This lifting action will result in the seat and disc faces being scored. In order to avoid this problem, valve manufacturers recommend installing globe and angle valves, with fluid flow from upper side in high temperature applications like super-heated steam. Such a flow direction shall keep pressure above the disc, forcing it into the seat and creating a tighter seal.
Check Valve	Check valves differ significantly from gate and globe valves. Check valves are designed primarily to prevent backflow. Backflow means the reversal of flow within pipe due to some process upsets. There are many designs of check valves, however, the two most common types are the Swing Check and the Lift Check Valves Check valves do not require hand wheels or any other device to control the flow of fluid. It uses either gravity or pressure of fluid to operate the valve. The swing check valves are often installed as a companion to gate valve, in lines where back flow need to be checked. As the name implies, this valve has a swinging gate that is hinged at top and opens as fluid is allowed to flow through the gate valve. When the gate valve disc is in open position, a clear flow path is created through the gate valve and swing arm of the swing check valve also shall be in full swing, providing a clear path for fluid with minimal turbulence and pressure drop within the valve. Pressure shall always be under the disc for the valve to function properly. When flow reverses, the pressure and weight of the fluid against the disc shall force the disc against the seat, stopping all backflow. Because of the above safeguard, check valves are often regarded as sort of safety device.

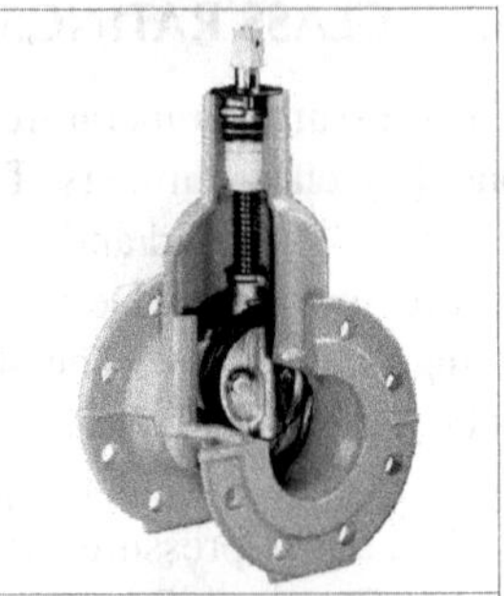

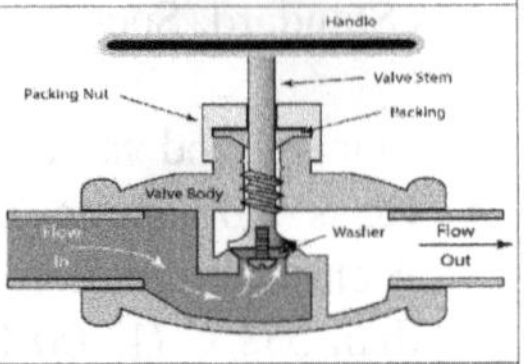

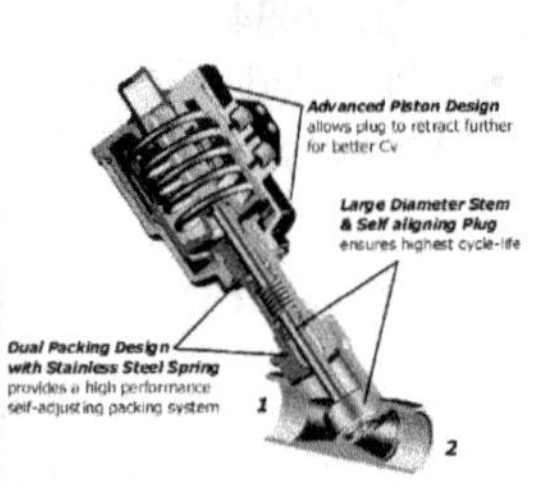

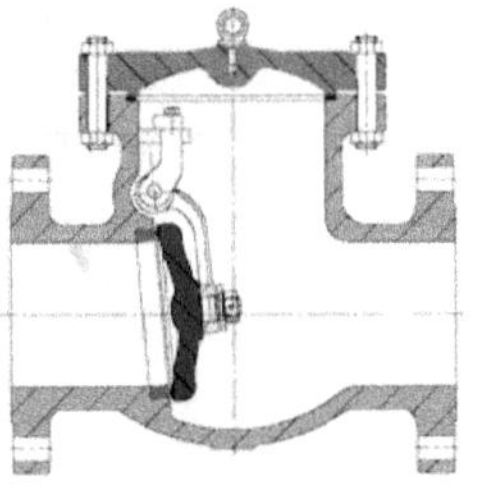

Swing Check Valve

FIGURE 8.1 Common types of valves and operators (*Gate*: VseSdelki, n.d.; *Globe*: Quora, n.d.; *Angle*: Fenghua Fly Automation Co., n.d.; *Check (1)*: Waterworld, n.d.; *Lift (2)*: Waterworld, n.d.; *Ball*: NewsLiner, 2023; *Plug*: Piping Engineering, 2013; *Butterfly*: Sölken, n.d.; *Relief Valave*: Piping Engineering, 2013; *Operator (1)*: Garlock, n.d.; *Operator (2)*: Teksal Safety, 2022; *Actuator (1)*: Kent Introl, n.d.; and *Actuator (2)*: VirtualExpo Group, n.d.).

(Continued)

Check Valve	The lift check valve is often installed along with globe valve. Lift check valve has a body style similar to globe valve. As flow enters the valve, disc is lifted up off the seat to allow flow to pass. Just as in the case of globe valve, there is significant turbulence and pressure drop associated with lift check valves. There are two types of lift check valves, *horizontal* and *vertical*. Both of these valves use either a disc or ball and force of gravity to close the valve in the event of a reverse flow. Horizontal lift check valve has a seat that lies parallel to flow, requiring an S-type body style that mandates valve to be installed in horizontal position only and have flow that enters from below seat and exit above seat. Flow entering valve raises disc or ball off the seat permitting fluid to pass through closure device. Vertical lift check valve is designed to work automatically on flow that is traveling in an upward direction only. Similar to horizontal lift check, vertical lift check valves use a disc or ball that raises off the seat when fluid flows upward through valve. When flow stops, gravity shall reseat the disc or ball preventing backflow. This check valve requires outlet end of valve to be installed in check valve *"up"* position always. Some manufacturers refer to lift check valves that employ the use of a ball as a *ball check* valve.	Lift Check Valve
Ball Valves	Ball valve is an inexpensive alternative to other valves and is extensively used on account of this feature. Ball valves use a metal ball with a hole bored through center, sandwiched between two seats to control flow. Ball valves are also capable of throttling gases and vapors and are especially useful for low flow situations. These valves are quick opening type and provide a very tight closure on hard to hold fluids (gases). Ball valves do not use a hand wheel but instead use a wrench to control flow, a 90° turn of which opens or closes the valve. This simple design yields a non-sticking operation that produces minimal pressure drop when valve is in its full-open position.	
Plug Valves	Unlike other valves, plug valve uses either a hand wheel or wrench to operate. Plug valves provide a tight seal against hard to hold commodities and requires a minimum amount of space for installation. Unlike ball valve, plug valve uses a tapered wedge rather than a ball to create a seal. This wedge, or plug, has an elongated opening, which when placed in *"open"* position, allows fluid to pass through valve. Plug is the only movable part of valve and its tapered shape assures positive seating Plug valves are designed with etched grooves along tapered plug to permit a lubricant to seal and lubricate internal surfaces as well as to provide a hydraulic jacking force to lift plug within the body, thus permitting easy operation. The clear and open passageway through valve body provides little opportunity for scale or sediment to collect. In fact, plug seats get cleaned well, as plug is rotated and foreign debris are wiped from plug's external surfaces. These valves, however, do require constant lubrication to maintain a tight seal between plug and body.	
Butterfly Valves	Butterfly valve has a unique body style unlike other valves. Butterfly valves uses a circular plate or wafer operated by a wrench to control the flow. A 90° turn of wrench moves wafer from fully open position to fully closed position. Wafer remains in the stream of flow and rotates around a shaft connected to the wrench. As the valve is being closed, wafer rotates to become perpendicular to direction of flow and acts as a dam to reduce or stop flow. When wrench is rotated back to original position, wafer aligns itself with direction of flow and allows fluid to pass through valve. Butterfly valves have minimal turbulence and pressure drop. They are good for on-off and throttling service and perform well when controlling large flow amounts of liquids and gases. However, these valves do not normally create a tight seal and must be used in lowpressure situations or where some leakage is permissible.	

FIGURE 8.1 (*Continued*) Common types of valves and operators (*Gate*: VseSdelki, n.d.; *Globe*: Quora, n.d.; *Angle*: Fenghua Fly Automation Co., n.d.; *Check (1)*: Waterworld, n.d.; *Lift (2)*: Waterworld, n.d.; *Ball*: NewsLiner, 2023; *Plug*: Piping Engineering, 2013; *Butterfly*: Sölken, n.d.; *Relief Valave*: Piping Engineering, 2013; *Operator (1)*: Garlock, n.d.; *Operator (2)*: Teksal Safety, 2022; *Actuator (1)*: Kent Introl, n.d.; and *Actuator (2)*: VirtualExpo Group, n.d.).

(Continued)

Relief Valves

Relief valves have a purpose quite different from that of all types of valves discussed previously. They are designed to release excessive pressure that builds up in equipment and piping systems, to prevent major damage to equipment, and more importantly, injury to workers. Relief valves shall release elevated pressures before they become extreme and uses a steel spring as a means to automatically open when pressures reach unsafe levels. These valves can be adjusted and regulated to "*pop off*" when internal pressures exceed predetermined settings. Once internal pressures return to operational levels, relief valve closes.

Another valve that performs the same basic function as relief valve is the *pressure safety valve*. Although similar in design and appearance, the two valves operate differently. Relief valves are used in piping systems that service liquids and are designed to open proportionally, meaning, as pressure of fluid increases so does the opening of valve. This means that at higher pressures, opening will be larger. The pressure safety valve, however, is used with higher pressure fluids such as steam and gas. Pressure safety valves designed to open completely when internal pressures exceed the setting for which internal spring has been set. As with the relief valve, once internal pressures return to operational levels, valve will close itself.

BONNET VENT

BELLOWS

Control Valves

Control valve is an automated valve that can make precise adjustments to regulate and monitor any fluid flowing through a piping system. Most common valve body style used as a control valve is the globe valve. Although many other body styles are used, globe valve provides most effective means to regulate and control flow. Control valves use signals received from instruments positioned throughout piping system to automatically make adjustments that regulate fluid flow within pipe. Though control valves can perform many functions, they are typically used to control the flow of a commodity within a pipe or to limit its pressure.

Control valves shall be arranged within a run of pipe so that they can be easily operated. To achieve this, "*control valve manifolds*" are configured. Control valve manifolds make control valves readily accessible to plant workers.

Valve Operators

A *valve operator* is a mechanism used to operate valve in situations where standard hand wheel is insufficient to operate valve. Manual operators, like levers, gears, or wheels are used to ease operation of valve. Bevel, spur, and worm gears supply hand wheel with a greater mechanical advantage to open, close, or throttle commodity within pipe. If a valve is installed at a height that is out of a worker's reach, a *chain operator* is often used. Chain operator is a sprocket-like attachment bolted to a valve's hand wheel. A looped chain is passed through the sprocket and is hung down to a height that is accessible to a worker. This allows a worker to operate valve without using a ladder or moveable scaffold.

Actuators

Automatic operators known as actuators use an external power supply to provide necessary force required to operate valves. Automatic actuators use hydraulic, pneumatic, or electrical power as their source for operating valves. Hydraulic and pneumatic actuators use fluid or air pressure, respectively, to operate valves needing linear or quarter-turn movements. Electric actuators have motor drives that operate valves requiring multiple turn movements.

Automatic actuators are often provided on control valves that requiring frequent throttling or those valves found in remote and inaccessible locations within a piping system. Another common application for automatic actuators is on control valves of large diameter pipes. These valves are often so large that a worker simply cannot provide the torque required to operate valve. Also, in an effort to protect workers, control valves located in extremely toxic or hostile environments are outfitted with automatic actuators. Additionally, in emergency situations, valves that must be immediately shut down are also operated automatically.

FIGURE 8.1 (*Continued*) Common types of valves and operators (*Gate*: VseSdelki, n.d.; *Globe*: Quora, n.d.; *Angle*: Fenghua Fly Automation Co., n.d.; *Check (1)*: Waterworld, n.d.; *Lift (2)*: Waterworld, n.d.; *Ball*: NewsLiner, 2023; *Plug*: Piping Engineering, 2013; *Butterfly*: Sölken, n.d.; *ReliefValave*: Piping Engineering, 2013; *Operator (1)*: Garlock, n.d.; *Operator (2)*: Teksal Safety, 2022; *Actuator (1)*: Kent Introl, n.d.; and *Actuator (2)*: VirtualExpo Group, n.d.).

9 Sequence of Mechanical Works after Award of Work

9.1 WORK PROGRESSION ON AWARD OF CONTRACT

The usual practice in construction of large process plants is to entrust entire Engineering, Procurement and Construction (EPC) work of a project to a renowned EPC contractor. In almost all cases, process engineering, equipment design and piping engineering, civil, electrical and instrumentation engineering and related procurement of bulk materials (from all disciplines) are carried out at headquarters (HQ) of the EPC contractor. Since piping construction activity is to take place at the plant site, setting up a proper workshop facility on site with adequate inspection and test facility is of utmost importance and is a time-consuming process. The required area for the workshop, storage of raw materials and finished piping spools are often provided by the client. Activities related to piping works start with the receipt of materials on site, provided engineering activity is fairly completed, meaning piping isometrics have reached "Approved for Construction" (AFC) status. Plant piping starts with piping spool fabrication at the temporary workshop facility set up near to plant site and followed by installation of piping spools on site, welding of field joints, installation of intervening equipment, inspection and testing of the same up to commissioning of piping.

As mentioned earlier, during the time period required to set up temporary workshop facilities and to receive piping materials on site, engineering works related to the project are carried out at HQ of the EPC contractor. With regard to piping, the development of following engineering documents is essential to proceed with procurement of materials and consumables. They include but are not limited to development of process flow diagram (PFD), process and instrumentation diagram (P&ID), unit plot plan (UPP), equipment and storage layout, pipe rack layout up to isometrics, followed by material take-off (MTO), piping material specification (PMS), commodity specification (CS) and material requisition (MR). With MR, enquiries are floated for bulk materials and bids received pass through Technical & Commercial Bid Evaluations (TCBE) and thereafter purchase orders (PO) are placed from HQs. In delivery instructions, all vendors shall be directed to deliver materials to nearest port to project site. The PO for piping materials shall essentially contain the following documents, usually as attachments:

1. Material take-off (filtered to provide the list of specific items)
2. Commodity specification for listed items

DOI: 10.1201/9781003328124-9

3. Delivery instructions
4. Commercial terms

Because of the above, every construction contractor shall have a well-established work execution plan, wherein bifurcation of responsibility between HQ and site team shall be clearly defined. In case the EPC contractor is headquartered in a different place or country, it is often an accepted practice to engage local subcontractors to carry out site construction activities. Assuming that the entire works at site shall be carried out by the EPC contractor at site through local subcontractors, a typical model of responsibility matrix, usually followed by most of EPC contractors, is provided below.

9.2   RESPONSIBILITY MATRIX (HEAD OFFICE AND SITE)

Please note that matrix below does not cover all activities related to process plant construction; however, an earnest attempt was made to cover all aspects applicable to piping under this matrix.

Though mechanical activities related to project start with receipt of piping materials, followed by piping spool fabrication in temporary workshop facility on site, the actual work starts with setting up of the workshop, wherein many machineries are required based on scope of fabrication envisaged. This includes specialized welding equipment for high productivity, spool positioners, rotators, etc., that too in required numbers. Once the facility is set up satisfactorily, upon arrival of piping materials and consumables, spool fabrication can start. The book assumes that the site workshop is fully geared up for fabrication of pipe spools in a wide range of diameters and all sorts of configuration and using the range of materials required for the project.

Apart from readiness of manufacturing facility, a lot of documentation needs to be at AFC status from various agencies involved, such as Consultant or Client as applicable. As it takes time to complete site workshop, this work is carried out from head office of the contractor, which includes preparation and approval of a host of design documents as elaborated elsewhere along with a long and complete list of piping isometrics and related construction documents. Apart from completing the design and piping isometrics, preparation of other technical documents, placement of POs for equipment (if any required), raw materials, consumables and issue of various site sub-contracts and so on are also to be in position by the time materials arrive on site. However, prior to start delving into those requirements, it is considered pertinent to discuss about the site organization required to execute such a large plant construction project, involving huge amount of piping to the tune of 100,000 to 200,000 inch diameter of piping usually involved in a medium scale project.

Sl. No.	Activity	Responsibility	
		HQ	Site
1	Receipt of Contract	X	
2	Preparation Detailed Project Design Basis (PDB)	X	
3	Preparation Process Flow Diagram (PFD) with Heat and Mass Balance	X	
4	Preparation Process and Instrumentation Diagram (P&ID)	X	
5	Preparation Material Selection Diagram (MSD)	X	
6	Development of Piping Material Specification (PMS)	X	
7	Material Take-Off (MTO) for bulk items	X	
8	Piping Material Commodity Specifications (PMCS)	X	
9	Material Requisition (MR) along with Vendor Document Requirement (VDR)	X	
10	Technical & Commercial Bid Evaluation (TCBE)	X	
11	Purchase Orders (PO) for Materials Including Consumables (Bulk)	X	
12	Preparations of Piping Key Plans	X	
13	Preparation of Isometric Drawings	X	
14	Receipt and Storage of Piping Materials		X
15	Receipt and Storage of Consumables and Other Miscellaneous Items		X
16	Inspection of Materials at Site		X
17	Welding Procedure Qualification Requirements		X
18	Welding Procedure Qualification and Documentation		X
19	Welder Qualification		X
20	Other QA/QC Documentation Including ITP	X	X
21	Manufacturing Pipe Spools		X
22	NDT of Pipe Spools		X
23	Inspection and Testing of Spools		X
24	Heat Treatment of Spools as Applicable		X
25	Surface Preparation and Coating of Spools		X
26	Storage of Spools till Installation on Site		X
27	Field Installation, Assembly and Welding		X
28	NDT, Inspection and Testing of Field Welds		X
29	Preparation of Test Packages		X
30	Final Hydrostatic Testing		X
31	Final Leak Testing		X
32	Punch Listing of Lines		X
33	Handing Over for Commissioning		X
	Contracts/Procurement		
34	All Main Contracts	X	
35	Equipment Erection Contracts	X	X
36	DT and NDT of piping	X	X
37	Scaffolding	X	X
38	Site Transport and Handling	X	X
39	Surface Preparation and Coating	X	X

(*Continued*)

Sl. No.	Activity	Responsibility	
		HQ	Site
40	Supplementary Purchases from Vendors		X
41	Local Purchases	X	
	Construction Documentation (Piping Only)		
42	Material Test Certificates		X
43	WPS/PQR and WQT Documents		X
44	Other QA/QC Documents	X	X
45	Isometric Drawings (AFC)	X	X
46	Design Change Notes (by Field Engineering Team – FET)		X
47	Red Line Copy of Isometrics (by FET) for Preparation of As-Built Isometrics		X
48	NDT Reports and Summary		X
49	PWHT Summary and Reports		X
50	Hydrostatic Test Reports (Test Pack Wise)		X
51	Any Other Data for As-Built Drawings and Documents		X
52	Report of Inspection and Tests		X
53	Field Test Reports	X	X
54	Handing Over Reports	X	X
55	As-Built Isometrics, Key Plans and P&IDs	X	X
56	Preparation of Construction Record Book	X	X
57	Preparation of Project Record Book (PRB)	X	X

X	Recommended	X	Optional between HQ and Site

9.3 ORGANIZATION SETUP ON SITE

9.3.1 INTRODUCTION

EPC or simple construction companies with proven track record in erection of plant piping and equipment, by default, shall have an established methodology in force to cover all activities right from submitting quotation for the piping, fabrication of spools, its erection, welding, inspection and testing up to handing over, that too within usually stringent contractual time frames. Effective and efficient progression of work requires well-defined systems within the organization which can only be achieved through a robust organizational structure within.

Manufacture and field erection of plant piping is predominantly a site activity even if piping spools are manufactured in workshops nearby. The closeness of temporary manufacturing facility to actual construction site would be an added advantage in carrying out site modifications very quickly. For any large EPC or contracting company, piping manufacture and installation shall form only a part of their activities and the overall organization structure of the company shall be oriented towards those corporate goals. As a part of this corporate organization, a well-defined site organization structure is necessary to carry out all site activities according to plans and procedures laid out for the purpose. Though activities of the parent contracting company differ widely depending on their spectrum of operation, site-related activities in the manufacture of piping spools, installation, field welding, inspection and testing of piping till commissioning remains more or less the same and hence site organizational structure for this kind of work remains almost the same universally.

In order to complete the entire scope of work under piping in any process plant construction with reasonably good quality (with an intention to avoid reworks), it is essential that strong organization structure is set up on site with experienced and knowledgeable people manning all key positions. Organization on site almost resembles that of any piping spool fabrication shop. However, due to bifurcation of works between the HQ and site for piping, as indicated in responsibility matrix, initial part of the work is carried out at the contractor's HQ and all construction activities are to be completed using site workshop and with handling facilities available on site. Therefore, a clear demarcation in scope of works proposed to be carried out by both the groups shall be laid out to avoid confusion during execution of the job with regard to responsibility for each stage as described in Section 9.2. The organization structure recommended below is based on the following presumptions:

1. It is presumed that the EPC contractor is responsible for full engineering, procurement, construction and commissioning of the project.
2. Out of the entire piping scope, manufacture of pipe spools, its inspection and testing and documentation can be taken up either by the EPC contractor themselves or subcontractor who is experienced in construction activities.
3. Due to geographical advantages, piping spool fabrication, its installation on site, welding, inspection and testing, etc. shall often be sub-contracted to local contractors under direct supervision of the EPC contractor.
4. Billing is done from HQ (of contractor) based on duly certified progress report from site (originated by subcontractor and modified by EPC contractor based on reported production according to contract payment terms) and approved by consultant or client as applicable.
5. Payment to large procurements shall be done directly by the HQ (of contractor) based on material receipts on site based on site inspection reports approved by contractor and client.
6. Site purchases shall be limited to those items that are essentially required to proceed with day-to-day activities on site.

9.3.2 ORGANIZATION CHART

The organization in Figure 9.1 gives an idea about various disciplines of work involved in site construction, their relative positions and number manning each disciplines and specialist required under each section shall be decided based on quantum of work involved under each discipline and so also the intricacies of the job in hand. Large projects may have more discipline heads such as safety manager and so on. The organization structure mentioned is just the minimum required for a median scale project.

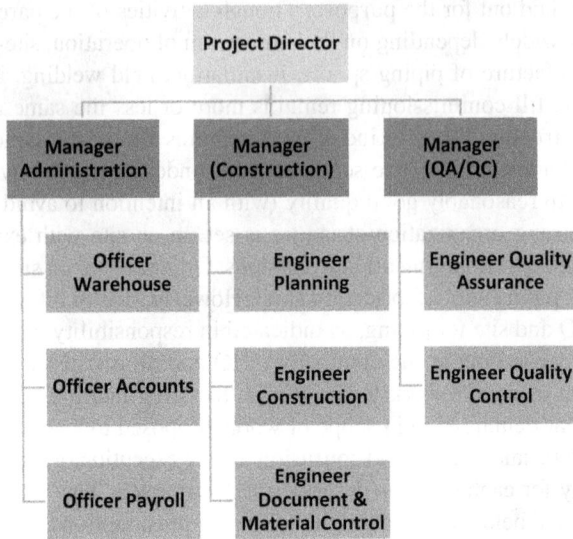

FIGURE 9.1 Organization chart.

9.4   DESIGN, DRAWINGS AND DOCUMENTS APPROVAL

As explained in Section 9.2, preparation and approval of all related technical documents required to start work on site shall be the responsibility of HQ group. Upon receipt of contract documents, engineering sections shall commence design process followed by preparation of documentation (deliverables and non-deliverables) as listed in Section 9.5 at a minimum. Table in Section 9.5 indicates deliverables related to piping and associated works only and hence is not a comprehensive list covering the entire scope of the project. The preliminary documents so prepared shall be reviewed by a group (joint review within contractor) consisting of engineers from design, planning, QA/QC and production departments with regard to all aspects related to construction. The comments of such review meetings shall be properly addressed by engineering (to make changes in drawings if required) as well as by other groups during various phases of erection. Each drawing/document shall have a unique identification number according to document numbering system adopted by the contractor. Comments of reviews are captured in respective drawings and document and same document under new revision number is issued. Revision are indicated using characters A, B and C – or numbers 0, 1 and 2– after the unique document number. All documents shall also contain a table indicating revisions history with a brief description of salient revision and reasons for the same, to provide an overview about the revisions made on the document since its initial issue. In addition, areas or regions where revisions were implemented shall be highlighted by clouding (for drawings) or by straight line in margin (for text documents) with Rev marking on right margin for ease in locating revision zones and to spot changes easily. Earlier versions of documents shall be marked as either "superseded" or as "obsolete" and these documents shall be with-drawn from respective end users of these documents. However, all end users shall have a system of maintaining superseded documents for verification of changes made. This is essentially required to resolve commercial or other related legal matters that may arise during the construction phase, but shall be available only with those autho-rized to do so, usually the concerned section heads. In this regard, a proper document control methodology shall be in position within EPC organization for documentation pertaining to the project and essentially shall address document revision, retrieval and maintenance/destruction of obsolete revisions of documents.

By this time the piping documents developed shall be ready for submission for approvals of various agencies involved like client, consultant statutory authorities, etc., as applicable. These documents are treated as deliverables under contract and a proper tracking system shall be in position for proper monitoring and control of approval process by various agencies concerned. For easy tracking and control of revisions of documents, it is also advisable to develop a proper document numbering system for all documents being generated for each work undertaken by the contrac-tor. In case there is a guideline for doing so in the contract, the numbering shall be as per client instructions or else, EPC contractor is free to adopt one according to their own document numbering system.

It is preferable to send documents to all reviewing agencies simultaneously for their approval. However, few back and forth transmissions might be required to reach AFC status by all agencies concerned. When submitted for approval, comments on drawings are usually given as a mark-up in red colour, whereas that for text docu-ments is provided as a comment sheet or as mark-up at the option of the reviewer.

In case the comments offered are within contractual requirements, the contrac-tor is liable to incorporate it without any financial implications and to submit as next revision of document for final approval of the concerned. Whereas if the com-ments fall outside contract, the contractor is eligible for extra claim, and hence to be notified to client through "Compliance/Non-Compliance Sheet" often enclosed with revised document which describes briefly what really was done with the com-ments offered during review. As mentioned earlier, incorporation of comments (with or without reservations) shall be the responsibility of originating section of that particular document. Therefore, the respective departments shall be respon-sible for maintaining a history of revisions implemented as well. Further, they shall also ensure that revised documents have reached all end users, immediately after release of new revision and to be ensured that old revision is no more under circula-tion with any of the sections involved in engineering, procurement, construction or QA/QC of the construction team.

9.5 LIST OF DOCUMENTS

During review of above-mentioned documents, it shall be noted that comments offered by different agencies shall be strictly within agreed terms of the contract, or in accordance with applicable codes or statutory regulations (which shall be a part of

Sl. No.	Description of Document	Approval			S	AB
		E	C	O		
1	**Piping Engineering**					
1.1	Process Design Basis	X	X	X		X
1.2	Process Flow Diagram (PFD) with Heat and Material Balance	X	X	X		X
1.3	Material Selection Diagram (MSD) and Guidelines (MSG) separate or together	X	X	X		X
1.4	Process and Instrumentation Diagram (P&ID)	X	X	X		X
1.5	Piping Key Plan	X	X			X
1.6	Piping Isometric Drawings (Isometrics)	X	X			X
1.7	Piping Material Specification (PMS)	X	X	X		X
1.8	Commodity Specification (Part of PMS or Separate)	X	X			X
1.9	Material Take-Off (MTO)	X	X			X
1.10	Material Requisition (MR)	X	X	X		X
1.11	Vendor Data Requirements (Part of MR)	X	X			X
1.12	Technical Bid Evaluation (TBE)	X	X			X
1.13	Purchase Order (Piping Bulk Items)	X	X			X
1.14	Vendor Data Requirements (Part of Purchase Order)	X	X			X
1.15	Packing/Delivery Instructions (Part of Purchase Order)	X				X
1.16	Hydrostatic Test Packages	X				X
2	**Construction Documents**					
2.1	Pipe Spool Manufacturing Procedure	X	X		X	X
2.2	Pipe Spool Identification and Storage Procedure	X	X		X	X
2.3	Pipe Spool Transportation Procedure					
2.4	Pipe Spool Installation and Alignment Procedure	X	X		X	X
2.5	Site Welding, Inspection and Testing Procedure	X	X		X	X
2.6	Static and Rotary Equipment Installation Procedure	X	X		X	X
2.7	Inline Instruments and Equipment Installation Procedure	X	X		X	X
2.8	Alignment, Levelling and Grouting Procedure	X	X		X	X
2.9	Punch Listing Procedure	X	X		X	X
2.10	Scaffolding Procedure	X	X		X	
2.11	Overall Construction Quality Plan	X	X	X	X	
3	**QA/QC Documents**					
3.1	Piping Quality Plan	X	X		X	X
3.2	Inspection and Test Plan (Piping Spool Fabrication)	X	X		X	X
3.3	Inspection and Test Plan (Piping Spool Installation and Welding)	X	X		X	X
3.4	Welding Procedure Specification Summary with Coverage	X	X		X	X
3.5	Welding Procedure Qualification Records	X	X		X	X
3.6	Welder/Welding Operator Qualification Test Records	X	X		X	X
3.7	Welding Consumables Control Procedure	X	X		X	X
3.8	Welding Inspection Reports	X	X		X	X
3.9	NDT and Other Test Procedures for Pipe Spools and Field Welds	X	X		X	X

(Continued)

SI. No.	Description of Document	Approval			S	AB
		E	**C**	**O**		
3.10	NDT Summary and NDT Reports	X	X		X	X
3.11	PWHT of Pipe Spool and Field Welds (as Applicable)	X	X		X	X
3.12	Hydrostatic Test Procedure for Piping Spools and Piping	X	X		X	X
3.13	Pneumatic Test Procedure for Reinforcement Pads (as Applicable)	X	X		X	X
3.14	Hardness Test Procedure	X	X		X	X
3.15	Punch List	X	X		X	X
3.16	Inspection and Test Plan (Mechanical Erection of Equipment)	X	X		X	X
3.17	Leak Testing Procedure	X	X		X	X
3.18	Procedure for Calibration of Welding Equipment	X	X		X	X
3.19	Procedure for Calibration of Electrode Oven	X	X		X	X
3.20	Surface Preparation and Coating Procedure (for Pipe Spools and Erected Piping)	X	X		X	X
3.21	Inspection and Test Plan (Surface Preparation and Painting/Coating)	X	X		X	X
	NDT Procedures					
3.22	Liquid Penetrant Test Procedure (Shop and Field Welds)		X		X	X
3.23	Magnetic Particle Test Procedure (Shop and Field Welds)		X		X	X
3.24	Ultrasonic Test Procedure (Shop and Field Welds)		X		X	X
3.25	Radiographic Test Procedure (Shop and Field Welds)		X		X	X
3.26	Visual Examination Procedure (Shop and Field Welds)		X		X	X
4	**Construction Records**					
4.1	Material Summary for Piping Materials (Cross-Reference between PO and MTC)				X	X
4.2	Material Test Certificate				X	X
4.2.1	Pipes				X	
4.2.2	Fittings				X	
4.2.3	Flanges				X	
4.2.4	Fasteners				X	
4.2.5	Gaskets				X	
4.2.6	Valves				X	
4.2.7	Strainers and Other Small Equipment				X	
4.3	Weld Joint Fit-Up/Welding/Visual Inspection Report				X	
4.4	Welding Summary				X	
4.5	Weld Joints NDT Reports				X	X
4.6	Weld Joints NDT Summary				X	
4.7	Weld Joints PWHT Reports				X	X
4.8	Weld Joints PWHT Summary				X	
4.9	Pad Air Test Reports (if Applicable)				X	
4.10	Spool Hydrostatic Test Reports				X	X
4.11	Spools Release Report				X	X
4.12	Pipe Spool Erection Report (Field)				X	X

(Continued)

Sl. No.	Description of Document	Approval E C O	S	AB
4.13	Field Joint Fit-up/Welding/Visual Inspection Reports		X	
4.14	Field NDT Reports		X	X
4.15	Field Pad Air Test Reports		X	X
4.16	BOM Check Report (Test Pack wise)		X	X
4.17	Hydrostatic Test Report (Test Pack Wise)		X	X
4.18	Signed off ITP for Pipe Spools/Test Packs		X	X
4.19	Surface Preparation and Painting Reports		X	X
4.20	Adhesion/Other Coating Inspection Test Reports		X	X
4.21	Signed off ITP for Surface Preparation, Painting		X	X
4.22	Punch List		X	X
4.23	Handing Over Report		X	X

Colour Legends

Essentially Required	Desirable
Not Directly Related to Piping	**E** EPC Contractor (HQ)
C Consultant	**O** Owner/Client
S Site (Contractor or Subcontractor)	**AB** As-Built

the contract). If not, in all probability, the contractor would swing back with an extra claim. Depending on quality of "Revision A" documents with respect to compliance to code, technical specifications of project and statutory requirements, it is possible that a document could reach AFC status in one round of review itself. Whereas, in reality, first submission of documents is usually done just to meet contractual obligation of submitting deliverables and because of this reason, usually a lot of omission can be seen which may call for a few more submissions to achieve AFC status. Strictly speaking, most POs placed shall insist that contractors start work only after obtaining approval from client/consultant. Strict compliance to this requirement may lead to time over-runs and hence in most cases, contractors are permitted to proceed with procurement (especially of long lead items) and other similar actions even before approval of necessary engineering documents at the risk and cost of the contractor.

9.6   PREPARATION OF MTO AND CS

Based on documents developed by engineering group (listed from 1.1 to 1.7) in the table under Section 9.5, a detailed MTO is developed, which gives the quantity of each of the item to be procured with ±10% accuracy, in the case of a preliminary MTO. While MTO provides quantity of each piping item, it does not specify technical requirement fully well. This is covered by document called Commodity Specification developed to specify all technical requirements applicable to each piping element. In addition to applicable international specification for the piping element, it specifies additional requirements (applicability of specific supplementary requirements spelled out in guiding specification) and any other specific requirements spelled out by the consultant or client above the code requirements. Though made as a very brief document, CS shall indicate all technical requirements for each of the components by referring to clauses and requirements of applicable code for manufacture. The basis for preparation of CS is also from the same set of documents used to develop MTO.

9.7   MATERIAL REQUISITION

MR is the document sent to various vendors to obtain quotations for piping bulk items like pipes, fittings, flanges, valves, etc. MR essentially contains MTO, CS and other commercial and delivery conditions along with a Vendor Document Requirement (VDR) as attachments. This document is sent to known and reliable manufacturers/ vendors (often based on approved vendor list of owner or consultant as mentioned in contract) across the world for obtaining competitive offers.

9.8   TECHNICAL AND COMMERCIAL BID EVALUATION

Soon after closing date for quotations, technical and commercial evaluation of bids submitted is prepared by procurement team. In order to reduce workload, commercial evaluations are carried out only for offers found technically acceptable. In commercial evaluation, delivery schedules, logistics, price etc. are considered in line with construction schedule. When bid evaluation is included as a deliverable, only TBE needs to be submitted to client or consultant and many contractors try to limit it to three TBEs from prospective vendors.

9.9   PURCHASE ORDER

Based on techno commercial evaluations, POs are placed on various vendors for supply of piping components. All enclosures of MR along with other communications transpired between contractor and vendor after floating of enquiry also shall form a part of the PO or referred in it. In many cases, unpriced PO is included as a deliverable to client/consultant for record purposes.

9.10   VENDOR DATA AND ITS APPROVAL

Once PO is placed with a vendor, thereafter vendor is liable to submit specified vendor data at stipulated timings indicated in Vendor Data Requirements (VDR) schedule, which is part of PO. Documents listed in VDR shall be submitted to EPC contractor and from there to consultant/client for the review and approval of contractor/consultant/client as agreed in main contract.

When vendor data review by consultant/owner is specified in main contract, by right, the same shall be sent for review of consultant/client after approval by EPC contractor. Since this process takes more time, to cut short review cycle time, documents with contractor's comments are forwarded to consultant/client simultaneously.

Thereafter, contractor consolidates comments received from client/consultant in addition to their own are sent back to vendor to get it addressed by vendor without any cost implications as long as it is within contractual requirements. This method often adopted in EPC contracts helps considerably in saving document review and approval cycle time.

9.11   MATERIALS PROCURED FROM STOCKISTS OR TRADERS

When final MTO is taken after approval of all documents, or due to modifications to be made at site, small quantities of materials may be required at a short notice. In such instances, procurement from local stockists and vendors is considered unavoidable. The following minimum requirements shall be complied to while doing so, in order to ensure that the material procured is genuine and traceable.

- Since stockists shall be holding bulk materials in their stock, they may not provide original Material Test Certificate (MTC) for small quantities of materials procured from them. In that case, it shall be ensured by buyer that original MTC is available with the stockist.
- The Third-Party Inspector (TPI) or client representative assigned to stockist's yard shall inspect materials physically prior to dispatch. Further TPI shall be instructed to make an endorsement in copy of MTC provided by stockiest or in Inspection Release Note (IRN) issued by TPI stating that original MTC was sited with the stockiest during inspection and verified MTC against stampings on materials.
- Material shall be traceable to MTC through proper stamping or stencilling on raw materials like pipes, fittings, and flanges as required in applicable specifications and manufacturing practices and endorsed accordingly.
- TPI shall ensure that origin of materials is from renowned sources with proven records.
- TPI shall ensure that additional testing/certification requirements (as specified in TPS) are also certified in the MTC. If not, this shall be reported as a discrepancy and buyers' confirmation shall be sought prior to release of materials.
- In case it is agreed by buyer that the discrepancies can be covered by additional testing at construction site, still discrepancy shall be reported in IRN, with a reference to buyers clearance to release materials with reported discrepancy.

9.12   WORK CONTRACTS

In case site installation of pipe spools and other equipment is directly taken up by contractor, at a minimum, the following sub-contracts would be required to complete construction as per specification. It is presumed that such contractors shall be available within the country in which construction site is situated. In such instances, it would be cheaper to use them rather than using contractor's own team to be mobilized from elsewhere, most likely from HQ.

The usual practice is to have following sub-contracts organized at site from cost, convenience and time frame considerations.

9.12.1 SUB-CONTRACT FOR NDT

All NDT work required to be carried out on piping, like Radiographic Testing (RT), Ultrasonic Testing (UT), Magnetic Particle Testing (MPT) and Liquid Penetrant Testing (LPT), are usually covered in this contract. However, some of the contractors would wish to carry out low-end NDT like MPT and LPT by their own staff. This is because of the fact that these tests may often be required at short notice during shop fabrication of spools and during erection, wherein availability of technicians from NDT subcontractor might not be possible within short notice. In such instances, it shall be the responsibility of main contractor to provide NDT technicians with requisite qualifications to carry out, interpret and document these NDT as required in applicable codes and specifications.

9.12.2 SUB-CONTRACT FOR SCAFFOLDING

Yet another sub-contract required on site is for scaffolding, which is often sub-contracted to scaffolding contractors in the locality. The decision in this regard is taken based on reliability and capability of contractors in the locality and also based on logistic and cost considerations involved in mobilizing own scaffolding materials from contractors' stock.

9.12.3 SUB-CONTRACT FOR SURFACE PREPARATION AND EXTERNAL PAINTING

All plant piping requires surface preparation and external painting as per client specification. Rarely some piping (large diameter) may require some special internal lining as well. Usually, a package consisting of surface preparation and application of painting is sub-contracted to an expert contractor who is specialized in this field. Subcontractor's applicators, especially those carrying out specialized coating/lining applications, need to be trained and qualified as per manufacturer's recommendations.

9.13 LOCAL CONTRACTS/PURCHASES AT SITE

Apart from above main contracts directly connected with piping fabrication and installation, for smooth progress of piping works on site, many more contracts also need to be in position. They include contracts such as scrap removal, handling, stacking and storage of pipes and fittings, handling of the same, house-keeping, trucking, maintenance support for equipment, supply of consumables like oxygen, acetylene, argon, etc. and supply of other consumables for day-to-day activities on site, which also can cast impediments in piping production. The project/construction manager on site shall be responsible for all such local contracts and purchases done from site. It shall be the strategy and decision of EPC contractor to carry out many works on their own or to sublet, based on availability of manpower with EPC contractor on site, cost of such services locally and execution plan and strategy.

10 Fabrication of Piping Spools

10.1 GENERAL

Piping fabrication work shall be carried out in accordance with standards, codes, specifications and special requirements of the project. In addition, works shall be carried out in accordance with good engineering practice by qualified and experienced trade personnel. Apart from physical completion of spool fabrication and its installation, the entire work shall be properly documented. These records need to be continuously updated as necessary in line with progression of work for ensuring quality of entire piping, in accordance with applicable codes and specifications. Further, all materials used in piping shall be in accordance with drawings and the piping class specified and shall be traceable.

10.2 LINEAR DIMENSIONAL TOLERANCE FOR PIPING

The general tolerance normally permitted in piping fabrication is given in table below.

	Linear Tolerance	
Description of Dimension	For Length Less Than 1.5 m	For Length 1.5 m and Longer
End to End		
Centre to End	±1.5 mm	±3.0 mm
Centre to Flange Face		
Flange Face to Flange Face		

Note: The above tolerances apply to each stated dimension in the drawing and are not cumulative.

10.3 WORKSHOP REQUIREMENTS

Adequately covered workshop facility with side cladding is absolutely essential to protect welding works from wind and other adverse weather conditions that have a direct impact on weld quality. In addition, ample storage, handling, machining, welding, inspection and testing facilities also shall be set up to ensure safe, efficient and continuous pre-fabrication of piping spools, preparation for welding, assembly and testing of high-quality piping spools under all weather conditions. Many times, surface preparation and painting of the piping spools also need to be completed prior to erection, for which yet another closed shop is often called for.

DOI: 10.1201/9781003328124-10

In addition to the above, specific area needs to be assigned for non-destructive testing (NDT). In case NDT such as radiography testing (RT) is planned during regular production shifts, then area specified for NDT shall be located at a safe distance from piping spool fabrication area to conduct RT without interrupting piping spool fabrication. Such area shall comply with all statutory and safety requirements set forth by the Atomic Energy Board of the country, wherein the process plant is to be constructed.

10.4   STAINLESS AND OTHER HIGH-ALLOY STEEL

When piping with stainless and other high-alloy steels is involved, direct contact between carbon steel and stainless steel or alloy steel is not permitted. Tools containing carbon steel and grinding discs containing carbon steel particles shall not be used on stainless and other high-alloy steels.

Tools used for fabrication of stainless steel shall be clearly identified. Tools to be used only for fabrication of stainless steel piping and piping components shall be stored separately to avoid accidental switching with tools previously used on carbon steel fabrication work.

10.5   PROGRESS CONTROL

Progress control and planning documentation shall be maintained to plan, control and report all facets of piping spool fabrication and shall include but not be limited to the following:

- Preparation of piping spool drawings
- Material availability per spool
- Surplus and/or shortages per spool
- Shop fabrication progress per spool
- Coating/lining of spools, if required

Similarly, inspection and testing of spools also can be included in progress control if desired:

- Non-destructive examination
- Weld repair and rework
- Post-weld heat treatment

Engineering changes also can be included in progress reporting:

- Revisions to isometric drawings

Coating Application

- Hot dipped galvanizing
- Surface preparation and coating

Procedures and documentation for control, reporting and recording of spool fabrication shall be submitted to owner's representative for approval prior to commencement of work.

10.6   SPOOL SIZES

Length, height and width of completed spools shall be within limits of road transport and handling constraints unless specifically requested otherwise by the owner's representative.

10.7   GALVANIZED PIPING

Galvanized piping and fittings NPS 80 (3″) and larger shall be fabricated as flanged spools and then hot dip galvanized. Pipe spools shall be pressure tested prior to galvanizing. Pipe fittings shall be abrasive blasted internally prior to fabrication and galvanizing. Piping spools shall be easily identifiable by drawing and spool number.

10.8   PAINTING OF SPOOLS

Shop fabricated spools, except galvanized and stainless steel spools, shall generally be shop painted. Prior to welding, remove supplier-applied coating for a margin of at least 25 mm from bevel on either side of joint (pipe to pipe or pipe to fitting). Stainless steel piping shall not be painted. Flange gasket faces shall be protected against damage and paint deposits during entire surface preparation and painting process. Pipe spools shall remain identifiable at all times, even during surface preparation and painting phases.

10.9   SHOP FABRICATION OF SPOOLS

10.9.1   RELEASE OF ISOs IN AFC STATUS

Piping spool fabrication work starts at workshop facility near construction site upon release of "Approved for Construction" (AFC) isometrics and receipt of all materials listed in isometrics. Yet another perquisite is to establish welding procedure specification (WPS) through a properly documented Procedure Qualification Record and subsequent qualification of welders/welding operators in required numbers and positions as required for carrying out welding in shop and site.

10.9.2   MATERIAL RELEASE

To start spool manufacture, all piping elements mentioned in spool drawing (isometric) shall be available at stores after satisfactory inspection and acceptance. Therefore, to start spool manufacture, release of AFC drawings to shop shall be followed by release of authorization from planning team to draw materials from stores with a consolidated material take-off (MTO) applicable to isometrics released for construction. The MTO so issued shall only contain piping materials already received at stores and accepted and no way shall contain any materials that are quarantined or rejected under any circumstances.

10.9.3 MATERIAL TRACEABILITY

All pressure retaining parts used in piping shall have positive material traceability, and specific test certificates for each of the pressure retaining components shall be retrievable at any point of time, right from construction phase up to scrapping of the process plant after its service life. The QA/QC wing of the contractor shall develop a fool-proof system for material traceability and this document shall form part of the project record book (PRB). Chapter 15 provides report formats that can be used for developing positive documentation pertaining to piping project, including material traceability. PRB is the authentic record pertaining to piping during the service life of plant and essentially required for carrying out modifications or alterations during service and also for carrying out root cause analysis in case of failures.

10.9.4 SIZING OF PIPE

Most of the time, pipes for piping project come in double random lengths. For fabrication of piping spools, often pipes need to be cut to smaller sections (or rarely joined together) to obtain required dimensions for piping spool to match dimensions in the drawing and for proper matching on site during installation. Sizing of pipes is done by a square cut initially (using either oxyacetylene flame or band saw or by cutting wheel depending on feasibility and cost). Depending upon the material of pipe and its wall thickness, suitable cutting methods are used as in below table.

	Pipe Cutting Process		
Pipe Material	Using Grinding Wheel (Cutting Wheel)	Oxyacetylene Flame	Plasma Arc Cutting
Carbon Steels	Yes	Yes	Not required
Low Alloy Steel	Yes	Yes	Not required
Stainless and Other High-Alloy Steels	Yes	No	Yes

10.9.5 TYPES OF WELD JOINTS REQUIRED IN PIPING

The following four basic types of joints are often required in piping (Figure 10.1).

In actual use, many times several variations of above typical joints may have to be used to cater to diverse needs. On closer scrutiny, it can be seen that the four types of joints constitute two basic types only, groove-type weld (butt weld) and the fillet type.

Groove welds are welds filled in grooves that are cut on the inside of two pieces of metal that are positioned next to each other, whereas in fillet type, weld fills in space on the outside of pieces of metal that are positioned at an angle to each other.

- **Typical Butt Weld Joint Edge Preparations**
  For pipes cut to size as mentioned in Section 10.9.4, proper edge preparation needs to be made for joining it with either another pipe or with a fitting to make the required pipe spool.

Two types of edge preparations are often adopted in piping joints, single bevel or double bevel, the commonly used configurations are shown in Figure 10.2.

Butt welding considered under piping fabrication is limited to girth or circumferential welds between pipes or between pipe and a fitting as the

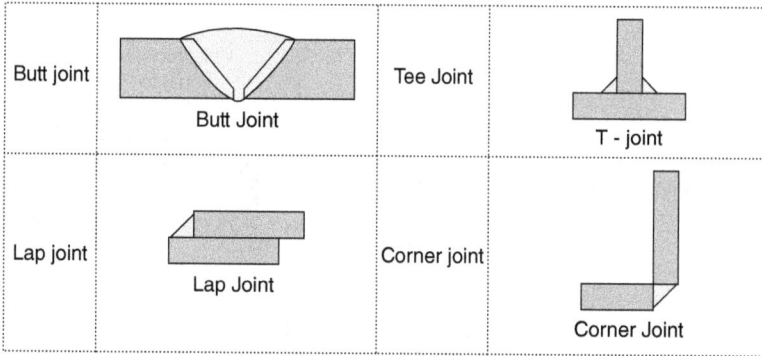

Butt joint	Butt Joint	Tee Joint	T - joint
Lap joint	Lap Joint	Corner joint	Corner Joint

FIGURE 10.1 Types of weld joints required in piping (Sölken, n.d.).

Bevel	Sketch	Process used
		Wall thickness Range
Square Bevel		Feasible by oxy acetylene, plasma arc and to follow by griding to remove oxidized metal from surface or by Grinding
		For nominal wall thickness of 3 mm (1/8") up to 6 mm max based on welding process
Plain Bevel	**Plain Bevel** 37 1/2° ± 2 1/2° / 1 less than 7/8" / .060 ± .010 land / .030	Feasible by oxy acetylene, plasma arc and to follow by griding to remove oxidized metal from surface
		Wall thickness less than or equal to 22 mm (7/8"). May be used even below 3 mm as well
Compound Bevel	**Compound Bevel** 37 1/2° ± 2 1/2° / 10°±1° / 3/4" transition / t greater than 7/8" / .060 ± .010 land / 4:1 load out (14°) / .030	Feasible through maching, with difficulty, feasible through grinding as well.
		Wall thickness greater than 22 mm (7/8")

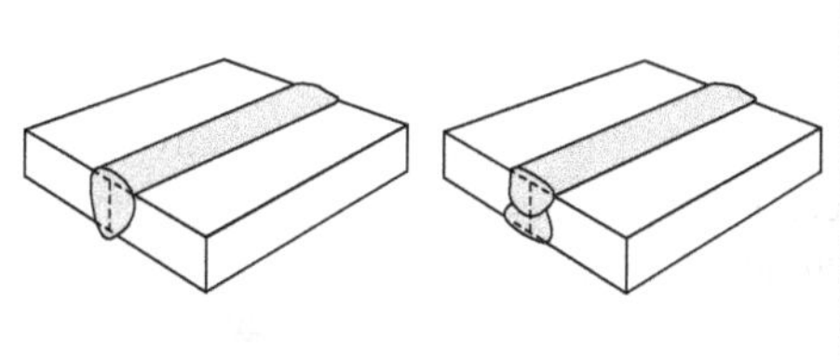

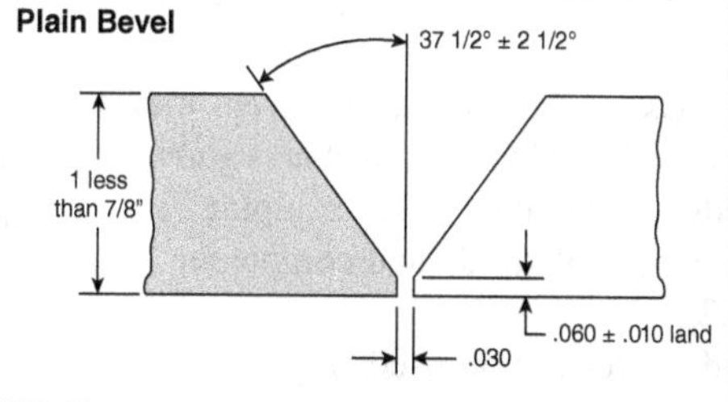

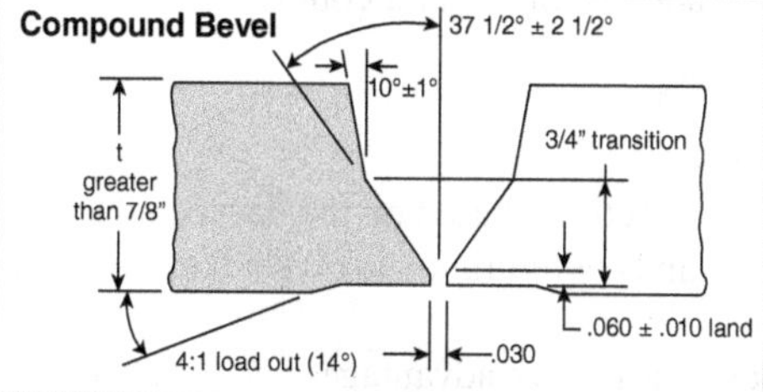

FIGURE 10.2 Typical edge preparations (Fabricating & Metalworking, 2020).

(Continued)

Bevel	Sketch	Process used
		Wall thickness Range
"J" Preparation	"J" Prep — 20° ± 2 1/2° — t less than 7/8" — 3/16" R — .060 ± .010 land — .020 land extension — 4:1 lood out (14°)	Feasible only through maching
		Wall thickness less than or equal to 22 mm (7/8"). May be used even below 3 mm as well
Compound "J"	Compound "J" Prep — 20° ± 2 1/2° — 10° ± 1° — 3/16" R — 3/4" transition — t greatar than 7/8" — .060 ± .010 land — .020 land extension — 4:1 lood out (14°)	Feasible only through machining
		Wall thickness greater than 22 mm (7/8")

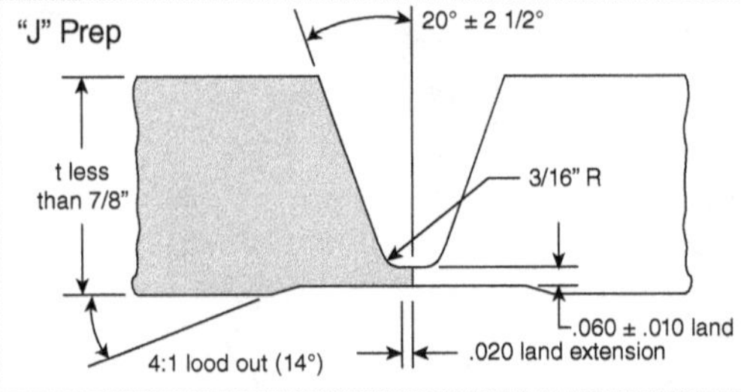

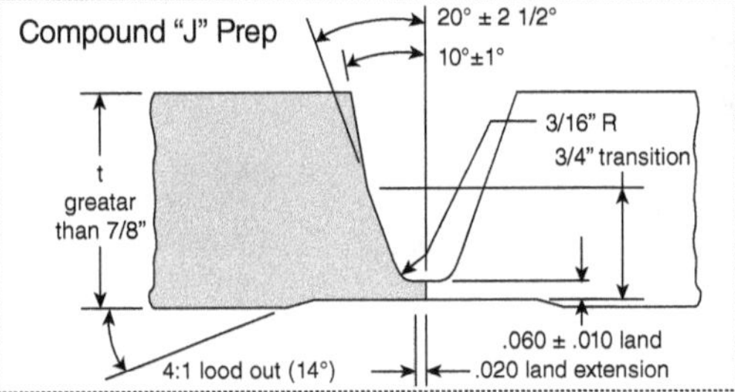

FIGURE 10.2 (*Continued*)   Typical edge preparations (Fabricating & Metalworking, 2020).

case may be. The welds on pipes and fittings (longitudinal seams) are considered out of scope of this book and not addressed for restricting the coverage within piping construction.

It is well known that a butt joint is the most universally used method of joining in piping, where very good quality weld is required, which is ensured through NDT techniques.

Generally, when thickness (t) of material at weld joint exceeds 4.8 mm (3/16"), weld edge needs to be chamfered at approximately 37.5°, leaving an upright portion (root face) on the square cut edge, which is often termed as weld bevel. As indicated, the most used bevels are the "plain bevel" from wall thicknesses (t) 4.8–22 mm, and the "compound bevel" for wall thicknesses above 22 mm.

Yet another type of edge preparation used in piping is the "J" groove preparation, which also includes plain "J" and compound "J" preparations as indicated in Figure 10.2.

Yet another type of edge preparation used in piping is the "J" groove preparation, which also includes plain "J" and compound "J" preparations as indicated below.

- **Edge Preparation Methods**
  The possible edge preparation methods based on the bevel configuration are also indicated in Figure 10.2 and configuration itself shall explain why other methods are not practical.

  Each configuration above has its own advantages and disadvantages. The very purpose of having single or double "J" or double "V" is to reduce deposited weld metal and associated problem. However, as indicated in

Figure 10.2, above-mentioned bevels are not possible through conventional edge preparation processes like gas cutting or plasma arc cutting. Predominant considerations in selection of edge preparation are based on the following aspects:

- Low weld deposit
- Proper access and electrode manoeuvrability for root, hot and filler pass welding
- Feasibility of the proposed configuration
- Cost of carrying out above all

- **Permissible Butt Weld Edge Preparations**
  Considering the above aspects, in general, most commonly used edge preparations are straight cut, single bevel and compound bevels. For butt-welded piping, ASME B 31.3 permits use of bevels as specified in ASME B 16.25 or as established in WPS or as provided in Clause Figures 328.4.2 (a)/(b). Consolidated edge preparations (from ASME B 16.25 and B 31.3) for butt welding of piping components are provided in Figure 10.2 based on wall thickness of pipes to be joined.

- **Commonly Used Butt Weld Edge Preparations**
  As mentioned earlier, a major portion of welding works in process plant piping takes place at temporary workshop facilities set up at a location near to plant construction site. Because of the same, full range of machining facilities as in a well-established workshop may not be available. However, minimum machining facilities shall be available in the site workshop to carry out bevelling of pipe segments to make a spool. Therefore, from cost and productivity points of view, edge preparations used in piping are often restricted to a few simple configurations which can be easily made on site, which also could ensure required quality for the weld. Further, adopted weld bevel preparations shall be in accordance with those specified in code (ASME B 31.3) and also shall meet any other requirements included in technical requirements of the contract, by way of consultant/client specifications (Figure 10.3).

  Standard dimensions of edge preparations are as given above, with V angles and usually permitted tolerances. Compound bevel is specifically designed for higher wall thicknesses with an intention to reduce weld metal volume to the extent possible, thereby reducing chances of defects, and to minimize stresses in weld and probable distortion associated with.

  As shown on RHS of Figure 10.3, during pre-fabrication of piping spool, while fitting up weld joints using appropriate bevel a gap of approx. 3–4 mm is provided to obtain proper weld penetration. However, this can vary according to welding process and electrode sizes used in production.

- **Typical Socket Weld Joint**
  Bevel preparation is required only for butt weld joints. For socket weld and threaded connections, square cutting is good enough; however, it needs

Edge Preparation for Pipe Thickness up to 22 mm	
Plain bevel Wall thickness (t) X* mm to 22 mm 37.5° ± 2.5° 1.6 mm ± 0.8 mm (Root face) Less than X* = Cut square or slightly chamfer, at manufacturer's option.	Plain Bevel 37.5° ± 2.5° t = 7/8 max 1/16 - 1/8 1/16 ± 1/32

Edge Preparation for Pipe Thickness above 22 mm	
Compound bevel Wall thickness (t) >22 mm Note: Radius R is not defined 9° to 11° R 37.5° ± 2.5° 1.6 mm ± 0.8 mm (Root face) 19 mm ± 2 mm	Compound Bevel 10° ± 1° 10° ± 1° R 37.5° ± 2.5° t > 7/8 3/4 1/16 - 1/8 1/16 ± 1/32

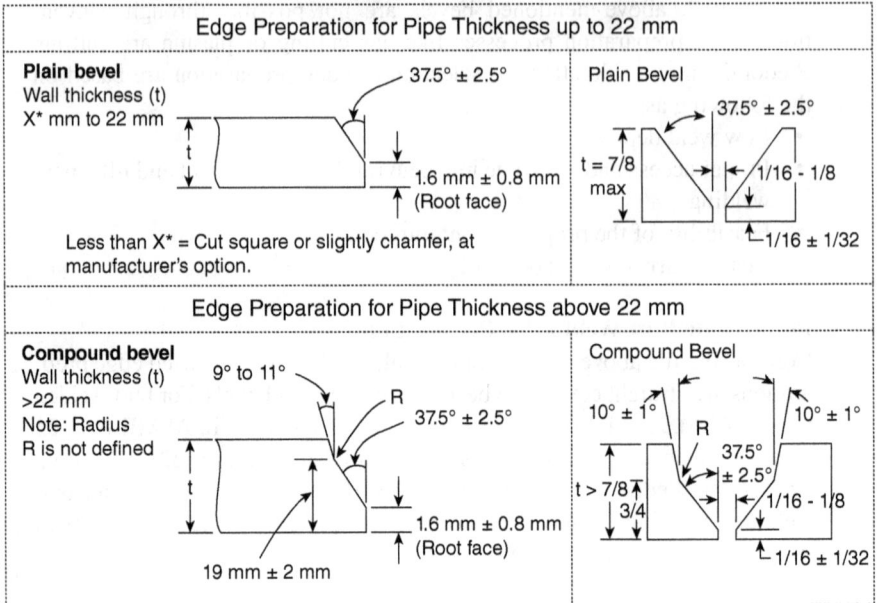

FIGURE 10.3 Typical edge preparations based on pipe thickness (Sölken, n.d.).

some more preparation before fit-up. In case square cutting of pipes is done using oxyacetylene cutting or plasma arc cutting, cut edge shall be rough and oxidation of edges is quite possible. In such cases, roughness of cut edge shall be removed by grinding and edge shall be made free from burrs by grinding or other suitable methods. For socket-welded joints, square cut edge of pipe is simply inserted into the socket of connecting fitting or threading is done as required to connect the mating pipe to fitting.

Socket joints are generally used in pipe sizes NPS 2 and smaller, and in systems where slip-on flanges are proposed (usually for low pressure no hazardous services). Figure 10.4 depicts a typical socket-welded pipe system.

While assembling the joint before welding, pipe shall be inserted into socket to maximum depth and then withdrawn slightly (say by 1.6 mm (1/16″)) from contact point between end of pipe and shoulder of socket. Purpose of providing a small clearance as above between socket and pipe is to reduce residual stress at root of the weld that could occur during solidification of weld metal, and to allow for differential expansion of mating elements.

The principal disadvantage of a socket weld also arises from above aspect, corrosion by stagnant fluid that gets entrapped in the crevice and narrow space between pipe OD and ID of shoulder of fitting. It may be noted that clients usually request for butt welding for all sizes of piping for some specific services, wherein corrosion due to stagnant fluid is quite significant.

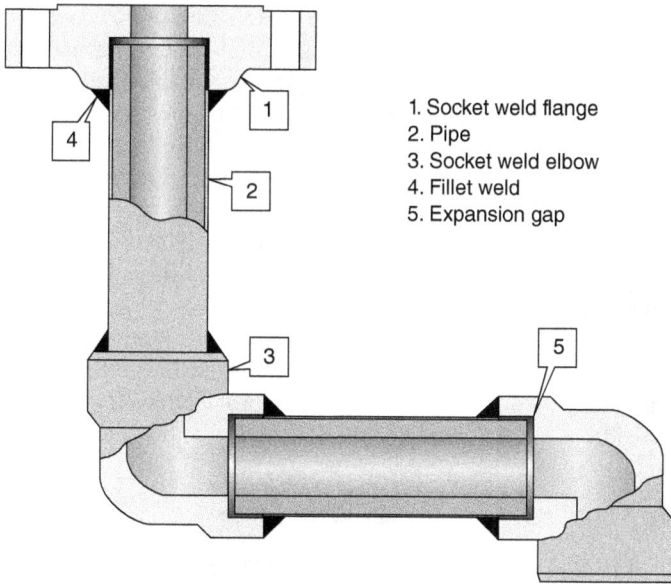

FIGURE 10.4 Typical socket weld joint (Sölken, n.d.).

Bevel preparation is required for butt weld joints alone. For socket weld and threaded connections, square cutting is good enough. However, square cut edge needs to be further prepared. In case square cutting of pipes is done using oxyacetylene cutting or plasma arc cutting, cut edge shall be rough and oxidized metal may be present at cut edge. In such cases, the roughness of cut edge shall be removed by grinding and edge shall be made free from burrs by grinding or other suitable methods. For socket-welded joints, square cut edge of pipe is simply inserted into socket of connecting fitting or threading is done as required to connect mating pipe fitting.

- **Threaded Connections**
  For threaded connections as well, edge preparation is similar to that required for socket-welded fittings.

- **Standard Edge Preparation on Pipes, Fittings and Flanges**
  Ends of components to be welded shall be bevelled to required geometric shape (bevel) recommended based on material of construction, wall thickness of components to be joined and welding process selected. Irrespective of the type of bevel preparation (joint design), prepared edges shall comply with the following:
  - End preparation is acceptable only if the surface is reasonably smooth and true, and free of slag from oxyacetylene or arc cutting, usually associated with thermally cut surfaces. Discoloration if any remaining on a thermally cut surface is not considered to be detrimental oxidation.

- In case of butt-welded fittings, usually items are supplied with required edge preparation. For all smaller sizes, it is done using machining and for larger sizes by grinding or any other suitable method or combination thereof. Whatever be the method of preparation, edges shall match design requirements and if any defects are found in edge preparation, same shall be rectified appropriately prior to joining of the same.
- For socket-welded fittings, socket shall be on the fitting and no preparation shall be required, except for cleaning of surface and adjacent location of the weld.
- Similarly, for threaded connections as well, pipe edge preparation required is same as that used in socket weld joints. Prior to external threading of pipe end, the fitting is provided with mating internal thread.

10.9.6 FIT-UP OF BUTT AND SOCKET WELD JOINTS

Weld joint, either butt or socket weld, shall be fit up as per joint configuration applicable as mentioned in Section 10.9.5. Mismatch between components joined is a very common defect found during fit-up inspection of butt weld joints.

The tolerances given in Figures 10.5 and 10.6 may be considered as the maximum permitted during fit-up (ASME B 31.3 is silent in this regard).

- **Valves**

    While welding butt or socket-welded valves to piping, there is a possibility of distortion of valve seat due to welding heat. To avoid that, valve stems shall be kept in fully open position prior to commencement of welding. This practise may be followed for all types of valves and sizes, except for large swing check valves.

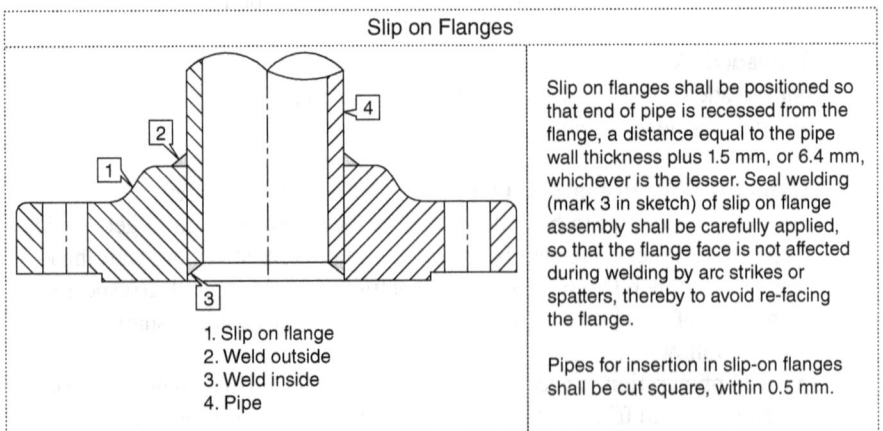

Slip on Flanges

Slip on flanges shall be positioned so that end of pipe is recessed from the flange, a distance equal to the pipe wall thickness plus 1.5 mm, or 6.4 mm, whichever is the lesser. Seal welding (mark 3 in sketch) of slip on flange assembly shall be carefully applied, so that the flange face is not affected during welding by arc strikes or spatters, thereby to avoid re-facing the flange.

Pipes for insertion in slip-on flanges shall be cut square, within 0.5 mm.

1. Slip on flange
2. Weld outside
3. Weld inside
4. Pipe

FIGURE 10.5 Typical slip-on flange joint (Sölken, n.d.).

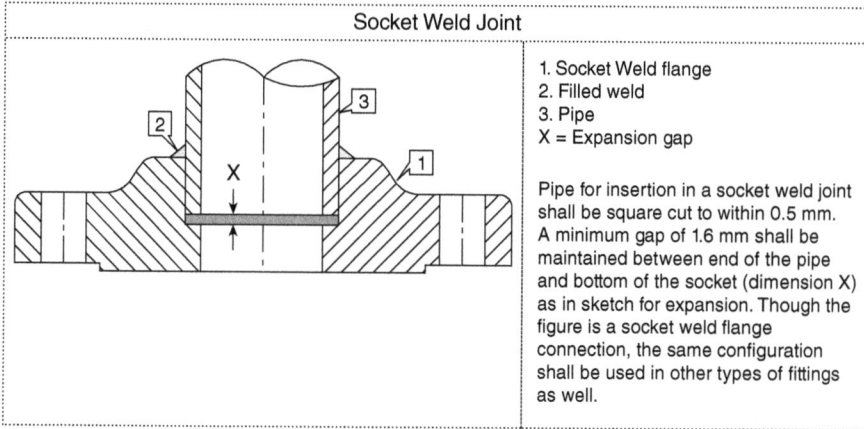

FIGURE 10.6   Typical socket weld flange joint (Sölken, n.d.).

10.9.7   THREADED CONNECTIONS

A threaded fitting has pipe threads machined into its bore. The fitting is screwed into matching threads on pipe end. Threaded fittings are easy to install and useful in areas where maintenance mandates assembling and dissembling process frequently. Therefore, threaded connections are used under special circumstances only where the main advantage is that they can be attached to pipe without welding.

Sometimes a seal weld is also applied in conjunction with threaded connection. Though threaded connections are still available in most sizes and pressure ratings, threaded fittings today are used almost exclusively in smaller pipe sizes.

It has been the general industry practice to avoid threaded connections as much as possible in flammable and toxic services. Similarly, they are avoided in service where crevice corrosion, severe erosion or cyclic loading is expected.

For threaded connections in piping, applicable specification referred in ASME B 31.3 is ASME B 1.20.1 – Pipe Threads – General Purpose (Inch). Some specific applications call for other types of threading as well. While making threads, inside ends of threaded pipes shall be deburred by reaming. All threaded connections shall be gauge-checked or chased after galvanizing. Threaded connections shall not be seal welded.

Threaded joints in piping system shall be made up using PTFE pipe tape or thread seal compound applied on male thread. Typical sketch of threaded flange connection is shown below. External threading shall always be made on pipe end and internal threading shall be on fitting to be connected to pipe and threading requirements are exactly the same as the one shown in Figure 10.7 for threaded flange connection.

10.9.8   FLANGED CONNECTIONS

Unless otherwise indicated on drawings, boltholes of all flanges shall straddle or off-set to vertical and horizontal centre lines. Maximum angular deviation of bolt holes shall not exceed 1.5 mm measured across bolt pitch circle.

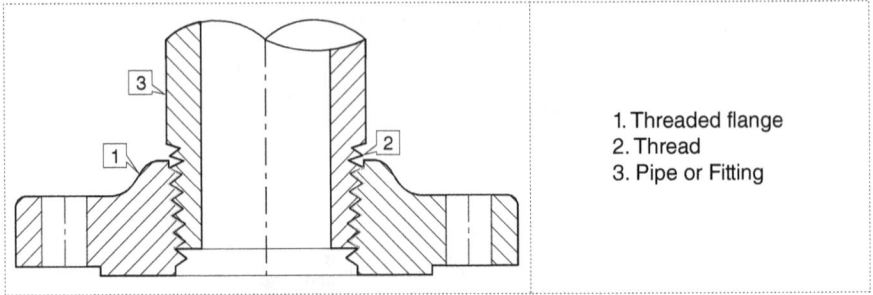

1. Threaded flange
2. Thread
3. Pipe or Fitting

FIGURE 10.7   Typical threaded flange connection (Sölken, n.d.).

Flange faces shall be square to pipeline to which they are fitted. Maximum deviation of flange face alignment measured at flange outside diameter from design plane shall not exceed the following, when measured in any radial direction:

Pipe Diameter Nominal	Maximum Deviation (mm)
DN up to 100 mm	1.0
DN 150–500 mm	1.5
DN 600–900 mm	2.5
DN 950 mm and over	3.0

Flanged connections to equipment supplied with raised face flanges shall have raised faced mating flanges and flat-faced flanges shall have flat faced mating flanges.

Shop fabrication of flanged spool pieces for connection to existing pipework or equipment shall have the mating flange tack welded to spool and an additional allowance of 50–100 mm of pipe shall also be provided for taking care of any probable adjustments required during fit-up at field.

10.9.9   BRANCH CONNECTIONS

Branch connection requirements shall be in accordance with that specified in ASME B 31.3. Figure 328.4.4 of ASME B 31.3 shows preparations for branch connections, and Figure 328.5.4 of B 31.3 shows typically acceptable branch connections. Locations for such branch connections and reinforcement shall be as indicated in respective drawings issued for the purpose. Reinforcement material shall be made from run pipe material as specified by relevant piping class and subject to same specification requirements as that of piping to which it is attached.

All cut edges of pipes shall be carefully bevelled and accurately matched to form a suitable preparation for welding and to permit full penetration of welds between branch and run pipe all around weld between run pipe and branch.

All reinforcement pads for pressure openings, or each segment of built-up type reinforcement pads for pressure openings, shall be provided with 6 mm NPT threaded hole for testing and venting. The vent hole shall be sealed after completion of pressure test of piping spool with grease or silicon sealant to prevent ingress of moisture.

Branch connections, vent nozzles, trunnions and other attachments including reinforcing pads shall not be welded over or near longitudinal or circumferential welds in the piping. The minimum distance from a longitudinal or circumferential weld to next weld shall be 50 mm measured between the heat-affected zones (HAZs). For reinforcing pads, the minimum distance measured between HAZs of weld in pipe and fillet weld of pad shall be 25 mm.

10.9.10   COLD BENDING

Pipe NPS 40 and smaller shall be bent only where cold bending is indicated in piping drawings. In all other cases, butt weld, socket weld or threaded elbows shall be used depending on what type of connection is specified in piping material specification (PMS) for specified piping class for line under question. Cold bending shall be carried out using pipe bending machines or presses using proper guides and dies to avoid flattening of pipe at bends. Unless otherwise specified, centre line radius of bends shall be five times nominal pipe diameter. Butt welds in arc portion of the bend or for addition of pulling legs shall not be permitted.

Bending shall not reduce pipe wall thickness below the minimum wall thickness required by design considerations like pressure and temperature plus any corrosion allowance.

Other requirements are as follows:

- No bending shall be performed at metal temperatures less than 4°C (40°F).
- Cold bending may be performed using hydraulic or mechanical bending machines. Bending machines shall be qualified by test for pipe minimum wall thickness and ovality.
- Mandrel and die used in bending stainless steel piping shall be free of zinc.
- In case of welded pipes, pipe longitudinal welds shall not be located within 30° of the plane of bend measured axially from pipe centreline.
- Necking as determined by reduction of outside circumference shall not exceed 4%.
- Creased or corrugated bends are not permitted.
- After bending, finished surface shall be free of cracks and substantially free from buckling and shall be ensured by visual inspection.
- Depth of wrinkles on inside of bend as determined from crest to trough shall not exceed 1.5% of nominal pipe size.
- Flattening or ovality of a bend, difference between maximum and minimum diameters at any cross section, shall not exceed 8% of nominal outside diameter for internal pressure and 3% for external pressure.
- Wall thinning in piping shall not exceed:
  - 10%: Bend radius of 5 pipe diameters and larger
  - 21%: Bend radius of up to 3 pipe diameters

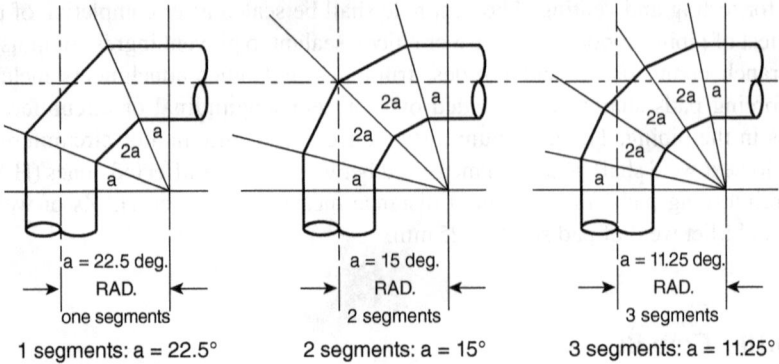

FIGURE 10.8   Mitre bends (GlobalSpec, 2010).

10.9.11   MITRE BENDS

Mitre or segmented bends are manufactured (as described in Section 304.2 of ASME B 31.4) by butt welding together segments of pipe, shaped to produce the required bend. Wherever possible, segments are to be taken from same length of pipe. However, the use of segmented bends shall be limited to applications detailed in piping specification or drawings (Figure 10.8).

10.9.12   MISALIGNMENT TOLERANCE

All piping fit-ups shall be within tolerance specified in the table below.

Components with Equal and Unequal Inside Diameters (More Than 1 mm Difference)	
Nominal Pipe Size	Misalignment (mm)
DN 150 and smaller	1
DN 200–300	2
DN 350 and larger	2.5

Note: Misalignment should be minimized wherever possible by rotating pipe/fitting for best fit and/or by grinding the bore as required.

10.9.13   PREPARATION FOR WELDING

Before carrying out any welding, the concerned shall be aware of specific details as to how it will be carried out. This implies that WPS shall be established well before this event. ASME B 31.3 refers to ASME Section IX for welding procedure and welder/welding operator qualifications along with any other additional requirements specified in client specification, which forms part of Engineering, Procurement and Construction contract.

Welding procedure qualification demonstrates that methodology proposed through WPS is capable of meeting acceptance criteria established in codes when properly applied.

Upon establishing WPS, next step is to qualify required number of welders or welding operators and welding equipment to carry out specific welding procedure. Here again, the relevant sections of ASME B 31.3, ASME Section IX and the client requirements shall establish the qualification criteria.

Satisfactory result of above two steps reaffirms that both welding procedure, and individuals and equipment intended to be used in piping production are adequately qualified to perform the welding works with required quality.

Apart from the above, sufficient number of weld joints shall be ready for welders to start work as also supply of consumables like filler rods, electrodes, fluxes, wires, shielding/trailing gases, etc. In addition, it shall be ensured that power sources for welding can offer uninterrupted supply during the course of welding.

10.9.14 CLEANING

Apart from weld bevel, adjacent portion of bevel (internal and external surface) also shall be cleaned thoroughly to make it free from paint, oil, rust, scale or other material that would be detrimental to either weld or fused base metal when welding heat is applied. If such items are not cleaned, they could mix with weld metal at elevated temperatures and result in poor quality welds. Cleaning to a distance of 15–25 mm on either side of bevel on both internal and external surface is considered adequate for this purpose.

10.9.15 PREHEATING

Preheating is used, along with heat treatment, to minimize detrimental effects of high temperature and severe thermal gradients that are inherent in welding. Necessity for preheating and temperature for preheat shall be specified in engineering design or in WPS, generally as required in design code, client specification and good engineering practices.

The following are the specific benefits of preheating:

- Dries base metal and removes surface moisture, which, if present, would result in porosity in weld metal.
- Reduces temperature difference between base metal and weld, eventually affecting weld in following manner:
  - Reduces cooling rate of weldment
  - Results in reduction of residual stresses
  - Reduces cooling/shrinkage stresses in weld metal
  - Manifested as lower hardness in weld metal and HAZ.
- Helps to maintain molten weld pool for a longer time to permit maximum fluxing and separation of impurities.
- Helps drive off absorbed gases (such as hydrogen) which could contribute to weld porosity.

While releasing fitted up joint for welding, concerned engineer/supervisor shall ensure that required preheat is applied to the joint, prior to start of weld. Preheat temperature shall be established over a minimum distance of 50–75 mm on either side of the weld. It is also essential to ensure that preheat temperature measured is the true "through thickness temperature" and not skin temperature.

10.9.16 WELDING

- **Welding of Butt Welds**

  Prepared edges are fitted according to joint configuration proposed in design within usual tolerances permitted. Fitted up joints shall be inspected for bevel angle, root face and root gap considering welding processes proposed, especially for root pass. When root penetration is critical, often, GTAW process is used both in CS and SS materials, which provides uniform and just enough penetration of root, provided root face and gap are fairly uniform over circumference. Excessive root face may result in poor penetration. As a standard practice, root and hot passes were carried out using same welding process. From productivity point of view, SMAW would be a better option compared to GTAW and in that case as well, root and hot passes are done using SMAW prior to using some other faster welding process like FCAW or GMAW for filler and capping passes. When root is welded using SMAW, low-diameter electrodes are used to prevent occurrence of burn troughs. Welding rods of 2.5 mm or at the most 3.15 mm diameter are often used for this purpose. The filler passes are made with larger diameter electrodes like 4 or 5 or even 6.0 mm. When it comes to the cover pass, lower diameter rods are used again to give a good finish for the weld. The above requirement shall apply only to SMAW process, which is extensively used in any piping spool fabrication work.

  As far as weld joint is considered in its entirety, apart from general guidelines, as far as possible, parameters mentioned in qualified WPS shall be adhered to, as specific WPS supersedes all common guidelines mentioned previously. It is presumed that all those aspects were considered based on each one's merits, demerits and past experience with the kind of fluid handled (the specific pressure and temperature) and a pragmatic approach is taken while establishing WPS.

- **Edge Preparations (Equal Thickness)**

  As mentioned earlier in table under Section 10.9.5, butt-welded joints in piping systems are primarily of the single-V configuration and are welded from pipe outside surface. Larger diameter pipes, which can be accessed from inside will often be welded from both sides using a double-V type of joint preparation (Figure 10.9).

- **Combination "V" Preparations**

  Joint preparation, procedure used and welding technique used by welder jointly ensure complete fusion between edges of components being joined.

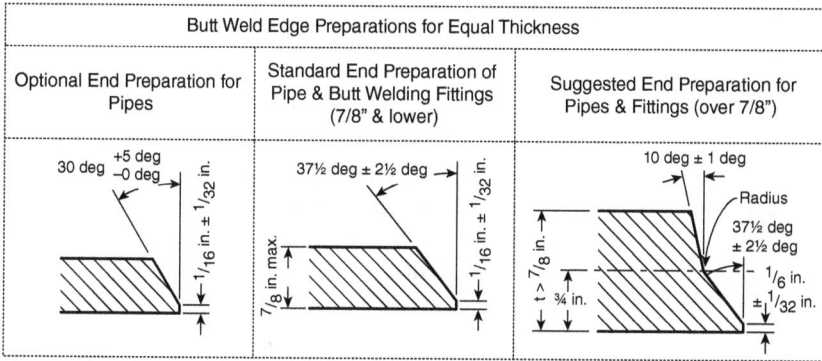

FIGURE 10.9   Edge preparations for equal thickness on both sides (ASME, 2022b).

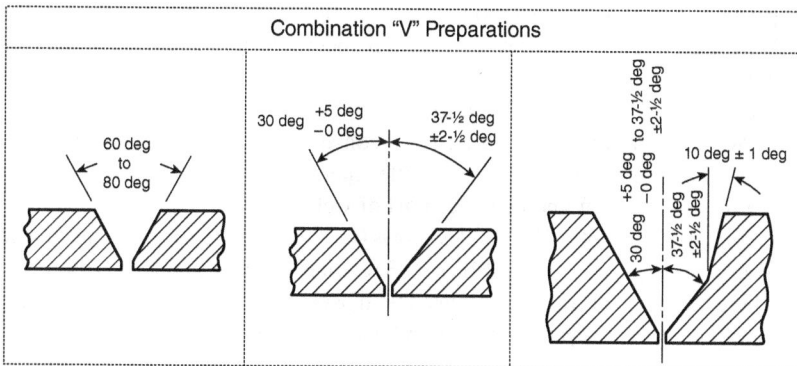

FIGURE 10.10   Combination "V" preparations (ASME, 2022b).

Apart from joint designs shown in ASME B 31.3, the following combinations of those joint designs are also acceptable for mating ends with equal thickness. Though these combinations of "V" preparations are not shown in ASME B 31.3, they are indicated in related pipeline codes such as ASME B 31.4 and 31.8 (Figure 10.10).

- **Edge Preparations (Unequal Thickness)**

  In piping systems, many times components with heavier thickness such as valves, castings, header sections, equipment nozzles need to be welded to pipe sections which are lower in thickness.

  In such instances, heavier sections are machined or ground down to match thinner pipe wall and the excess thickness tapered, either internally or externally, to form an acceptable transition zone. Limits imposed by various codes in this regard are fairly uniform.

  The external surface of heavier component is tapered at an angle of 30° maximum for a minimum length equal to 1½ times the pipe minimum wall thickness and then at 45° for a minimum of 1½ times the pipe minimum wall thickness.

Internally, either a straight bore followed by a 30° slope or a taper bore at a maximum slope of 1–4 for a minimum distance of 2 times the pipe minimum wall is required.

The surface of weld can also be tapered to accommodate differing thickness. This taper shall not exceed 30°. It may be necessary to deposit weld metal to ensure that these limits are not violated.

Figure 10.11 presents some of the acceptable design for unequal wall thicknesses based on other related codes like ASME B 31.4 and B 31.8.

- **Joints with Backing Strip**

Joints with backing strip are also commonly required in piping. This requires component ends to be prepared accordingly to accommodate a backing ring. A backing ring is a material in the form of a ring whose primary function is to support or hold molten weld metal in weld puddle. Where component ends are trimmed, inspector shall verify that remaining net thickness of finished ends is not less than the minimum required wall thickness for service condition for which line is designed.

As an alternative to backing ring, consumable inserts are also used. A consumable insert is a pre-placed filler metal that is completely fused into the root of the joint and becomes part of weld.

Figure 328.3.2 of ASME B 31.3 shows typical backing rings and consumable inserts (see Figure 10.12). ASME B 31.3 requires that backing rings be removed where the resultant crevice associated with backing rings is subject to corrosion, vibration or severe cyclic condition (ref. para. 311.2.3 of ASME B 31.3)

- **Consumable Inserts**

In case of orifice flanges, only weld-neck flanges with internal bore matching that of the pipe alone can be used. If backing rings are used in such cases, the inspector shall verify their removal and confirm that butt weld is ground flush at the root from inside pipe. When it is impractical to remove backing ring, consideration shall be given to welding without backing rings or else through use of consumable inserts (Figure 10.13).

- **Socket Weld Joints**

Fit-up requirements for socket weld joints are described in Section 10.9.5 in detail. In order to maintain clearance in socket weld joints as shown in sketch under Section 10.9.5, it is a common practice to use non-metallic washers of 1–2 mm with the same OD and ID of pipes to be inserted inside socket during fit-up. Thereafter, fillet weld of required size is deposited, using low-diameter electrodes (2.5 or 3.15 mm diameter) using SMAW process. Even in case of thin fillet sizes on thin-walled pipes, a minimum of two passes of welds shall be deposited, to avoid presence of through holes in welds. Moreover, use of low-diameter electrodes helps in providing an undercut-free fillet weld. The weld metal shall be deposited only on cleaned

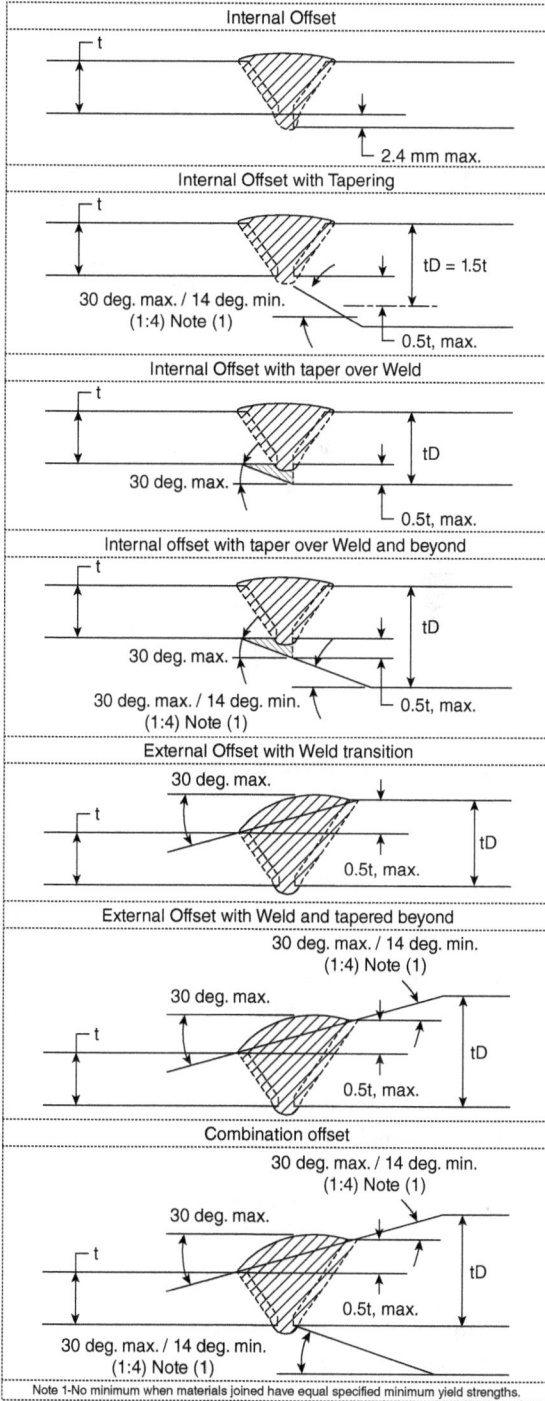

Internal Offset

t

2.4 mm max.

Internal Offset with Tapering

t

tD = 1.5t

30 deg. max. / 14 deg. min.
(1:4) Note (1)

0.5t, max.

Internal Offset with taper over Weld

t

tD

30 deg. max.

0.5t, max.

Internal offset with taper over Weld and beyond

t

tD

30 deg. max.

30 deg. max. / 14 deg. min.
(1:4) Note (1)

0.5t, max.

External Offset with Weld transition

30 deg. max.

t

tD

0.5t, max.

External Offset with Weld and tapered beyond

30 deg. max. / 14 deg. min.
(1:4) Note (1)

30 deg. max.

t

tD

0.5t, max.

Combination offset

30 deg. max. / 14 deg. min.
(1:4) Note (1)

30 deg. max.

t

tD

0.5t, max.

30 deg. max. / 14 deg. min.
(1:4) Note (1)

Note 1-No minimum when materials joined have equal specified minimum yield strengths.

FIGURE 10.11 Edge preparations for unequal thickness (ASME, 2022b).

Butt Joint with Bored Pipe Ends and Solid or Split Backing Ring (Note 1)	Butt Joint with Taper Bored Pipe Ends and Solid Backing Ring (Note 1)
5 mm ($^3/_{16}$ in.) >t_m 3 mm to 5 mm ($^1/_8$ in. to $^3/_{16}$ in.) 19 mm (¾ in.)	5 mm ($^3/_{16}$ in.) >t_m 3 mm to 5 mm ($^1/_8$ in. to $^3/_{16}$ in.) 19 mm (¾ in.)
Non-metallic Removable Backing Ring (Refractory)	Note 1- Refer to ASME B 16.25 for detailed dimensions of welding edges

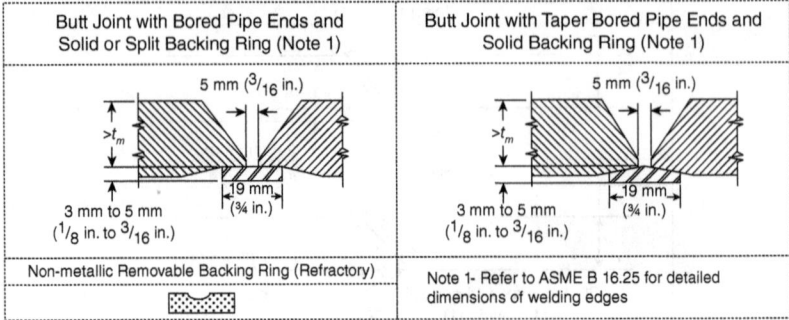

FIGURE 10.12 Joints with backing strip (ASME, 2022a).

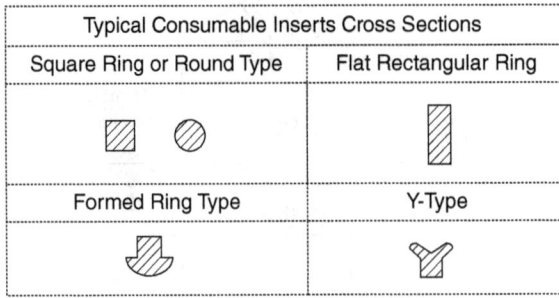

Typical Consumable Inserts Cross Sections	
Square Ring or Round Type	Flat Rectangular Ring
Formed Ring Type	Y-Type

FIGURE 10.13 Consumable inserts (ASME, 2022a).

surface and it is always recommended to clean at least 25 mm (1″) width on either side of the weld as well, prior to start of welding.

Similar to threaded joints, socket welds are also restricted to maximum size of NPS 40 (1½″) in hazardous services. For sour service applications, socket welds are often avoided even for lower pipe sizes below NPS 40 (1½″). Similarly, socket-welded joints are generally avoided in piping services where crevice corrosion, severe erosion or cyclic loading is also expected.

- **Fillet Weld Sizes**
  A fillet weld consists of an angular weld bead that joins components positioned normally at a 90° angle to each other. After deposit of weld metal, a fillet weld may be concave to slightly convex in shape as shown in Figure 10.14.

  Either concave or convex in profile, the ideal fillet is one with equal legs, which is often not possible. In reality, the following would be the resultant weld shape after depositing it to required size specified in design (Figure 10.15).

  The size of a fillet weld is stated as a leg length of the largest inscribed right isosceles triangle as shown in sketches in Figure 10.16 for various resultant weld contours.

  In piping systems, fillet welds are used for slip-on flanges, socket welds and for welding attachments to piping components (e.g. reinforcing pads, supports).

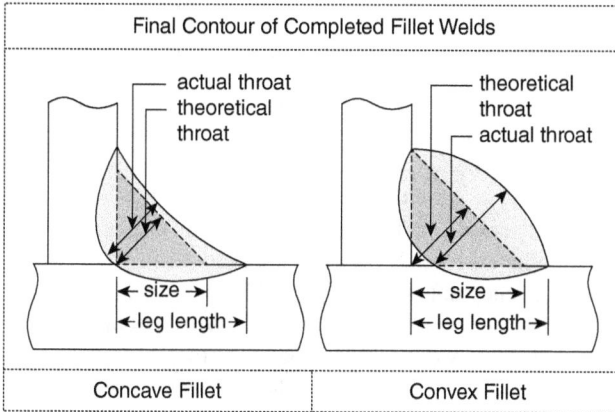

FIGURE 10.14 Fillet weld sizes (Kobe Steel, LTD, n.d.).

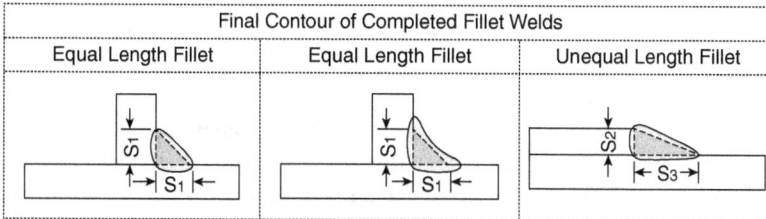

FIGURE 10.15 Practical fillet weld configurations and sizes (Kobe Steel, LTD, n.d.).

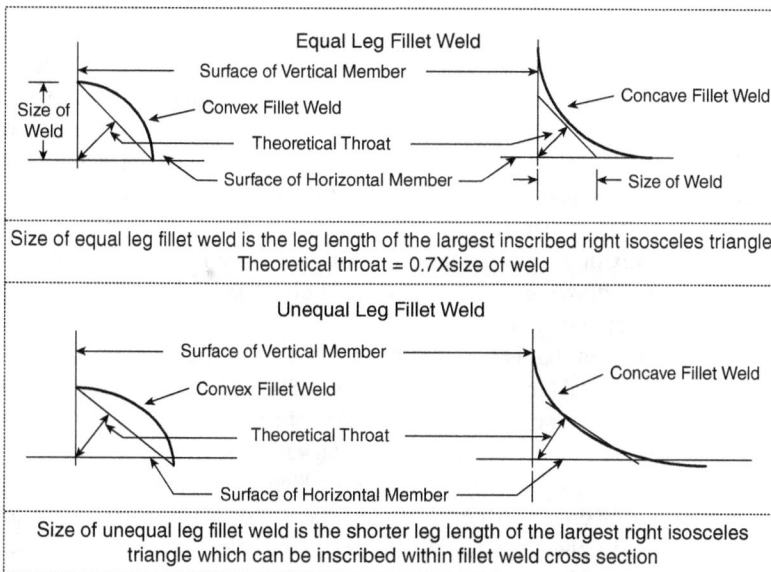

FIGURE 10.16 Fillet weld measurement (Quora, n.d.).

10.9.17 VISUAL INSPECTION

- **Butt Welds**

  All welds require visual inspection to ascertain contour, reinforcement, undercut and other surface opening defects on weld. In piping, most of the times, visual inspection is possible from outside only. Being a powerful tool in ensuring quality of weld, visual inspection of welds, supplemented by DPT (in specific cases), shall be used to eliminate presence of surface opening defects. Whatever be the contour of weld (even when weld reinforcement is within acceptable limits), it shall be free from any abrupt changes. In other words, weld shall merge smoothly with parent metal so that concentration of stress at weld edge can be minimized.

  Weld ripples shall also be reasonably good, so that if a radiograph from that joint is taken, it can be clearly interpreted. In case weld surface is irregular or ugly looking, though there are no clear-cut requirements in applicable code sections, at the discretion of inspection engineer, these welds shall be given a dressing up or cosmetic grinding. When weld requires dressing as mentioned, it shall be done prior to radiography of that joint, if radiography is required for the line, either full or random spot is done, based on code and other technical requirements of the contract.

  Selection between full and random spot radiography is principally based on fluid service and its pressure–temperature ratings. Often client specifications make radiographic requirements more stringent than basic minimum code requirements and these requirements are elaborated in Chapter 11.

- **Socket Welds**

  Socket welds are basically a fillet weld. Code requires these welds to be inspected visually, for contour, leg sizes, undercut and other surface opening defects. In addition to the above, other tests (like LPT, MPT, etc.) as specified in code/client specification also need to be done to ward off leakages during hydrostatic test, leading to rework and retesting. In some critical applications, client specifications often insist for LPT or MPT as an additional requirement. For any fillet weld, the essential visual inspection shall be to ensure size of fillet to drawing/specification. Profile of weld may be convex or concave, but minimum throat size needed to be ensured as required in drawing. Further, this weld also shall be free from surface defects as mentioned above.

  Another aspect that is often being overlooked by many inspection engineers is the unequal leg size of fillet welds. ASME 31.3 does not specify any acceptance criteria for difference in leg size of fillet welds. As a thumb rule, a maximum of 2 mm difference between legs is considered acceptable, with the lowest leg having the minimum specified leg size, derived from the fillet size as shown in drawing. It is always practical and easy to measure leg size of a fillet weld rather than the throat size, which can be translated to weld throat size as required in applicable drawing.

- **Threaded Connections**

  Threaded connections are often visually inspected for improper tightening of mating components resulting in crossed threading, which will not ensure leak tightness. In addition, it shall also be ensured that a sufficient number of threads have already been engaged. In case these conditions are met, threaded connections can be released for further processing. In some cases, seal welding may be required (as per specification), which is recommended to be carried out after the satisfactory testing of the line, either hydrostatic or pneumatic as applicable.

  For threaded joints, following aspects need to be confirmed by the inspector:
  - Maximum joint size
  - Seal welding
  - Thread engagement

- **Maximum Joint Size**

  In hazardous services, the inspector shall confirm that the maximum size of threaded connections is no more than NPS 40 (1½") for standard fittings and valves. NPS 50 (2") joint connections may be used when required for maintenance, minor field modifications of existing piping systems and to match threaded specialty devices such as scraper signals and access fittings for corrosion monitoring.

  In non-hazardous services, the inspector shall ensure that the maximum size of threaded connections used is no more than NPS 80 (3") for standard fittings and valves.

- **Seal Welding**

  ASME B 31.3 defines seal welding as a weld intended primarily to provide joint tightness against leakage in metallic piping. Some client specifications require seal welding for threaded joints if used in flammable and toxic services.

  Where seal welding is required, seal weld shall be a fillet weld on diameter of the female part, and it shall be smooth with slight concavity as allowed by codes, to male part covering all exposed threads without undercut. The inspector shall make sure that no polytetrafluoroethylene (PTFE), (commercially known as teflon), tape or joint compounds are used in threaded connections requiring seal welding. Limitations on seal welding are given below:

  Seal welding is often required for all threaded joints up to first block valve in the following services and applications:
  - All hydrocarbons
  - Boiler feed water, condensate and steam systems utilizing ASME Class 300 and higher flange ratings
  - Toxic materials such as chlorine, phenol and hydrogen sulphide
  - Corrosive materials such as acid and caustic
  - Oilfield chemicals (e.g., corrosion inhibitors, emulsifiers, electrolytes)
  - Piping which is subject to vibration, whether continuous or intermittent

Seal welding is not required for the following services and applications:
- Thermowells
- Bar stock plugs downstream of a seal welded block valve
- Special devices such as access fittings and scraper signals
- Joints which require frequent disassembly and are located downstream of a seal welded block valve, for example, sample connections
- Instrument piping downstream of primary instrument isolation valve
- Pipe union ring threads and joints with elastomer o-rings
- Threaded joints, downstream of a seal welded valve, which discharge directly to an open drainage system or to the atmosphere
- Extended body valves with integrally reinforced welding end

- **Thread Engagement**
  A minimum thread engagement shall be maintained to ensure integrity of threaded connection and to preclude possibility of leakage. Minimum length of engaged threads on pipe shall meet requirements of ASME B 1.20.1 for taper pipe thread. The minimum number of engaged pipe threads according to pipe size is provided in the table below.

Thread Engagement Requirements for Taper Pipe Threads	
Nominal Pipe Size	Number of Threads Engaged
1/2" and 3/4"	6
1" through 1–1/2"	7
2" through 3"	8
4" and above	10

10.9.18 Dimensional Tolerances for Pre-fabricated Piping Assemblies

Dimensional control of pre-fabricated piping spools shall be performed in a systematic manner, to ensure that the final installation will be correct in accordance with dimensions shown in drawings (within permitted tolerance). Pre-fabricated spools for installation shall be 100% dimensionally controlled so that rework on spools during installation and site assembly can be reduced to a considerable extent.

Tolerances on linear dimensions (intermediate or overall) are illustrated in figures given in the table below. It may be noted that these tolerances are not accumulative. Similarly, angularity tolerances across the face of flanges and rotation of flanges are also provided in the table below.

When closer tolerances other than those given in the table below are required, these shall be as specified on the isometric drawing in question (Figures 10.17–10.19).

In plant piping projects, piping contractor is primarily responsible for achieving tolerances as above. Even if above tolerances are achieved during piping spool

Linear Dimensions (Length in m)	Tolerance in mm	
	ΔL	T
≤6	±5	±1.5
>6	±10	±1.5

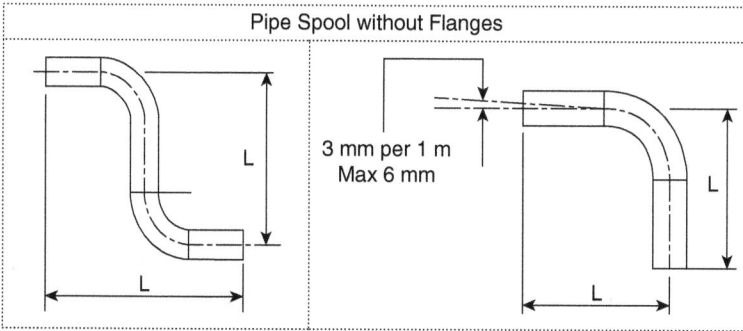

Pipe Spool without Flanges

3 mm per 1 m
Max 6 mm

FIGURE 10.17   Dimensional tolerance for piping spools without flanges (KLM Technology Group, 2011).

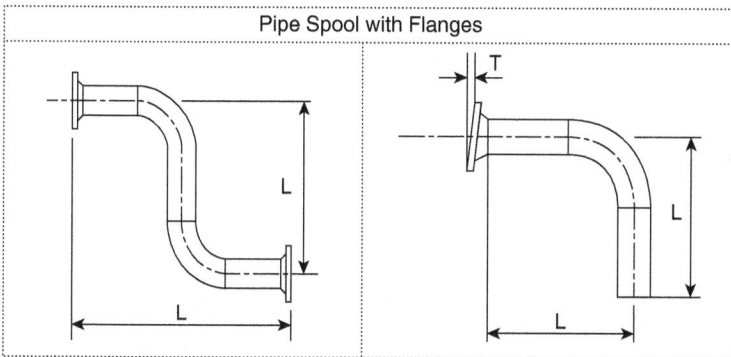

Pipe Spool with Flanges

FIGURE 10.18   Dimensional tolerance for piping spools with flanged ends (KLM Technology Group, 2011).

manufacture, variation in dimensions can be expected at site due to minor variations in structural members supporting piping or that of equipment to which piping is to be connected. Contractor shall be responsible to accommodate such variations as well, without any additional claims. To take care of such variations "Field Weld" locations are pre specified in drawings, with some extra length of pipe (usually of 100 mm) over and above the dimensions required in drawing is provided. Depending on errors expected, this extra length is provided to spool on one side of each field weld. During erection, pipe end with extra length shall be cut to obtain proper joint at site. Isometrics usually have the field welds marked

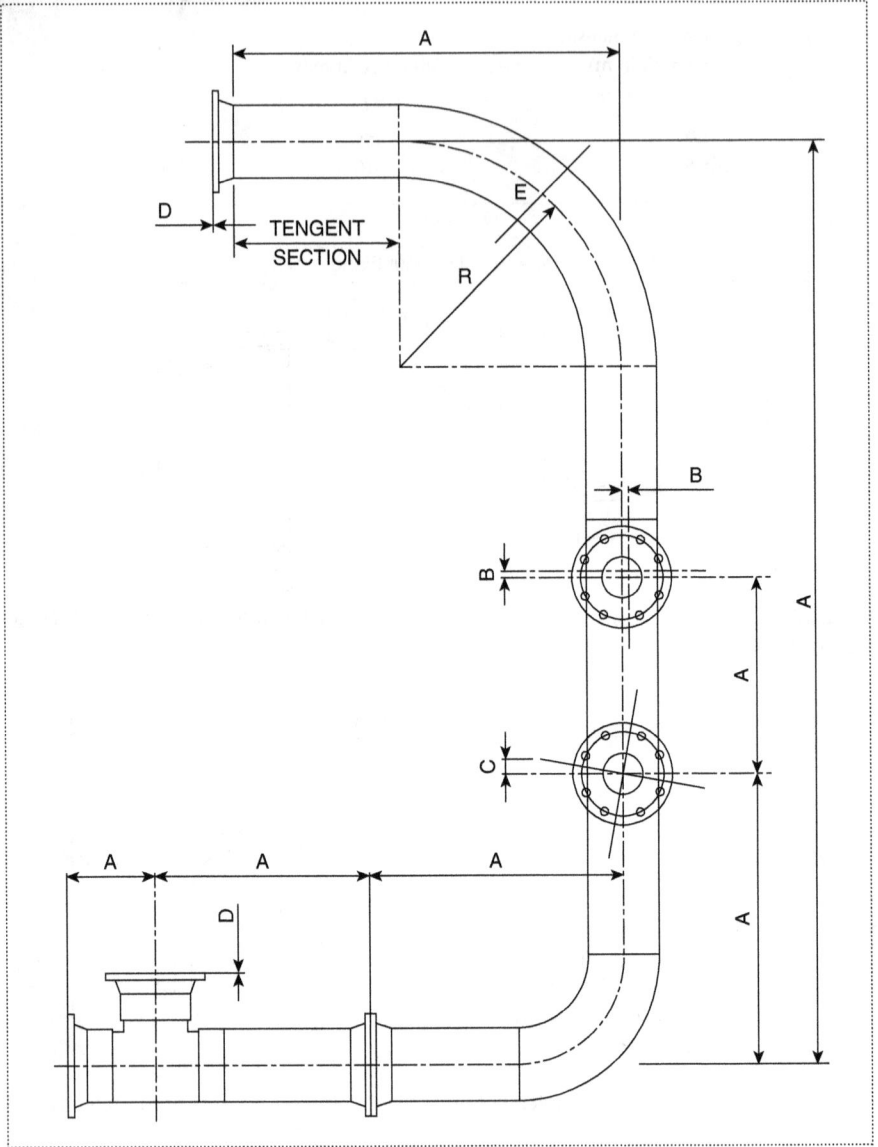

FIGURE 10.19 Tolerance on various dimensions of a completed piping spool.

on them so that there shall not be any ambiguity in this regard. Responsibility to provide adequate number of filed joints at site to maintain flexibility as well as to minimize rework at site, shall also fall under the scope of work of piping contractor. Further, in case of any omissions or errors in drawings supplied by client,

Dimension	(in mm)	Description	
		Linear Dimension	
A	±3	Maximum permitted from indicated dimensions for face-to-face, centre-to-centre location of attachments	
B	±1.5	Maximum lateral translation in any direction from the indicated position	
		Flange Alignment	
C	±1.5	Maximum rotation from the indicated position measured across as shown	
D	±1.0	Out of alignment from the indicated position measured across any diameter as shown	
		Bend Tolerance	
R	±6.4	For bending radii up to 600 (24″) NB	Bending radius "R" shall be as specified on the isometric in terms of multiple pipe diameters
	±2.8	For bending radii above 600 (24″) NB	
Angle	±1/4°	The angular dimensions of bends shall be within ±¼°	
E	−5%	Flattening measured as difference between the maximum and minimum OD at any cross section maximum	For pipe with external pressure

responsibility to report the same also shall fall under the contractor's scope of work as they are expected to carry out detail engineering of piping.

10.9.19   RELEASE FOR NDT OF COMPLETED SPOOLS

After satisfactory completion of visual inspection of all butt, socket and other fillet (attachment welds) welds of spool, spool can be released for NDT (RT, UT or MPT) as required. NDT requirements as per code and the common additional requirements for sour environment envisaged in oil and gas are dealt with in detail in Chapter 11. It is in fact better to use applicable format given in Chapter 15 to release all weld joints (butt and socket weld) for NDT. The said format (spool inspection release) is capable of identifying weld joint to be subjected for NDT in case of spot NDT required as per code as well as client specification.

10.9.20   REPAIR OF DEFECTS

Repairs of welded joints, defective or damaged pipe and fittings or any other pressure part shall not be carried out without approval of the owner's representative and shall be shown as a hold point in the inspection and test plan (ITP).

Repair procedure for welds shall be approved by all concerned (client/consultant/third party inspector (TPI) prior to commencement of work.

If repairs or modifications are carried out after heat treatment has been completed, the areas affected by repair or modification shall be heat treated again and this also requires concurrence of all concerned.

10.9.21   VERIFICATION OF COMPLETION OF SPOOL AND RELEASE FOR POST-WELD HEAT TREATMENT (PWHT)

Upon completion of all butt, socket and other attachment welds, each spool is inspected for completeness. All welding works on spool including any supports or pads shall be completed, except for those to be done after installation on site. Basically, this completion check is carried out based on isometrics and bill of materials (BOM) provided therein. Spools that are complete in all respects except for installation of supports and other similar parts in position are released for PWHT if applicable, otherwise directly for hydrostatic testing.

10.9.22   PIPE SPOOL PWHT

It is not mandatory that pipe spools need to be post-weld heat treated entirely. Heat treatment in piping can be restricted to local PWHT of welds alone or the entire spool can be heat treated in a furnace. When heat treated as a spool, protection of flange faces during PWHT cycle plays a vital role. Irrespective of methodology (whether furnace of local PWHT), PWHT charts with reports shall be available for review and acceptance.

10.9.23   HARDNESS TEST OF WELDS (IF REQUIRED)

In sour service piping, often hardness of weld is restricted. In such cases, hardness survey on weld HAZ and base metal (optional) is to be recorded after PWHT of joints. Method adopted to record hardness shall be as required in applicable procedure, typical format for which is provided in Chapter 15.

10.9.24   HYDROSTATIC TEST OF SPOOL (AS APPLICABLE)

Upon completion of all works up to hardness survey, spools (if spool hydrostatic testing is required in contract) are released for hydrostatic testing, especially when they have flanges at both ends. Hydrostatic testing of spools is carried out by interconnecting many spools judiciously (to avoid air entrapment) based on configuration of spools so connected. Hydrostatic testing of spools is generally carried out based on a written down procedure, duly approved by all concerned. Format used for recording hydrostatic testing is provided in Chapter 15. Advantage in carrying out spool hydrostatic test is that weld joints in spools that are covered by hydrostatic test can be released for full surface preparation and painting including weld joints.

10.10   SURFACE PREPARATION AND PAINTING

Spools after completion of mechanical works or after completion of hydrostatic test (as specified) shall be released for surface preparation and painting as required in contract. When hydrostatic testing is not carried out on spools, weld area shall be left uncoated to facilitate inspection of weld joints during hydrostatic testing of the line after site installation and welding.

In case it is possible to carry out hydrostatic test of spools, then weld joints except field welds can be completely prepared by blasting and can be coated fully as required. Filed joints alone shall be left uncoated till final hydrostatic testing of the completed line on the pipe rack.

10.10.1 Inspection during and after Completion of Surface Preparation and Painting

Entire surface preparation and painting of piping shall be under the strict quality control and various stages of inspection and tests are to generally follow an ITP sample provided as Appendix A.

10.11 RELEASE OF PIPE SPOOLS AND TAGGING

After satisfactory completion of surface preparation and painting, open ends are properly protected to avoid entrapment of debris during storage, till installation of spool on site. Thereafter each piping spool is provided with a unique tag traceable to specific spool as described in isometric drawing. Care shall be taken during handling of spools not to damage applied coating and flange faces during storage, transportation and installation of spools on site.

10.11.1 Tagging and Numbering

- All completed spools shall be marked on their outside with letters and numbers defining their respective line numbers/isometric numbers.
- Markings by welding shall not be allowed. Markings using permanent ink markers are permitted over masking tapes stuck on painted surface which can be removed after installation.
- Drawing number and spool number shall be hard stamped on the rim of the flange of a flanged fabricated spool. For non-flanged fabricated spools, markings shall be hard stamped on the bevel edge of fitting for thick-walled fittings and at least 50 mm away from bevel edge for thin-walled fittings. All markings shall be hard stamped using a low stress die.
- In addition, metal tags stamped with drawing number and spool number shall be securely tied to each spool.
- Stampings on spools shall be suitably masked prior to blasting.
- After painting, drawing number and spool number shall be stencilled on the outside surface of the spool with a permanent ink marker, that too over masking tape. Marker used shall be compatible with paint on the spool and shall not pose any health hazard. For stainless steel spools, markers shall be chloride free even when used over masking tape which can be peeled off later.

10.12 TEMPORARY STORAGE OF PIPE SPOOLS

In case site installation work is delayed on account of any unforeseen issues, then spools produced need to be stored somewhere temporarily till site is ready for installation. Therefore, adequate space shall be available for storage of spools and a proper scheme for storage of spools shall be devised for easy retrieval of spools from storage yard.

10.12.1   GENERAL STORAGE REQUIREMENTS

- All piping components shall be stored in a clean area away from fabrication and construction activities and handled in such a way that no damage or mixing of spools occurs.
- Spools shall be stored on pallets and not on ground.
- End caps shall be ensured at both ends and on branch pipes.
- Spools with threaded ends also shall be protected by end caps.
- Hooks or chains shall not be used for lifting.
- Spools shall not be rolled off transport vehicles, dropped onto the ground or dragged over ground.
- Flange facings shall be protected from damage either by providing end caps or temporary wooden blanks.
- Covers shall be securely fastened to flange facings during handling, transportation and storage on site.
- Partly installed piping components and spools shall be protected at all times from ingress of moisture or foreign matter, by covering and taping.

10.12.2   STAINLESS STEEL MATERIALS

Storage and handling of stainless steel piping components shall be as follows:

- Stainless steel materials shall be stored on non-metallic pallets.
- End caps shall be provided at all open ends of spool.
- All flanges and flanged connections shall be sealed with blinds to prevent ingress of water, moisture and foreign matter.
- Threaded ends shall be capped with plastic cap and sealed.
- Stainless steel piping and components shall be stored in separate areas away from storage areas for carbon steel and other materials to avoid direct contact between carbon steel and stainless steel.
- Steel wire slings shall not be used for handling and transportation of stainless steel pipes. Canvas or nylon slings shall be used.
- Spools shall be cleaned with "acetone" and then rinsed with demineralized water to remove any contaminants, deposits of foreign materials found on spool surface.

10.12.3   LINED STEEL PIPES

Storage and handling of lined steel pipes requires special arrangements and shall be as follows:

- Pipes and piping components shall be handled in such a way that lining applied is not damaged.
- Lined pipe shall be stored under cover to protect it from high atmospheric temperatures (40°C and over).
- Provide 20 mm bolted plywood flange at all open ends of the spool.

11 Non-Destructive Testing (NDT) of Piping

11.1 INTRODUCTION

Radiography is one of the most powerful non-destructive testing (NDT) techniques used in process plant construction projects to control quality of welds (mainly girth welds) in piping for hydrocarbon or similar hazardous services. Depending upon severity and criticality of service, percentage of NDT varies from random spot (5%) to 100%. Percentage of NDT required for piping is decided based on guidelines provided in applicable standards and specifications for design, construction and testing as called for in the contract. Based on past experience, with regard to specific fluid and in accordance with general industry guidelines, client/consultants have the prerogative to increase this percentage, above the minimum specified in applicable standards and codes for design and construction. Almost all consultants and clients across the world have developed their own requirements over and above code requirements, especially for oil and gas sector, wherein sour/lethal environment is often encountered. Basic considerations in arriving at an enhanced NDT (especially for RT) percentage are criticality of service conditions (like pressure, temperature, lethality, etc.) and consequential environmental hazards in the event of failure. Though the easy solution to above considerations would be to opt for 100% radiography for all piping girth welds, the direct/indirect cost tags associated with this decision call for a judicious and rational level of spot NDT. Further, percentage so decided shall ensure required quality for welds in question with reasonable confidence level. In random inspection, degree of randomness and number of inspections finally decide the confidence level.

Considering the sour environment often encountered in surface production facilities or other similar services, increase in percentage NDT for piping welds is often called for by clients/consultants over and above code requirements based on their experience in respective industries and these restrictions are provided against all piping classifications in ASME B 31.3.

It may be worth noting that welds made on pipes and fittings during their manufacture (which include the long/helical seam welds on pipes and long seam welds on pipe fittings) are excluded from the coverage and construction girth, branch, socket and seal welds alone are considered by ASME B 31.3 and hence in this chapter as well.

DOI: 10.1201/9781003328124-11

11.2   BASIC ASME 31.3 NDT REQUIREMENT

A brief narration of the requirements of NDT as in ASME B 31.3 is provided below as a quick reference.

		Visual	RT/UT/MPT/LPT	Supplementary
Category D Fluid		Random inspection at the discretion of the inspector. No specific minimum provided	No radiography	
Normal Fluid		**Visual** — Minimum 5% of welds both groove and fillet for each welder's and welding operator's work	**RT/UT/MPT/LPT** — Not less than 5% of circumferential butt and mitre groove welds shall be examined fully by random radiography or by random ultrasonic examination	
Category M Fluid		**Visual** — All groove and fillet welds	**RT/UT/MPT/LPT** — Not less than 20% radiography or ultrasonic examination of circumferential butt and mitre welds and of fabricated lap and branch connection welds	For circumferential butt welds and other welds, it is recommended that extent of examination be not less than one shot in each 20 welds for each welder or welding operator. Unless otherwise specified, acceptance criteria are as stated in Table 341.3.2 for radiography under normal fluid service for the type of joint examined.
Severe Cyclic Conditions		**Visual** — All groove and fillet welds	**RT/UT/MPT/LPT** — All circumferential butt and mitre groove welds and all fabricated branch connection welds shall be examined by 100% radiography or by 100% ultrasonic examination. Socket welds and branch connection welds which are not radiographed shall be examined by magnetic particle or liquid penetrant methods	Progressive sampling examination Hardness examination Examinations to resolve uncertainty
Elevated Temperature Service		**Visual** — All groove and fillet welds	**RT/UT/MPT/LPT** — Not less than 5% of circumferential butt and mitre groove welds shall be examined fully by random radiography or by random ultrasonic examination. Socket welds and branch connection welds in P-No. 4 and P-No. 5 materials that are not radiographed or ultrasonically examined shall be examined by magnetic particle or liquid penetrant methods.	

(Continued)

	Visual	RT/UT/MPT/LPT	Supplementary
High-Pressure Piping (K)	All groove and fillet welds	All girth and branch connection welds shall be 100% examined by radiography. Ultrasonic testing in lieu of RT permitted when $Tw \geq 13$ mm (1/2″) and with owners' approval	UT/MPT/LPT also may be specified as required with applicable acceptance criteria. Hardness test. Examinations to resolve uncertainty
	Visual	**Radiography**	**Supplementary**
High-Purity Piping (U)	Same as normal fluid	A weld coupon examination in accordance with para. U344.8 may be used in lieu of the 5% random radiography/ultrasonic examination required in para. 341.4.1(b)(1) when orbital welding is employed in fabrication	Same as normal fluid
	Visual	**RT/UT/MPT/LPT**	**Supplementary**
High-Purity Piping in Category M (UM)		The 20% random radiography/ultrasonic examination required in para. M341.4(b)(1) applies. *(2)* The in-process examination alternative permitted in M341.4(b)(2) applies, except a weld coupon examination in accordance with para. U344.8 is also an acceptable substitute when specified in the engineering design or by the inspector.	

TABLE 341.3.2
Acceptance Criteria for Welds and Examination Methods for Evaluating Weld Imperfections

Category D Fluids		Normal and Category M Fluids			Severe Cyclic Service		Weld Imperfection Type	Examination Method			
Girth and Mitre Groove	Fillet	Branch Connection	Girth, Mitre Groove & Branch Connection	Fillet	Girth, and Mitre Groove	Fillet		Visual	RT	MPT	LPT
A	A	A	A	A	A	A	Crack	X	X	X	X
C	N/A	A	A	A	A	A	Lack of Fusion	X	X	—	—
C	N/A	B	B	N/A	A	N/A	Incomplete Penetration	X	X	—	—
N/A	N/A	N/A	E	N/A	D	N/A	Internal Porosity	—	X	—	—
N/A	N/A	N/A	G	N/A	F	N/A	Internal Inclusions	—	X	—	—
I	H	H	H	H	A	A	Undercutting	X	X	—	—
A	A	A	A	A	A	A	Surface Porosity or Exposed Slag Inclusion	X	—	—	—
N/A	N/A	N/A	N/A	N/A	I	I	Surface Finish	X	X	—	—
K	N/A	K	K	N/A	K	N/A	Concave Root Surface (Suck-up)	X	—	—	—
M	M	M	L	L	L	L	Weld Reinforcement or Internal Protrusion	X	—	—	—

(Continued)

TABLE 341.3.2 (Continued)
Acceptance Criteria for Welds and Examination Methods for Evaluating Weld Imperfections

Symbol	Criteria — Measure	Acceptance Value Limits
A	Extent of imperfection	Zero (no evident imperfection)
B	Depth of incomplete penetration	≤1 mm (1/32″) and ≤0.2T w
	Cumulative length of incomplete penetration	≤38 mm (1½″) in any 150 mm (6″) weld length
C	Depth of lack of fusion and incomplete penetration	≤0.2T w
	Cumulative length of lack of fusion and incomplete penetration [Note (7)]	≤38 mm (1½″) in any 150 mm (6″) weld length
D	Size and distribution of internal porosity	**See BPV Code, ASME Section VIII, Division 1, Appendix 4**
E	Size and distribution of internal porosity	For T w ≤6 mm (1/4″), limit is same as D; For T w >6 mm (1/4″), limit is 1.5 * D
F	Slag inclusion, tungsten inclusion or elongated indication	
	Individual length	≤T w/3
	Individual width	≤2.5 mm (3/32″) and ≤T w/3
	Cumulative length	≤T w in any 12T w weld length
		≤2T w
G	Slag inclusion, tungsten inclusion or elongated indication	
	Individual length	≤2T w
	Individual width	≤3 mm (1/8″) and ≤T w/2
	Cumulative length	≤4T w in any 150 mm (6″) weld length
H	Depth of undercut	≤1 mm (1/32″) and ≤T w/4
I	Depth of undercut	≤1.5 mm (1/16″) and ≤[T w/4 or 1 mm (1/32″)]
J	Surface roughness	≤500 minutes Ra in accordance with ASME B 46.1
K	Depth of root surface concavity	Total joint thickness, incl. weld reinf., ≥T w

(Continued)

TABLE 341.3.2 (Continued)
Acceptance Criteria for Welds and Examination Methods for Evaluating Weld Imperfections

Symbol	Criteria			
	Measure	Acceptance Value Limits		
		For T_w, mm (in.)	Height, mm (in.)	
L	Height of reinforcement or internal protrusion [Note (8)] in any plane through the weld shall be within limits of the applicable height value in the tabulation at right, except as provided in Note (9). Weld metal shall merge smoothly into the component surfaces.	≤6 (1/4)	≤1.5 (1/16)	
		>6 (1/4), ≤13 (1/2)	≤3 (1/8)	
		>13 (1/2), ≤25 (1)	≤4 (5/32)	
		>25 (1)	≤5 (3/16)	
M	Height of reinforcement or internal protrusion [Note (8)] as described in L. Note (9) does not apply.	Limit is twice the value applicable for L above		

Notes

(1) Criteria given are for required examination. More stringent criteria may be specified in the engineering design. See also paras. 341.5 and 341.5.3.

(2) Branch connection weld includes pressure-containing welds in branches and fabricated laps.

(3) Longitudinal groove weld includes straight and spiral seam. Criteria are not intended to apply to welds made in accordance with a standard listed in Table A-1 or Table 326.1. Alternative Leak Test requires examination of these welds; see para. 345.9.

(4) Fillet weld includes socket and seal welds, and attachment welds for slip-on flanges, branch reinforcement and supports.

(5) These imperfections are evaluated only for welds ≤5 mm (3/16″) in nominal thickness.

(6) Where two limiting values are separated by "and", the lesser of the values determines acceptance. Where two sets of values are separated by "or", the larger value is acceptable. T_w is the nominal wall thickness of the thinner of two components joined by a butt weld.

(7) Tightly butted unfused root faces are unacceptable.

(8) For groove welds, height is the lesser of the measurements made from the surfaces of the adjacent components; both reinforcement and internal protrusion are permitted in a weld. For fillet welds, height is measured from the theoretical throat, Fig. 328.5.2A; internal protrusion does not apply.

(9) For welds in aluminium alloy only, internal protrusion shall not exceed the following values:

    a) 1.5 mm (1/16″) for thickness ≤2 mm (5/64″)

    b) 2.5 mm (3/32″) for thickness >2 mm and ≤6 mm (1/4″)

    For external reinforcement and for greater thicknesses, see the tabulation for symbol L.

(Source: ASME, 2022)

NDT Requirement for Welds in High-Pressure Piping (Table K341.3.2)

Criteria (A–E) for Types of Welds, and for Required Examination Methods [Note (1)]

Type of Imperfection	Method		Type of Weld		
	Ultrasonic or Visual	Radiography	Girth Groove	Fillet (Note 3)	Branch Connection (Note 4)
Crack	X	X	A	A	A
Lack of fusion	X	X	A	A	A
Incomplete penetration	X	X	A	A	A
Internal porosity	—	—	B	NA	B
Slag inclusion or elongated indication	—	—	C	NA	C
Undercutting	X	X	A	A	A
Surface porosity or exposed slag inclusion	X	X	A	A	A
Concave root surface (suck-up)	X	X	D	NA	D
Surface finish	X	X	E	E	E
Reinforcement or internal protrusion	X	X	F	F	F

(Continued)

Criterion Value Notes for Table K341.3.2 (Continued)

Symbol	Criterion — Measure	Acceptable Value Limits [Note (5)]
A	Extent of imperfection	Zero (no evident imperfection)
B	Size and distribution of internal porosity	See BPV Code, Section VIII, Division 1, Appendix 4
C	Slag inclusion or elongated indication. Indications are unacceptable if the amplitude exceeds the reference level, or indications have lengths that exceed individual length	6 mm (1/4") for $T_w \leq 19$ mm (3/4"); $T_w/3$ for 19 mm (3/4") $< T_w \leq 57$ mm (2¼"); 19 mm (3/4") for $T_w > 57$ mm (2¼")
	Cumulative length of above	$\leq T_w$ in any 12 T_w weld length
D	Depth of surface concavity	See table below
E	Surface roughness	≤ 12.5 μm (500 μ in.) Ra (see ASME B46.1 for definition of roughness average, Ra)
F	Height of reinforcement or internal protrusion [Note (6)] in any plane through the weld shall be within the limits of the applicable height value in the tabulation at the right. Weld metal shall be fused with and merge smoothly into the component surfaces.	See table below

D — Depth of surface concavity:

Wall Thickness, T_w, mm (in.)	Depth of Surface Concavity, mm (in.)
≤13 (1/2)	≤1.5 (1/16)
>13 (1/2) and ≤51 (2)	≤3 (1/8)
>51 (2)	≤4 (5/32)

and total joint thickness including weld reinforcement $\geq T_w$

F — Height of reinforcement or internal protrusion:

Wall Thickness, T, mm (in.)	External Weld Reinforcement or Internal Weld Protrusion, mm (in.)
≤13 (1/2)	≤1.5 (1/16)
>13 (1/2) and ≤51 (2)	≤3 (1/8)
>51 (2)	≤4 (5/32)

(Continued)

Criterion Value Notes for Table K341.3.2 (*Continued*)

Notes

(1) Criteria given are for required examination. More stringent criteria may be specified in the engineering design.

(2) Longitudinal welds include only those permitted in paras. K302.3.4 and K305. The criteria shall be met by all welds, including those made in accordance with a standard listed in Table K326.1 or in Appendix K.

(3) Fillet welds include only those permitted in para. K311.2.2.

(4) Branch connection welds include only those permitted in para. K328.5.4.

(5) Where two limiting values are given, the lesser measured value governs acceptance. T_w is the nominal wall thickness of the thinner of two components joined by a butt weld.

(6) For groove welds, height is the lesser of the measurements made from the surfaces of the adjacent components. For fillet welds, height is measured from the theoretical throat; internal protrusion does not apply. Required thickness t_m shall not include reinforcement or internal protrusion.

(*Source:* ASME, 2022)

11.3   ENHANCED NDT REQUIREMENTS FOR PIPING

At the outset, it may please be noted that enhanced NDT requirements as below are basically meant for sour environment often encountered in surface production facilities and easily can be extended to other similar services such as toxic and lethal. The abridged NDT requirements as per ASME B 31.3 are narrated in sections above to give a feel of the stringency incorporated in enhanced NDT requirements often called for in sour service. Though enhanced NDT helps tremendously in improving quality of welds and thereby the reliability of piping system, it involves additional cost. Therefore, it is all the more important to apply these requirements judiciously in every project, principally based on severity of service, in terms of pressure, temperature, corrosivity, cyclic stressing, etc., as envisaged during operation phase of the plant.

Since ambit of code (ASME B 31.3) covers a wide variety of industries, they are intended to provide general minimum requirements common to many different applications. Therefore, setting criteria for NDT for specific applications, considering severity and service conditions of fluids handled by each industry, is impractical for ASME. Hence, this responsibility falls under the prerogative of client/consultant, responsible to formulate contract/specifications for the project in addition to basic minimum design envelope envisaged by selected codes and specifications. Further, the utilities services associated with process plant may require only less stringent NDT requirements. Therefore, the prime responsibility to decide upon the NDT requirements for various services considering severity of services of the entire plant piping rests with client/consultant engaged to do basic engineering/detailed engineering for the project, based on industry practices and past failure/maintenance history of similar plants and other published literature.

With regard to plant piping designed and constructed as per ASME B 31.3, almost all clients/consultants shall have a well-defined piping material specification, wherein an array of piping material classes would be identified covering all types of piping envisaged for the plant. In many cases, this document shall also provide the additional requirements for NDT, against various piping classes developed for the project, specifying the minimum as per code (ASME B 31.3) or more as considered appropriate.

Table below provides enhanced requirements for NDT usually called for in oil and gas service for the benefit of inspection engineers. Since ASME B 31.3 specifies NDT requirements against various piping services, enhanced NDT requirements per table below are also developed in similar lines. To make it more specific and elaborate, types of NDT required against each type of weld against various piping services are provided. A few of the NDT methods are also indicated as interchangeable as well from operational convenience point of view. As mentioned earlier, the table below is developed for piping connected with surface production facilities wherein severe sour environment cannot be ruled out.

Recommended Minimum Random NDT Percentage for Piping Welds

Extent of NDT Proposed (%)

Sl. No.	Description of Service	RT			UT			MPT			LPT		VT
		SW	BRW	BW	SW	BRW	BW	SW	BRW	BW	BRW	BW	ALL
1.1	Category D Fluids	5	—	5	—	—	5	—	—	—	—	—	100
1.2	Normal Fluids	5	—	10	5	—	10	5	—	—	—	—	100
1.3	Category M Fluids	5	—	20	5	—	20	5	—	—	—	—	100
1.4	Severe Cyclic Conditions	10	100	100	10	100	100	10	10	10	10	10	100
1.5	Elevated Temperature Service	5	10	10	5	10	10	5	5	—	5	—	100
1.6	High-Pressure Piping (K)	10	100	100	10	100	100	10	10	10	20	10	100
1.7	High-Purity Piping (U)	10	—	30	5	—	30	5	5	5	5	5	100

Colour Legends for Percentage NDT

Either of two for same category of welds fully interchangeable

Interchangeable NDT with restraint for same category of weld joint (SW, BRW or BW) from RT to UT and MPT to LPT

Interchangeable NDT with prior concurrence

If applicable

UT may be permitted in case RT is not possible due to practical constraints

Abbreviations Used

RT	Radiographic Testing
UT	Ultrasonic Testing
MPT	Magnetic Particle Testing
LPT	Liquid Penetrant Testing
VT	Visual Testing
SW	Socket Weld
BRW	Branch weld
BW	Butt Weld
ALL	All types of welds, BW, BRW and SW

11.4   PROGRESSIVE RADIOGRAPHIC EXAMINATION

When defects (reparable) are observed in random NDT, ASME B 31.3 calls for progressive examination culminating to 100% examination of the lot as required in Clause 341.3.4, which goes as below.

When required spot or random examination reveals a defect, then

a. Two additional weld joints of the same welder (bonder or operator) from the originally identified lot shall be given the same type of examination
b. If items examined as required by (a) above are acceptable, the defective item shall be repaired or replaced and re-examined, and all items represented by these two additional welds shall be accepted, but
c. If any of the items examined as required by (a) above reveals a defect, two further welds of the same kind shall be examined for each defective item found by that sampling
d. If all the items examined as required by (c) above are acceptable, the defective item(s) shall be repaired or replaced and re-examined, and all items represented by the additional sampling shall be accepted, but
e. If any of the items examined as required by (c) above reveals a defect, all items represented by the progressive sampling shall be either
   1. Repaired or replaced and re-examined as required, or
   2. Fully examined and repaired or replaced as necessary, and re-examined as necessary to meet requirements of the Code

12 Preheat and Post-Weld Heat Treatment (PWHT) of Piping

12.1 PREHEATING

Purpose of preheating (along with subsequent post-weld heat treatment (PWHT)) is to minimize the detrimental effects caused by thermal gradient created during fast cooling of molten puddle of weld metal. Therefore, quantum of preheat applied plays a vital role in reducing thermal gradient and thereby reducing the detrimental effects in resultant weld.

ASME B 31.3 recommends preheat requirements for various piping covered by the code under Clauses 330, M 330, K 330 and U 330. The requirements spelled out therein apply to all types of welding including tack welds, repair welds and seal welds of threaded joints.

Apart from the above, applicable engineering specification for the project shall be listing out the additional restrictions required in preheating when special materials are required for construction. These restrictions can be enforced during welding by incorporating it in applicable Welding Procedure Specification (WPS) and through religious compliance to WPS during production welding ensured by rigorous quality checks.

12.1.1 PREHEAT REQUIREMENTS

General requirements for minimum preheat temperatures for materials envisaged in ASME B 31.3 are given in Table 330.1.1 against their P-Number Classification as in QW 422 of ASME Section IX and also based on material thickness at weld joint. The thickness considered in Table 330.1.1 is that of the thicker component measured at the joint. When ambient temperature falls below 0°C (32°F), preheat mentioned becomes mandatory. In case of unlisted materials, preheat to be deployed shall be indicated in WPS and shall be based on studies and experiments carried out on those materials previously.

12.1.2 VERIFICATION OF PREHEAT

Often preheat temperatures are verified using temperature indicating crayons (known under trade name tempilstik), or using thermocouple pyrometers, or even infra-red thermometers as available and feasible. It shall be ensured that required temperature

DOI: 10.1201/9781003328124-12

is attained at the joint prior to welding and is maintained during subsequent weld passes as well.

In case thermocouples are used to monitor temperature, they shall be attached temporarily to pressure-containing parts using capacitor discharge method. It may please be noted that such welding does not require any qualification. However, after removal of thermocouples, area where it was attached shall be visually inspected for any surface defects and if required examined by LPT or MPT techniques as considered appropriate and may even require some cosmetic grinding.

12.1.3 OTHER ASPECTS OF PREHEATING

Though codes do not specify any bandwidth for preheating, commonly suggested bandwidth extends to at least 25 mm (1″) beyond both edges of the weld. When weld is between dissimilar metals, with different recommended preheating temperatures, code specifies that the higher temperature specified in Table 330.1.1 of ASME B 31.3 shall be used.

Similarly, when welding is interrupted due to any reason, cooling rate needs to be controlled, meaning that sudden cooling is to be avoided to ward off detrimental effects in weld due to fast cooling. In such instances, blanketing of welded and preheat area would help immensely in reducing thermal gradient. In addition, welding shall be resumed only after applying preheat again as specified in Table 330.1.1 or as specified in WPS.

12.2 POST-WELD HEAT TREATMENT

As mentioned under preheat, PWHT is yet another tool used to avert or relieve detrimental effects of high temperature and severe temperature gradients associated with fast cooling of molten pool of weld metal, which acts as a casting made with steel mould, which allows fast cooling being a good conductor of heat. Further, PWHT helps immensely in reducing or relieving the stresses created in parent metal bending and forming process it undergoes (if any) in the process of manufacture. Clause 331 of ASME B 31.3 spells out the requirements and guidelines for PWHT applicable to piping welds.

12.2.1 GENERAL REQUIREMENTS

Table 331.1.1 of ASME B 31.3 provides temperature and soaking time requirement against various material groups (P Nos specified in QW 422 of ASME Section IX) and for thickness ranges.

PWHT temperature and soaking time to be used in actual production shall be as specified in respective WPS based on recommendation as in Table 331.1.1 and other considerations as required in technical requirements of the project. It may also be noted that the very same preheat/PWHT parameters shall be used while establishing the WPS through PQR test as well.

Other engineering requirements if any specified within technical specification shall be above code requirements and shall in no way relax any of the mandatory requirements specified in ASME B 31.3.

12.2.2 GOVERNING THICKNESS FOR DETERMINING PWHT TEMPERATURE AND SOAKING TIME

In piping, thickness of thicker part to be welded is taken as the governing thickness for selection of temperature and time from Table 331.1.1. When components are joined by welding, thickness to be used in applying heat treatment provisions of Table 331.1.1 shall be that of the thicker component measured at the joint. However, exemptions as below are also applicable in determining governing thickness, especially in the case of branch connections where an ambiguity is possible.

For branch connections, metal other than weld metal added as reinforcement (whether an integral part of a branch fitting or attached as a reinforcing pad or saddle) shall not be considered in determining heat treatment requirements, whereas heat treatment is required when thickness through weld in any plane through the branch is greater than twice the minimum material thickness requiring heat treatment, even though the thickness of components at joint is less than the minimum thickness.

Thickness through the weld in joint configurations shown in Figure 328.5.4D of ASME B 31.3 shall be computed using the following formulae (Figure 12.1):

Heat treatment is required only when thickness through weld in any plane is more than twice the minimum material thickness requiring heat treatment (even though thickness of components at the joint is less than the minimum thickness requiring PWHT) for the following types of joints:

- Fillet welds of slip-on and socket welding flanges and piping connections of NPS 50 (2″) and below
- For seal welding of threaded joints in piping of NPS 50 (2″) and below
- For attachment of external non-pressure parts such as lugs or other pipe-supporting elements in all pipe sizes

However, the above requirements have following exemptions based on material classifications:

- For P-No. 1 materials when weld throat thickness is 16 mm (5/8″) or less, regardless of base metal thickness
- For P-No. 3, 4, 5 or 10A materials when weld throat thickness is 13 mm (1/2″) or less, regardless of base metal thickness, provided that not less than the recommended preheat is applied, and the specified minimum tensile strength of base metal is less than 490 MPa (71 ksi)
- For ferritic materials when welds are made with filler metal addition, which does not air harden. Austenitic welding materials may be used for welds to

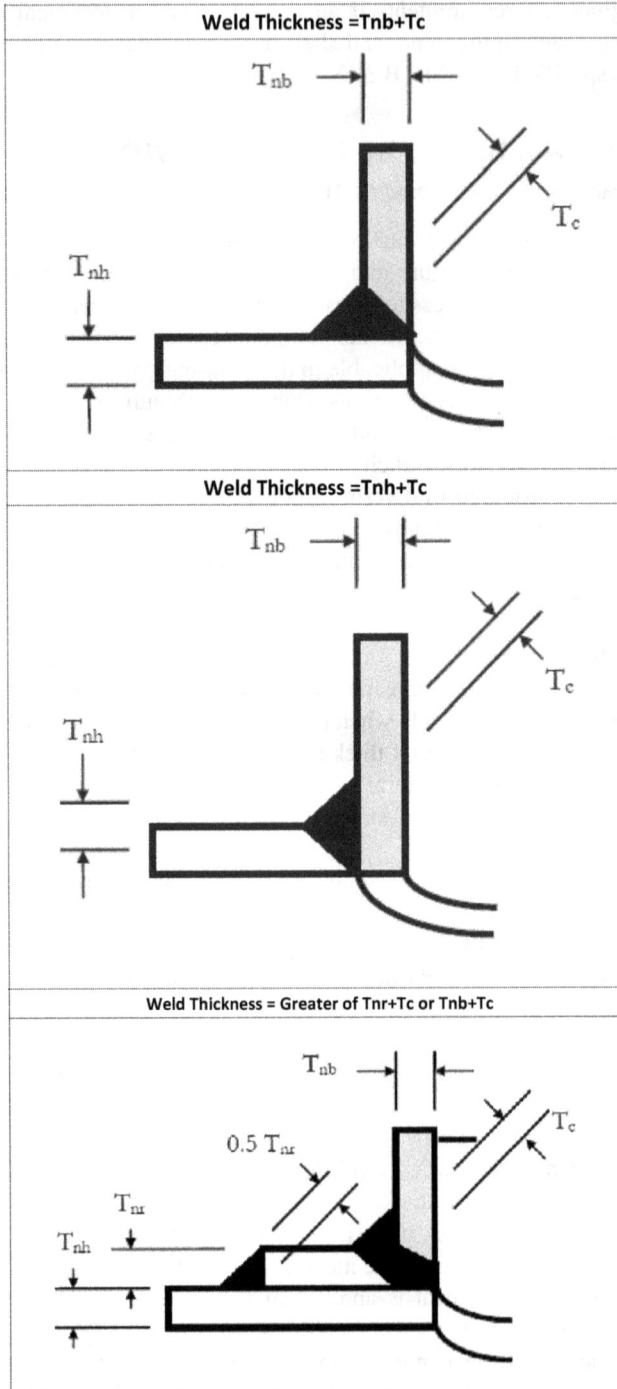

FIGURE 12.1 Governing thickness calculation (ASME, 2022).

(Continued)

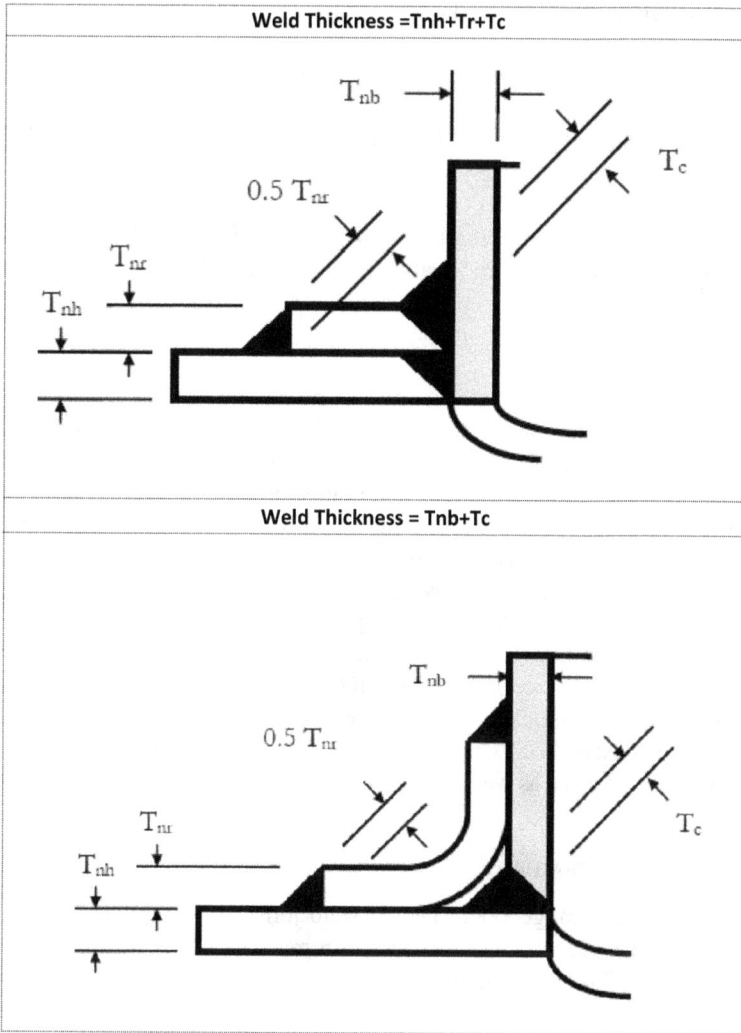

FIGURE 12.1 (*Continued*) Governing thickness calculation (ASME, 2022).

ferritic materials when the effects of service conditions, such as differential thermal expansion due to elevated temperature or corrosion, will not adversely affect the weldment.

12.2.3 HEATING AND COOLING METHODS

Heating method used shall be capable of producing required temperature and with fair uniformity while heating with proper temperature control and recording features.

Following types of heating methods may be used in PWHT of piping welds:

- Enclosed furnace, electrical, gas or diesel fired
- Local flame heating
- Electric resistance or electric induction heating
- Exothermic chemical reaction

Cooling method adopted also shall meet the requirements (cooling rate specified) and include cooling in a furnace, in air, by application of local heat or insulation or by other suitable means.

12.2.4 HEATING RATE

In PWHT, weld material is heated at a controlled rate to avoid detrimental effects caused due to thermal gradients related to uncontrolled heating.

Commonly used heating rates fall in the range of 60°C–200°C/hr. ASME B 31.3 does not indicate any specific heating rates and hence the requirements of ASME Section VIII Div (1) can be used as a guidance, the essence of which is given below.

When furnace heat treatment is used, temperature of the furnace shall not exceed 425°C (800°F) at the time of loading pipe spools. This implies that temperature control and recording shall start from this temperature onwards. Above 425°C (800°F), rate of heating shall not exceed 222°C/hr (400°F/hr) divided by the maximum metal thickness of the piping spools in inches placed in one charge of furnace. However, in no case shall this exceed 222°C/hr (400°F/hr). The same heating rate can be applied to other modes of PWHT as well.

12.2.5 SOAK TEMPERATURE

It is the temperature range where heated weld joints (or component) are held for some period of time, recommended based on material of construction and thickness at weld joint. Soaking as above helps immensely in reducing stress level within weld. For carbon steels, this falls in the range of 593°C–643°C (1,100°F–1,200°F). Table 311.11.1 of ASME B 31.3 provides required soaking temperature based on parameters like chemical composition, wall thickness of piping and minimum specified tensile strength of material of piping. In addition to this, any further stringency imposed by client/consultant specification also shall apply.

12.2.6 SOAK TIME

It is the specific time period for which welds or pipe spools are retained at sock temperature specified above, in order to have uniform heat throughout weld thickness. Pertinent aspect here is to predict ideal soak time required to homogenize heat throughout the weld material thickness, essential for reduction of undesirable stresses within weld. Normally codes require soak time based on metal thickness at weld in the range of 1 hour per inch thickness of weld joint. Just as in the case of soak temperature, Table 311.1.1 of ASME 31.3 also provides recommended soaking time.

During heating and soaking periods, if in furnace, furnace atmosphere shall be so controlled as to avoid excessive oxidation of surface of piping spools and flange faces. Flange faces shall be properly protected from flame impingement and any other possible damages.

12.2.7 COOLING RATE

Just as in the case of heating, cooling rate also needs to be controlled and monitored carefully in PWHT to make it more effective, by avoiding consequential detrimental effects caused by thermal gradient that may occur in uncontrolled cooling. Cooling starts after satisfactory completion of soaking. Since ASME 31.3 is silent on cooling rate as well, cooling rates proposed in ASME Section VIII Div (1) can be used as guidance, the gist of which is given below.

As in heating, above 800°F (425°C), cooling shall be done at a rate not greater than 280°C/hr (500°F/hr) divided by maximum metal thickness of pipe spools in inches placed in furnace under one charge, but in no case more than 280°C/hr (500°F/hr). Above cooling rate can be adopted for other modes of PWHT as well. From 425°C (800°F), spools can be cooled in still air.

Entire cycle from start of heating through soaking and cooling down to 425°C (800°F) as specified above constitutes a PWHT cycle. Though minimum temperature specified for start of PWHT cycle is 425°C (800°F), it is highly recommended that recording starts from 300°C onwards till it reaches 300°C back after soaking. This implies that still air cooling often starts below 300°C only.

12.2.8 TEMPERATURE VERIFICATION

Temperature of components heat treated are often verified by thermocouple pyrometers connected to recorders, both with valid calibration. Attachment of thermocouple is often done using capacitor discharge method of welding. All measuring devices used in PWHT shall have valid calibration at the time of performing PWHT.

12.2.9 HARDNESS TESTS

Hardness tests of production welds are often carried out to verify satisfactory heat treatment. Hardness limits specified in Table 331.1.1 apply to both weld and heat-affected zone (HAZ) tested as close as practicable to edge of the weld.

- Where a hardness limit is specified in Table 331.1.1, at least 10% of welds in each furnace heat-treated batch and 100% of those locally heat treated shall be tested.
- When dissimilar metals are joined by welding, hardness limits specified for base and welding materials in Table 331.1.1 shall be met for each material.

For sour service applications, many client specifications restrict final hardness values still further and may even ask for more hardness survey percentage.

12.2.10   PWHT OF WELD JOINTS BETWEEN DISSIMILAR MATERIALS

In the case of dissimilar metal welds within ferritic range, higher of the temperatures indicated in Table 331.1.1 (against respective P-No.) shall be used in PWHT. Whereas in the case of weld joints between ferritic and austenitic components, temperature applicable to ferritic material (as given in Table 331.1.1) shall be used if no other temperature is put forward by the applicable engineering specification.

12.2.11   DELAYED HEAT TREATMENT

Often PWHT of pipe welds is carried out after NDT and because of the same, weld joint shall be at atmospheric temperatures. Therefore, cooling down after welding to atmospheric temperature also needs to be properly controlled to avoid detrimental effects due to sudden cooling. Similarly, while heating it up again for PWHT, heating rate, soaking time and final cooling rate shall be within the range specified in code or in applicable specifications.

12.2.12   HEAT TREATMENT IN PARTS

In furnace heat treatment, when a piping spool cannot be accommodated into the furnace, ASME B 31.3 permits to heat-treat spool in more than one heat. An overlap of at least 300 mm (1 feet) between successive heats shall be ensured in that process and parts of the assembly protruding outside the furnace shall be protected from harmful temperature gradients through blanketing.

12.2.13   LOCAL HEAT TREATMENT

When entire pipe spool does not require PWHT, ASME B 31.3 permits local heat treatment of girth and other welds, such as branch connections, etc. In such cases, a circumferential band of run pipe and of branch welds shall be heated until the specified temperature range is achieved. It shall be ensured that entire section band all around the weld is heated to required level and blanketing shall extend much beyond (25–50 mm) the heating band to ensure gradual thermal gradient. Preheat and Post-Weld Heat Treatment requirements as required in Tables 330.1.1 and 331.1.1 of ASME B 31.3 are reproduced in Appendix E.

13 Installation, Welding, Inspection and Testing of Piping Spools

13.1 COMPLETED PIPING SPOOLS

It is presumed that piping spools ready for installation on site, either through field welding or bolting, are ready in all respects including hydrostatic testing and painting so that after installation and welding/bolting, painting applicable on site shall be limited to application of that to field weld joints and touch-up at damaged locations during installation.

13.2 OTHER MISCELLANEOUS COMPONENTS TO BE INSTALLED AND WELDED OR BOLTED

Engineering, Procurements and Construction (EPC) contractor shall usually be responsible to fabricate miscellaneous piping components like flash pots, seal pots, sample cooler, etc., and supporting elements like turn buckles, extension of spindles and interlocking arrangement for valves, operating platforms, etc., as required in the contract. These items shall also be ready on site after necessary pre-testing and inspection at respective manufacturing locations (either in own workshop or at sub-contractor's premises) ready for shifting to site and for installation in pipe rack or along piping.

13.3 CLEANING OF PIPING SPOOLS BEFORE INSTALLATION OR ERECTION

Before installation of pre-fabricated spools or pipe lengths or fittings, they shall be cleaned inside and outside by suitable means. Cleaning process shall include removal of all foreign matter such as scale, sand, weld spatter chips, etc. (if not painted already) by wire brushes, cleaning tools and blowing with compressed air or flushing out with water for inside surface. Some specific fluid services may even require special cleaning requirements, if indicated in piping material specification or isometrics or in line list.

13.4 PIPING ROUTING

No deviations from specified piping route indicated in drawings shall be permitted without consent of field engineering team. In case changes are made on site with the concurrence of field engineering team, design change notes shall accompany such changes and same shall be incorporated in "as-built" piping drawing to be prepared later for records.

Pipe-to-pipe, pipe-to-structure/equipment distances/clearances as shown in drawings shall be strictly followed, as these clearances may be required for free expansion of piping or equipment. No deviations (beyond permitted tolerance) from these clearances shall be permissible without approval of field engineering team of EPC contractor.

13.5 SLOPES

In case any slopes are specified for various lines either in drawings or piping and instrumentation diagram (P&ID), the same shall be maintained as required. Slope is provided for draining the line and is very critical in hydrocarbon or other inflammable or toxic liquid-carrying lines. If by any chance it is found unfeasible to provide slope as required, changes shall be made on site, in consultation with field engineering team of EPC contractor and all such changes made shall be incorporated through design change notes (DCN) or through revision of applicable drawings as appropriate for the situation. DCNs are often issued for small changes, whereas large changes affecting many systems may require total revision of affected drawings.

13.6 FIT-UP AND WELDING OF FIELD WELD JOINTS

Two types of field weld joints are possible, butt-welded groove and socket (fillet) welded joints. While butt and socket welds are pressure-retaining welds in piping, fillet welds are required for attaching structural members to the piping to provide rigidity/flexibility for piping as required, based on configuration and operational parameters, principally the operating temperature.

Bevel preparation for field weld joints is the same as that used in shop fabrication of spools, both for butt and socket welds. Difference if any felt during fit-up of the joint may call for huge force to align the field weld joint. Because of rigidity already gathered by spool due to its intricate configuration, bringing the field joint exactly matching the mating pipe or fitting of another spool may require some trial-and-error grinding operations as well as some stretching, mainly attributable to dimensional tolerances permitted in spool fabrication. It may further be noted that all spools are provided with 25–50 mm extra length to make site corrections to match the mating spool both dimensionally and to obtain a desired "V" preparation for the field joints. Excessive stress on spools shall not be applied to align the spools to have matching weld edges or joints, which may result in excessive locked-in stresses within weld joint. This may give rise to development of cracks in weld or heat-affected zone (HAZ) during service life of piping, especially in the case of piping designed for low-temperature service or for cyclic loading conditions.

Upon completion of fit-up, welding is to be carried out in most of the cases on top of pipe racks or at heights requiring temporary scaffolding. Since mobility of spools is totally absent in field welds, for welding a field joint (either butt or socket), welders with all position qualification as well as thickness range using applicable welding procedure specification (WPS) need to be deployed.

Field welds shall also essentially follow applicable WPS and upon completion of welding, it shall be subject to visual inspection, followed by non-destructive testing (NDT). Upon satisfactory clearance of all applicable NDTs for the joint, it can be released for post-weld heat treatment (PWHT) if applicable and thereafter for hydrostatic testing of the completed line. Surface preparation and painting of field welds are usually carried out after hydrostatic testing of completed line. Surface preparation and painting methodology adopted for field welds shall not adversely affect the already coated pipe spools in near vicinity.

13.7 FLANGE CONNECTIONS

While fitting up mating flanges, utmost care shall be taken to align pipe pools properly and to check trueness of flanges, so that faces of mating flanges can be brought together, without inducing much stresses in pipe spools and from there to connected equipment nozzles. Extra care shall be taken for flange connections to rotating equipment like pumps, turbines, compressors and static equipment like cold boxes, air coolers, etc. Flange connections to these equipment shall be checked for misalignment, excessive gap, etc., after final alignment of the static/rotating equipment is over. Fit-up of such critical joints, welded or bolted, is often inspected by all stakeholders like contractor, consultant and client as applicable (sometimes including representative from the connected equipment manufacturer as well). Temporary protective covers shall be retained on all flange connections of pumps, turbines, compressors and other similar equipment, until the piping is finally connected, to avoid any foreign material from entering these equipments.

Assembly of a flange joint shall be done in such a way that gasket between these flange faces is uniformly compressed. To achieve this, bolts shall be tightened in proper sequence. All bolts shall extend completely through their nuts but not more than 6 mm (1/4″) or by the specified number of threads above the nuts (usually three threads).

13.8 VENTS AND DRAINS

High-point vents and low-point drains are required in piping for venting and draining of lines, and this requirement is generally included in technical specifications for piping. Often, there will be standard designs approved for vents and drains and provided at highest and lowest points, respectively, even if it is not indicated in technical specifications for piping.

13.9 VALVES

Valves shall be installed with spindle/actuator orientation/position as shown in layout drawings and to be viewed from safe operability and maintenance points, especially with regard to those placed at heights on pipe racks or other similar places. In case of any difficulty in doing this or if the spindle orientation/position is not shown in drawings, site engineering team shall be consulted and to be installed as instructed. While installing unidirectional valves bearing "Flow direction arrow" on valve body, make sure that flow direction shown on valve is in line with that required by process. If direction of flow (using an arrow) is not marked on such valves, generally it can be assumed that the valve is bidirectional. However, it would be better to consult field engineering team prior to installation.

Additional works like extension of stem for better operability, locking and interlocking arrangements of valves (if called for), etc., shall be carried out as per drawings or as directed by field engineering team.

13.10   INSTRUMENTS

Installation of inline instruments such as restriction orifices, control valves, safety valves, relief valves, rotameters, orifice flange assembly, venturi meters, flowmeters, etc., also shall form a part of piping installation/erection work. Fabrication and installation of piping up to first block valve/nozzle/flange for installation of offline instruments for measurement of level, pressure, temperature, flow, etc. shall also form part of piping construction work. The limits of piping and instrumentation work will be shown in drawings/standards/specifications. Regarding package equipment ordered with specialized manufacturers like compressors, turbines, etc., piping within battery limits specified for that shall fall within the package vendor's responsibility. However, all piping related to process flow lines within battery limit shall follow the piping specifications applicable for the service as per contract. Orientations/locations of take-offs for temperature, pressure, flow, level connections, etc. shown in drawings shall be maintained within the specified tolerances for the same.

13.11   LINE-MOUNTED EQUIPMENT/ITEMS

Installation of line-mounted items like filters, strainers, steam traps, air traps, desuperheaters, ejectors, sample coolers, mixers, flame arrestors, sight glasses, etc., including their supporting arrangements shall also form part of piping erection work and they are also to be installed as specified in the P&IDs, isometric drawings, etc., as applicable.

13.12   BOLTS AND NUTS

Bolting to be used in plant piping shall be in accordance with the piping material specification. PMS shall specify clearly the type of bolting (with material specification), whereas material requisition provides diameter and length of each of the bolts, arrived at based on diameter and pressure rating of flanges to be bolted and also considering thickness of gasket to be used between them. Various types of coatings commonly applied to bolts are elaborated in Section 13.6, and this information shall be reflected in the piping material specification (PMS) against bolting specified in each piping class.

13.13   PIPE SUPPORTS

Pipe supports are designed and placed at pre-determined places to effectively accommodate dead/live weight of piping, thermal effects of piping system due to temperature of service fluid and vibrations induced by flowing fluid. Standard practice is to show location and type (design) of pipe supports in the drawings for line sizes NPS 50 (2″) and above, whereas for sizes NPS 40 (1½″) and below, it shall be provided based on general specifications for piping included as part of the contract. Even after providing supports as required in drawing, it may require additional supports on site depending on site condition and configuration of piping, which are often provided based on mutual agreement.

Temporary supports are permitted to facilitate safe installation of piping. Attaching temporary supports to pipes by welding is not permitted. All temporary supports shall be removed completely upon completion of field joint and thereafter piping shall have only the specific supports at pre-determined positions as contemplated in isometric drawing.

Piping shoes and sliding pipe support attachments shall be centred over the concrete or steel support beams before any field welding or bolting to the pipe is carried out.

13.14   INSTALLATION AND INTERCONNECTION OF PIPING SPOOLS AND OTHER INLINE EQUIPMENT

13.14.1   GENERAL REQUIREMENTS

- All piping shall be installed in accordance with drawings and specifications.
- Flanges or unions shall be installed to allow for easy installation and dismantling during maintenance of intervening equipment or valves.
- Changes to pipe routing (from the drawing) on site may often be required to avoid interference of existing piping or vessels, especially in brown field (or modification) projects. In such instances, all such changes shall be incorporated in "as-built" drawings by the field engineering team.
- Straight run pipes shall not be pulled through the pipe racks, unless supported on rollers. In case of coated pipes, installation methodology shall be so chosen that the coating is not damaged during installation.
- Extreme care shall be taken during handling and installation of expansion joints. Shipping rods (restraints) shall remain in place until installation and welding or bolting is completed. However, restraints shall be removed immediately after expansion bellow is installed. Proper care shall be taken to prevent ingress of moisture or dirt into the expansion joint during erection, which may affect the performance of bellow during service.
- A straight run of pipe shall contain only minimum number of welds. Use of off-cut pipes to form a long straight stretch of piping shall be avoided. Proximity of piping butt welds shall be a minimum of 50 mm or four times the thinnest wall thickness (at the joint), whichever is greater, measured between HAZs.

- Cold springing in piping, resulting out of forcefully held segments for fit-up, welding/bolting for making a joint is not permitted unless specified in drawings.
- Misalignment beyond acceptable tolerances in straight pipe runs shall not be permitted. Seam orientation of welded straight pipe and pipe to fittings shall be offset at circumferential welds. The longitudinal welds shall be staggered over the top centre line on both sides preferably by 30° on either side (left and right) of the centre line. Minimum distance between staggered joints shall be 50 mm or six times thinner pipe wall thickness (to be joined), the distance measured between HAZs, whichever is greater. Care shall be taken to ensure that longitudinal welds clear branch connections.
- Pipes passing through concrete walls or floors shall not be cast in, but shall be passed through cast in sleeves. Sleeve shall be of standard weight pipe having an internal diameter large enough to give 12 mm radial clearance to the passing flange, including its lagging, where applicable, and shall have a light infill if required (if air passage is to be restricted).
- Where pipework is to pass through floor plate and grid mesh, a 75 mm high collar having an internal diameter large enough to give minimum 25 mm clearance to passing flange shall be welded to the floor plate/mesh to prevent chaffing and for safety.
- Installation of piping with special types of connections like "Victaulic" couplings or similar shall be in accordance with the manufacturer's recommendations.

13.14.2 INSTALLATION OF VALVES

- Valves shall be installed with stem orientation as indicated on piping drawings. Hand wheels and levers shall be easily accessible for operation from grade or platform.
- Valves shall not be installed with their stems projecting into walkways. All valves located underground or in trenches shall be provided with valve boxes and extension stems as required in drawings.
- Relief valves shall be installed in an upright position and accessible from a platform or grade. Relief valves or scour valves discharging hazardous liquids or gases shall be piped to a safe and environmentally acceptable location.
- Chain wheel operators shall be provided for valves with hand wheels, placed at a height more than 2,050 mm above the operating level. Chains shall clear operating floors by 900 mm and shall not hang in access areas.

13.14.3 PIPING FOR ROTATING EQUIPMENT

- Pump and compressor piping shall be installed up to a break point between nearest pipe support and equipment. Remainder piping shall be site measured, fabricated and properly fitted between equipment nozzle and its break point. It is essential that such connections are made as accurately as possible to avoid any external loadings on equipment connections.

- After welding of field joints, piping shall be disconnected from equipment after installation to demonstrate and ensure that no stress has been transferred from piping to equipment, and then shall be reinstalled.
- Inlet lines of turbines and suction lines of compressors, when constructed out of carbon steel, shall be pickled (chemically treated) to ensure that internal surfaces of pipe works are free from rust. Presence of rust and other similar loose material adhered to pipe internals may prove detrimental to functioning of turbines and compressors, especially during startups.
- Following procedure shall be applied for alignment of flanges to rotating equipment:
  - Install, level and bolt down the equipment on foundation (on shims) and grout the equipment on foundation using no-shrink grout. Alignment of driver to equipment shall be made in accordance with requirements spelled out by the manufacturer of the equipment.
  - Align pipework to equipment flanges to within manufacturer's tolerance and in accordance with applicable engineering specifications applicable in this case.
  - Recheck alignment of coupling (between driver and equipment) during bolting and tensioning of pipe flanges to the equipment flanges. Bolting up and tensioning of flanges shall in no way affect the alignment of coupling between driver and driven equipment.

13.14.4 FIELD WELDING AND TESTS

Field weld joints are set up after installation of pre-fabricated piping spools in position according to dimensions and orientations is shown in the isometric drawing. Fit-up of field weld joints is similar to that of any other spool weld fit-up carried out in spool fabrication shop. As mentioned earlier, care shall be taken not to induce much strain in the joint during the process to meet dimensional tolerance or to reach the connecting equipment. Excessive strain during fit-up, coupled with weld shrinkage stresses may cause failure of the weld during operation, especially when the lines are subjected to cyclic loading or cryogenic conditions or both, which is a very dangerous combination.

Fit-up and welding process is exactly the replica of process adopted in spool welding. Upon completion of welding, visual inspection and NDT follow.

Upon satisfactory completion of NDT, local PWHT of weld joint using electrical pad heating is carried out if PWHT is required as per code or as a service requirement imposed by client. In such cases, hardness survey of weld and HAZs also shall be required often as a corollary requirement.

On completion of above activities, line is now ready to have pipe supports, hangers, etc., as per drawing. A specific line segment can be considered complete only when all such supports and intervening piping elements and inline equipment are in place and necessary welding/bolting is completed.

13.15 FLANGED JOINT BOLTING

13.15.1 FLANGED JOINT BOLTING TORQUE

- When bolting gasketed and flanged connections, gasket shall be uniformly compressed to minimum torques as per the table below:
- Stud bolts of diameter 40mm and larger shall be tightened by torque wrenches or other tightening methods which result in uniform tightening of flanges.
- Bolt torques for valves and lined piping shall be in accordance with the manufacturer's recommendations.

Recommended Bolting Torques

Bolt Size (mm)	Bolt Size (inch)	N (m)
M12	1/2	55
M16	5/8	109
M20	3/4	190
M24	7/8	300
M27	1	450
M30	1⅛	640
M33	1¼	910
M36	1⅜	1,180
M39	1½	1,560
M42	1⅝	1,900
M45	1¾	2,440
M48	1⅞	2,850
M52	2	3,660

- Flanged connections shall be tightened sequentially, diagonally opposite in a clockwise or counter-clockwise order so that even tightening of gasket across entire seating area is obtained.
- Flanged spools mating with equipment flanges shall be field fitted to required alignment to ensure that no stress or load is transferred onto equipment flange and the spool flange shall then be fully welded in accordance with applicable welding procedure specification.
- Before flanged pipe spools are connected to static equipment like pressure vessels, heat exchangers and other flanged equipment, inspection shall be carried out by loosening up to 10% of flanged or union joints to equipment to ascertain that no undue stress is exerted on equipment due to misalignment.
- Finish of flange faces (gasket seating area) shall be in accordance with requirements in piping material specification developed for the project. Moreover, surface of flange face shall be free from rust, weld spatter, scars, paint, dents, arc strikes, corrosion pitting and other imperfections as well to obtain proper pressure tightness during testing and prolonged service.
- All flange connections shall be made using fully threaded stud bolts and nuts. A minimum of one and a maximum of three complete threads shall protrude from the nut after completion of tightening (as required in table above).

- Insulating gaskets shall be installed when shown in piping drawings. Utmost care shall be taken not to damage bolt sleeves and gaskets while installing insulation gaskets. This type of gaskets is often used in dissimilar metal bolting, wherein the service fluid acts as an electrolyte.
- Bolts and nuts shall be protected by an anti-seize compound, in addition to coatings (if any) specified for bolts/nuts as in PMS.

13.15.2   FLANGE BOLT-UP PROCEDURE

Bolt tightening is an activity of paramount importance in plant piping. However, lack of knowledge, experience and guidance is often visible in this front as many of the readers would admit. There are various types of tools available to achieve specified torque value such as impact wrench, torque wrench or a stud tensioner (Figure 13.1). Selection of proper tool depends on stud bolt size, physical location of flanged joint and criticality of the flange. Procedure prepared for bolt tightening shall clearly identify proper tools to be used in each instant and this procedure shall be available with the group responsible for work and they shall be able to obtain specified tools for carrying out the tensioning in accordance with the following procedure.

Step 1	Align flanges and gasket.	Forced tightening is not allowed to overcome non-acceptable alignment tolerances. Clamp securely in place.
Step 2	Apply lubricant to stud threads over length and nut engagement and to face of nut which contacts flange.	Ensure that the nuts run freely down the thread of the studs.
Step 3	Install all studs and nuts hand tight. Ensure that studs pass freely through the flange holes. Position the nut on one end of the stud such that only crown of the stud projects beyond face of the nut.	The excess stud length should project beyond nut on the other side.
Note	*By doing the above, nut that is installed nearly flush with end of stud can be easily removed since threads are not coated, and normally have not been subjected to corrosion. The side of stud with flush nut should be chosen by taking into consideration factors such as whether one side has better access for maintenance personnel and/or tightening tools, for example, torque wrench or impact wrench, etc.*	
Step 4	Number each stud according to its position in the flange as shown in figures below for stud tightening sequence.	
Step 5	Tighten studs per sequence derived based on number of studs with an appropriate tool such as an air impact wrench or equal.	
Step 6	For joints containing RTJ or Spiral Wound Gaskets, repeat Step 5.	
Step 7	Tighten the stud bolts in stages to obtain final required torque from the specified torque table or from table given above. The first stage shall not be more than 30% of final torque. The final torque shall be within ±5% of the required torque value.	Apply the torque evenly to each stud following the stud bolt tightening sequence. The final torque must be within ±5% of the required values per Step 5 above.

FIGURE 13.1 Commonly used torque wrench.

Inspector shall verify that manufacturer's instructions are followed for this operation and also shall be aware of the limitation and maintenance done on these torque wrenches used to perform bolt tightening. In addition, the inspector shall keep track of torque wrench calibration to ensure that it is performed in accordance with manufacturer's recommendations, or as required by the piping engineering team.

Inspectors are expected to ensure the following steps with regard to bolt tightening of flanges wherein controlled torque is specified.

13.15.3 BOLT TIGHTENING SEQUENCE

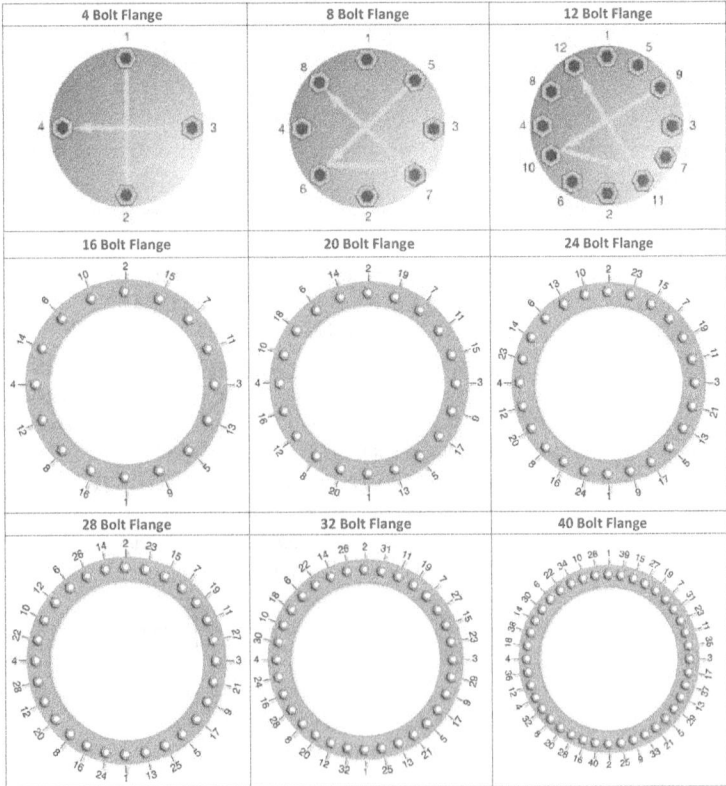

FIGURE 13.2 Bolt tightening sequence (Enerpac, 2022 and Westorc, n.d.).

(Continued)

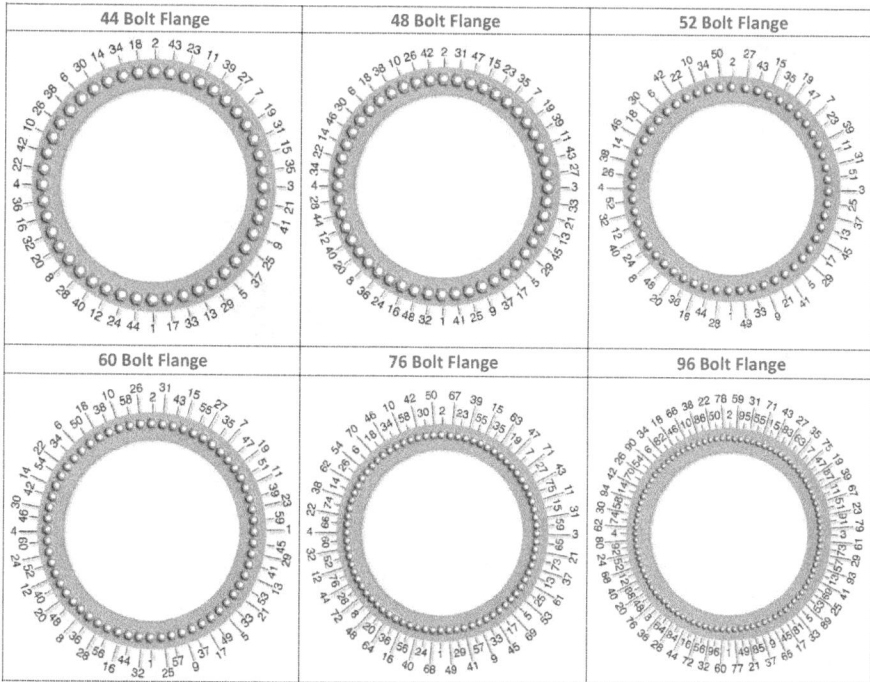

FIGURE 13.2 (*Continued*) Bolt tightening sequence (Enerpac, 2022 and Westorc, n.d.).

13.15.4 Recommended Torque Value Chart

Use this chart as a guideline to correct torque to be applied to standard size metric and imperial bolts in grades 8.8 (metric) and ASTM A193 grade B7 (imperial) or similar. Torque figures are calculated in both metric (Nm) and imperial (lbs ft.) values using a choice of three commonly used bolt thread lubricants. Always consider coefficient of friction applicable for chosen bolt lubricant. For calculating these values for grade 10.9 bolts add 40% and for grade 12.9 bolts add 70% to figures detailed against each of 8.8 grade metric stud bolt size. Remember these torque values are for guidance purposes only. Always check with equipment/bolt manufacturer for actual torque required and specified for bolted components within the particular equipment design.

Bolt Diameter (Note 1)		Nut AF Size (Note 2)		Bolt Tension (Note 3)		Torque for Specified Lubricant Nm (lbs ft.) w(Note 4)		
(mm)	(in.)	(mm)	(in.)	(kN)	(lbs force)	Moly (Friction = 0.06)	Cooper (Friction = 0.10)	Machine Oil (Friction = 0.15)
16		24		60	(15,760)	102 (75)	155 (114)	184 (136)
	5/8		1.1/16	67	(15,218)	99 (73)	150 (110)	179 (131)
	3/4		1.1/4	96	(21,652)	175 (129)	263 (194)	314 (231)
20		30		99	(24,730)	202 (148)	305 (224)	363 (267)
	7/8		1.7/16	134	(30,135)	277 (204)	423 (311)	501 (368)
24		36		159	(35,634)	346 (254)	526 (387)	621 (457)
	1		1.5/8	199	(44,981)	416 (307)	631 (464)	746 (549)
27		41		210	(47,153)	505 (372)	774 (569)	913 (672)
	1.1/8		1.13/16	261	(58,622)	629 (463)	965 (709)	1,081 (795)
30		46		255	(57,302)	687 (505)	1,051 (773)	1,177 (866)
	1.1/4		2	329	(74,068)	872 (642)	1,348 (991)	1,505 (1,107)
33		50		319	(71,695)	928 (682)	1,428 (1,050)	1,592 (1,171)
	1.3/8		2.3/16	407	(91,316)	1,169 (860)	1,820 (1,338)	2,026 (1,490)
36		55		374	(84,088)	1,198 (881)	1,837 (1,351)	2,067 (1,520)
	1.1/2		2.3/8	492	(110,369)	1,526 (1,122)	2,352 (1,729)	2,654 (1,951)
39		60		451	(101,353)	1,546 (1,140)	2,384 (1,753)	2,646 (1,946)
	1.5/8		2.9/16	584	(131,226)	1,949 (1,433)	2,388 (1,756)	2,650 (1,949)
42		65		517	(116,220)	1,967 (1,446)	3,037 (2,233)	3,386 (2,490)
	1.3/4		2.3/4	684	(153,888)	2,444 (1,797)	3,856 (2,835)	4,301 (3,163)
45		70		606	(136,153)	2,388 (1,756)	3,690 (2,714)	4,096 (3,012)
	1.7/8		2.15/16	793	(178,353)	3,021 (2,221)	4,781 (3,515)	5,317 (3,910)
48		75		680	(153,016)	2,881 (2,119)	4,444 (3,267)	4,933 (3,627)
	2		3.1/8	910	(204,621)	3,676 (2,703)	5,835 (4,291)	6,482 (4,766)

(continued)

Bolt Diameter (Note 1)		Nut AF Size (Note 2)		Bolt Tension (Note 3)		Torque for Specified Lubricant Nm (lbs ft.) w(Note 4)					
						Moly (Friction = 0.06)		Cooper (Friction = 0.10)		Machine Oil (Friction = 0.15)	
(mm)	(in.)	(mm)	(in.)	(kN)	(lbs force)						
52		80		819	(184,051)	3,690	(2,713)	5,718	(4,204)	6,347	(4,667)
56		85		944	(212,237)	4,578	(3,366)	7,081	(5,207)	7,860	(5,779)
	2.1/4		3.1/2	1,168	(262,571)	5,250	(3,860)	8,372	(6,156)	9,297	(6,836)
60		90		1,105	(258,590)	5,707	(4,196)	8,838	(6,499)	9,810	(7,213)
	2.1/2		3.7/8	1,458	(327,737)	7,223	(5,311)	11,561	(8,501)	12,827	(9,432)
64		95		1,250	(281,106)	6,835	(5,026)	10,595	(7,791)	11,760	(8,647)
68		100		1,435	(322,736)	8,328	(6,123)	12,961	(9,530)	14,387	(10,578)
	2.3/4		4.1/4	1,611	(362,012)	8,735	(6,423)	14,031	(10,317)	15,519	(11,411)
72		105		1,633	(367,081)	9,820	(7,220)	15,326	(11,269)	16,935	(12,452)
	3		4.5/8	1,931	(434,028)	11,343	(8,340)	18,273	(13,436)	20,226	(14,872)
80		115		1,870	(420,273)	12,264	(9,018)	19,251	(14,155)	21,272	(15,641)
	3.1/4		5	2,280	(512,574)	14,451	(10,626)	23,335	(17,158)	25,796	(18,968)
	3.1/2		5.3/8	2,659	(597,647)	18,081	(13,295)	29,257	(21,513)	32,311	(23,758)
90		130		2,424	(545,035)	17,654	(12,981)	27,881	(20,501)	30,809	(22,653)
	3.3/4		5.3/4	3,066	(689,249)	22,252	(16,362)	36,075	(26,526)	39,834	(29,290)
100		145		3,050	(685,695)	24,410	(17,949)	38,742	(28,571)	42,810	(31,478)
	4		6.1/8	3,503	(787,381)	27,024	(19,871)	43,883	(32,267)	48,444	(35,621)
	4.1/4		6.1/2	3,133	(704,242)	25,605	(18,830)	41,647	(30,633)	45,956	(33,791)
	4.1/2		6.7/8	4,146	(932,023)	30,423	(22,370)	49,538	(36,425)	54,641	(40,177)

Note:
1. Bolt Material Grades 8.8 (Metric) and ASTM A193/BS4882 Grade B7 (Imperial) or similar
2. AF size based on heavy series nuts
3. Bolt tension equates to a bolt stress of 70% of the minimum yield strength
4. Torque figures detailed are also based on 70% of the minimum yield strength of bolt

(Source: Sanger Metal, n.d.)

13.16   CAUSES OF FLANGE LEAKAGE

Common or primary causes of flange leakage are directly related to poor or unprofessional installation and tightening of flanges, summarized below:

13.16.1   UNEVEN BOLT STRESS

An incorrect bolt-up procedure or limited working space near one side of the flange can leave some bolts loose while others crush the gasket. This is especially troublesome in high-temperature services, when heavily loaded bolts relax during high operating temperature.

13.16.2   IMPROPER FLANGE ALIGNMENT

Improper flange alignment, especially nonparallel faces, causes uneven gasket compression, local crushing and subsequent leakage.

13.16.3   IMPROPER GASKET CENTRING

If a gasket is off-centre compared to flange faces, gasket will be unevenly compressed and more prone to leakage.

13.16.4   DIRTY OR DAMAGED FLANGE FACES

Dirt, scale, scratches, protrusions or weld spatter on gasket seating surfaces provide leakage paths or can cause uneven gasket compression that may result in leakage.

13.16.5   EXCESSIVE LOADS IN PIPING SYSTEM AT FLANGE LOCATIONS

Excessive piping system forces and moments at flanges can distort them during operation and cause leaks. Common causes of this are inadequate flexibility, usage of excessive force to align flanges during erection and improper location of supports or restraints/expansion loops.

13.16.6   THERMAL SHOCK

Rapid temperature fluctuations can cause flanges to deform temporarily, resulting in leakage.

13.16.7   IMPROPER GASKET SIZE OR MATERIAL

Using the wrong gasket size or material also can result in leakage.

13.16.8   IMPROPER FLANGE FACING

A rougher flange surface finish than specified for spiral wound gaskets can result in leakage.

From the above, it can be seen that most of the flange leakages are caused because of simple oversight during manufacture and installation of piping and hence can be remedied to a considerable extent through earlier mentioned precautions.

13.17   SAFETY OF FLANGE JOINT ASSEMBLY

Though flanged joint assembly is relatively simple, straightforward and effective, petrochemical industries have experienced tragic incidents due to failures of flanged joints. Hence safety of flanged connection is of paramount importance in process industry piping and it principally depends on the following aspects:

- Selection of flange type, material and rating.
- Selection of gasket material, type and quality.
- Choice of appropriate surface finish of flange based on ratings and materials selected for flange and gasket.
- Adoption of right bolting torque, sequence and procedure.
- Good workmanship of the team assembling flange joints.
- Careful inspection during above stages.

Flanges generally have a very high safety margin in its design. Experience over the years with flanged connections and leaking incidents has proven beyond doubt that human factor is the main reason for failures in flanged joints.

13.18   INSULATED PIPING

- For pre-insulated piping or piping with shop-applied insulation every precaution shall be taken that insulated parts remain weatherproofed at all times during storage, handling and erection to prevent moisture from entering behind or into the insulation materials.
- The above (moisture/water ingress through cladding into insulation) becomes extremely important in the case of pipes operating at sub-zero temperatures, wherein ingress moisture/water can expand anomalously causing breaking of cladding.

13.19   STAINLESS STEEL PIPING

- Special precautions shall be taken during installation of stainless steel piping in close vicinity of carbon steel piping. Direct contact between stainless steel and carbon steel shall not be permitted during any stage of construction. Carbon steel blinds, spades and caps also shall not be used for stainless steel pipe and components, even as a temporary arrangement.
- Stainless steel clamps and U-bolts shall be used for supporting stainless steel piping. Or else, proper positive padding shall be available between the two at all times.
- Stainless steel or polytetrafluoroethylene (PTFE) or other similar material spacer strips of adequate size shall be installed in areas where stainless steel piping is expected to rest on carbon steel supports.

13.20 PLASTIC PIPING

- The installation and testing shall be in accordance with the applicable international standards, engineering specifications and all materials used shall be as mentioned in the PMS developed for the project.
- As known, various types of joining methods are available in polyethylene (PE) pipe connections. However, based on service requirements, preferred joining methods are often stated in engineering specifications or the PMS, mainly based on previous experience. The commonly used connection methods include the use of compression fittings or butt weld fittings or using mechanical joint fittings and hence to be judiciously selected in accordance with the preferences specified in engineering documents and considering limitations or restraints on site.
- Backing rings shall be installed when plastic PE pipe flanges are connected to valves. PE spacers shall be used when wafer-type butterfly valves are used, to ensure free movement of the disc.
- PE piping shall be laid so that it is capable of operating between maximum and minimum service temperatures. There shall be sufficient excess length between anchors to allow for expansion and contraction. Adequate allowances shall ensure that no excessive residual stresses remain in the piping prior to or after startup of plant operations. Where excessive expansion is expected, adequate flexibility shall be built in to accommodate the deflections.
- Piping shall be suitably anchored in accordance with the standard drawings at pre-determined locations specified in isometric drawing. Anchoring plastic pipes by clamping with saddles or U-bolts is not acceptable.
- U-bolt guides with two lock nuts shall be loosely clamped to allow free movement of the pipe during expansion and contraction.
- Un-plasticized polyvinylchloride (UPVC) pipe connections of the socket type shall be made using solvent cement as recommended by the manufacturer. Threaded joints shall be made using PTFE (Teflon) tape or thread seal compound, applied only on male threads.
- In slurry service, pipes shall be flanged at least every 20 m if not mentioned otherwise. In solution service, pipes shall be flanged at least every 40 m, unless specified otherwise. Spools may be pre-fabricated in convenient lengths suitable for storage, transportation, lifting and installation.
- All fittings in slurry service shall be flanged. Spools in solution service may be pre-fabricated in convenient lengths. Piping shall be protected against chafing.

13.21 LINED PIPING (RUBBER, POLYURETHANE AND PTFE LINING)

- Flanged spools requiring any of above types of lining shall be fabricated and lined in accordance with relevant codes, specifications and applicator's recommendations. The internal lining applied shall cover the face of the flange as well. Rubber lining is expected to cover full face of flange and polyurethane lining shall extend up to the inside of bolt holes. Quite often, fittings like bends, elbows, tees, etc. shall have flanged ends. Flanged spools with internal lining shall be transported to field and installed in accordance with applicator's recommendations, relevant codes and of course as configured in isometric drawing.

- Makeup pieces if any required shall be site measured, fabricated, lined and then installed at field to ensure that no external loads are exerted on interconnecting equipment or mating flanges. For rubber-lined flanged joints, gaskets are not required. However, for frequently dismantled joints 1 mm PTFE full face gaskets shall be used to prevent bonding of rubber. For polyurethane and PTFE lined flanged joints, gaskets shall not be used.
- Torque used for tightening bolting of flanged joints in lined pipes shall be in accordance with applicator's recommendations.
- Repairs to damaged lining shall be done by cutting the affected area to the base metal and relining the affected area in accordance with lining repair procedure established along with original lining qualification procedure. In addition, repair of lining (if required) shall be a hold point in inspection and test plan (ITP) for application of lining and may require the presence of inspectors at all stages of repair work depending on extent of repair. The inspectors shall also have the authority to call for complete stripping of the already done lining and a full rework, if repairs are found to be unsatisfactory during any stage of repair.

13.22   UNDERGROUND PIPING

- All exterior surfaces of underground steel piping, fittings and valves shall be coated and wrapped as specified, after a proper surface preparation suitable for the intended coating and wrapping.
- The external surfaces of steel pipe and fittings to be coated and wrapped shall be clean, dry and free of any oil and grease, using appropriate methods to obtain the required cleanliness.
- Wrapping material commonly used is PVC pressure-sensitive tape or approved equivalent 0.5 mm thick. Widths shall be 50 mm for pipe diameter up to NPS 50; 100 mm for pipe diameter NPS 80 to NPS 200; and 150 mm for pipe diameter NPS 250 and larger.
- The tape shall be spirally wrapped on the steel pipe and fittings starting from the above-ground portion (80–100 mm above ground level) with 12 mm overlap shall extend over the entire underground portion and up to 80–100 mm above the ground level.
- Underground pipework shall be laid in trenches. Trenches shall be of suitable width and generally 300 mm deep except at vehicle traffic locations where the trench shall be 600 mm or deeper based on specific requirements. Pipe shall be laid on a sand bed and backfilled with suitable backfilling material approved by consultant or client representative.

- Piping and fittings after coating and wrapping shall be inspected visually and thereafter for defects such as holidays prior to lowering into trench.
- All finished piping shall meet the requirements of spark tests to be applied with Holiday detector.
- Recommended voltage for detector shall be approximately 400 V for each 100 μm thickness or as recommended by manufacture of the wrapping tapes.
- All pinholes, voids, holidays, air bubbles, cracks and other breaks shall be carefully marked. Repairs may be made immediately following the coating and wrapping operation or may be deferred, but in any case, shall be repaired and re-inspected prior to laying.
- All coating repairs shall be made with the wrap smoothly applied and without wrinkles or buckles.
- Repair shall be at least equal in effectiveness to the coating applied to principal part of the line.
- All repairs shall successfully pass test using electronic holiday detector.
- Thickness of coating and wrapping shall be checked with suitable instruments by the inspectors and if found insufficient, same shall be remedied satisfactorily.
- Coated and wrapped steel pipes shall be laid directly in trench bottom by lowering pipe carefully into trench using canvas or leather slings to ensure that wrapping is not damaged during laying process.
- Any damage to coating during laying/installation shall be repaired immediately. When repairing damaged coatings, wrapping in defective area shall first be removed and the pipe re-wrapped in accordance with coating and wrapping repair procedure. Visual and electronic spark testing is again required to ensure quality of wrapping at repaired area.
- After inspection (visual and electronic) and completion of testing, trench shall be backfilled and compacted in accordance with procedure in place for backfilling and compaction.
- Cathodic protection, if required, shall be provided as per the applicable engineering specification and standards referred in the contract.

13.23 GALVANIZED PIPING

- Usually, pipes and fittings NPS 50 and below are supplied galvanized with threaded ends or as specified in engineering specification developed for the project and are often field erected.
- Pipe sizes NPS 80 and larger in galvanized piping shall be shop fabricated and hot dipped galvanized in accordance with standards called for in engineering specification.
- If any modifications are required on galvanized spools of sizes NPS 80 and over during installation, spool shall be modified and hot dipped galvanized again. For minor modifications and repairs, spool may be cold galvanized. Procedure for both dip and cold galvanizing shall be submitted for approval of all, prior to start of galvanizing process and shall essentially include the surface preparation requirements as well.

13.24   MINING HOSE

- Hoses if any to be used in plant premises shall be stored in a cool dry location away from direct sunlight and high temperatures.
- Mining hoses shall be handled carefully. Nylon slings shall be used and properly placed while handling mining hoses in order to support the hose properly. Care shall be taken not to kink the hose during handling (rolling up and using). While rolling up the hose, bend shall never be below ten times the diameter of the hose.
- While dragging the hose for use, care shall be taken not to drag it over sharp or abrasive surfaces.

13.25   PRESSURE TESTING

13.25.1   Test Requirements

After satisfactory completion of all category "A" punch items (described in sections below), the lines are prepared for hydrostatic/pneumatic testing as applicable.

Pressure testing shall be carried out in accordance with requirements spelled out in engineering specifications developed for the project and in accordance with applicable codes and standards referred in contract.

Pressure testing is required for all piping designed to convey or contain process or utility fluids to operate either at a positive or negative internal pressure, and shall be tested generally as follows:

- Shop hydrostatic testing of piping spools is not generally mandated in piping codes. However, in the case of piping spool fabrication wherein spool painting is also required to be completed prior to installation, spool hydrostatic testing is required and this shall be captured in isometric drawings or any other document as appropriate. Clients often require spool painting in shops to have better surface preparation and painting quality than those done on site due to controlled atmosphere available within workshop area.
- Similarly, for galvanized piping spools also, shop hydrostatic testing is essentially required prior to galvanizing.
- Pressure piping shall be either hydrostatically or pneumatically tested (often both tests shall be required) as specified in engineering specification for the project.
- For hydrostatic testing, usual test pressure required is 1.5 times the normal working pressures, which will often be shown in the isometric drawing and also in test pack system.
- Open-ended vent, drain and other similar piping systems operating at atmospheric pressure are not considered as pressure piping and need not undergo pressure testing, but shall be leak tested by filling the lines with water.
- Piping systems to be tested shall be divided into sections often referred to as "test packs".

- Procedure for carrying out hydrostatic test of pipe spools and the completed test packs shall be reviewed and approved by all concerned prior to start of testing.
- Reinforcing pads provided at branch connections shall undergo pad air test at 35 kPa gauge (5 psig). All weld surfaces of such branch connections (inside and outside) shall be swabbed with a leak testing solution and to be inspected for leaks. After testing is complete, the vent hole on reinforcing pad shall be plugged with heavy grease or silicon sealant. Records of all reinforcing pad tests need to be maintained.
- Short pieces of pressure piping which were to be removed to facilitate installation of test blinds shall be tested separately.
- Flanged connections at points where blinds are used during pressure tests do not require separate tests after test blinds have been removed.
- Threaded and socket weld connections shall be inspected thoroughly after makeup to ensure tightness.
- Re-testing of any "cut in" or repair work into a line already tested (site modification after test) shall generally be carried out to same procedure and test pressure as original test.
- Testing shall be carried out in presence of all stakeholders and pressure test reports shall be signed off by all parties witnessing the same.
- Ensure that insides of all pipes, valves, fittings and other associated equipment are clean and free from loose foreign matter prior to commencement of pressure test.
- Normally, intervening equipment shall be isolated or removed from test section. However, where equipment (which could be damaged by foreign debris) is included in a test section, temporary inline strainers shall be installed to mitigate the issue.
- Bench testing of pressure relief valves, calibration and setting of relief valves though required prior to testing is not addressed here.

13.25.2 TEST PACK SYSTEMS

It is always preferable to carry out final hydrostatic test of pipelines as a single large piping system as practicable, including its associated equipment. This is not considered advisable when the difference among their design pressures exceeds 15% of the lowest design pressure of the system components.

Test pressure of entire system so connected shall be equal to test pressure applied to the system component of which the design pressure is the lowest among the system as mentioned below.

Where piping systems of different design pressures are fully welded together, then "cascade" pressure testing will be required (i.e. complete the pressure test on the higher pressure system prior to system closure welding and then repeat full pressure test for the combined systems at the lowest pressure).

Equipment like heat exchangers, pressure vessels and fired heaters may be included in the test system, provided system test pressure does not exceed shop test pressure of any of the included equipment.

The following equipment and components are not normally included in piping test pack system and hence these equipment need to be isolated from the system.

- Rotary equipment such as pumps, compressors and turbines
- Safety valves rupture discs, flame arresters and steam traps
- Pressure vessels with sophisticated internals
- Equipment and piping lined with refractory
- Storage tanks
- Filters, if filter element(s) is not dismantled
- Heat exchangers of which tube sheets and internals have been designed for differential pressure between tube side and shell side
- Instruments such as control valves, pressure gauges, level gauges and flow-meter (excluding thermocouples)

Apart from the above, the following are also often excluded from system hydrostatic testing of piping on site:

- Any package unit previously tested by manufacturer in accordance with applicable codes.
- Plumbing systems, which are tested in accordance with applicable plumbing codes.
- Lines and systems which are open to atmosphere such as drain, vents, open discharge of relief valve and atmospheric sewers.
- Instrument impulse lines between block valve in process or utility line and the connected instrument.

Though instruments are normally excluded from test pack system, process lead lines shall be tested up to first block valve together with piping system.

Any vents or bypasses downstream of instrument's first block valves shall be opened or the instrument shall be disconnected during test to ensure protection of that instrument.

Medium for hydrostatic testing shall be water (unless otherwise specified), except when there is possibility of damage due to freezing of water or when piping material is expected to be adversely affected by water. In such cases, any other suitable liquid can be used with client approval.

- Testing with kerosene or other inflammable fluids or compressed air shall be avoided as far as possible. If it is required in contract, it shall be carried out with prior approval of the client.
- In case flammable liquid is to be used for testing, its flash point shall be not less than 50°C and consideration shall be given to the test environment.
- Pneumatic testing shall be considered for the following:
- Gas, steam or vapour lines when weight of hydro test liquid would over-stress supporting structures or pipe wall.
- Piping with linings subject to damage by hydro test liquid.
- Instrument air headers shall be tested with dry oil free air. Commodity test may be used if the instrument air system is completed and the instrument air compressor is operational.

13.25.3   TEST PACK PREPARATION

When a specific process piping line is to be tested, P&IDs for the line shall be identified to determine extent or limit of test. From so identified P&IDs, EPC contractor subdivides into suitable test packs considering feasibility and operational convenience. EPC contractor is free to revise extent of each test pack provided; the extent covered by each test pack added together covers the entire scope of test originally identified for the selected line. Further, all concerned shall be notified of such changes made.

13.25.4   TEST PACK LIMITS

Test pack limits shall be clearly identified through a marked-up P&ID followed by concerned isometrics with proper mark-up of blinded locations and bypassed/dropped inline equipment/valves.

13.25.5   TEST PACK DOCUMENTATION

Test pack documentation shall essentially contain following documents depending on applicability:

- Pressure test report for piping with test pressure and medium
- Marked-up PIDs
- Applicable isometrics with mark-up showing extent of coverage
- Blind check list
- Welding history report and welding joint identification on isometrics
- NDT report for all applicable weld joints
- PWHT and hardness test report as applicable
- Copies of reinforcing pad pressure test certificates (if necessary)
- Post-test punch list (conducted jointly by contractor and client)

13.25.6   SPECIAL NOTES ON TEST PACK DOCUMENTATION

- Test packs shall include latest available revision of each isometric covering scope of test pack range.
- Test limits shall be clearly identified in P&IDs and isometrics by highlight or through mark-up.
- Size (thickness) and location of all test blinds, including those required at instrument connections (e.g. orifice flanges), need to be highlighted.
- Identify location for connection of fill and drain points for the system.
- Vent points and drain points to be marked up.
- Requirements for isolation or removal of inline equipment and instruments.
- Location and range of pressure gauges to be used. Minimum (2) calibrated gauges enquired, one located at an accessible low point of the tested line and one gauge located at highest points on test manifold.
- Any other special requirements such as chloride content if applicable.

13.25.7 TEST MEDIA

Most common test media used is water, and care shall be taken to ensure use of clean water for hydrostatic tests. Sea water is often prohibited in hydrostatic testing of plant piping. Suitable filtration arrangement shall be provided at filling point to avoid inclusion of foreign matter such as sand, rust, etc., in proposed test water.

For hydrostatic testing of carbon steel piping systems, test medium shall be potable water at ambient temperature with pH value between 6 and 7.

For hydrostatic testing of piping systems of austenitic stainless steel, test medium shall be demineralized water with a chloride content of maximum 1 ppm (some client specifications permit up to 50 ppm chlorides) and pH between 6 and 7. Water shall be drained immediately after completion of hydrostatic testing and system dried out to avoid concentration of chlorides, by dry air blowing. For hydrostatic testing of piping systems with high nickel content, water used shall be checked for possibility of generation of hydrogen sulphide (H_2S) during test.

A report on water analysis including chloride content and pH value of water shall be attached to test report at all times when austenitic stainless steel systems are hydrostatic tested.

While above being the requirements for carbon and alloy steel piping, saline water can be used for testing of PE or UPVC piping.

13.25.8 TEST PRESSURE

During hydrostatic testing of piping designed for internal pressure, minimum hydrostatic test pressure at any point in the system shall be calculated and maintained as follows:

- Not less than 1½ times the design pressure.
- For a design temperature above test temperature, the minimum test pressure shall be as calculated by the following equation:

$$Pt = 1.5*P*St/S$$

where
- Pt = minimum calculated hydrostatic test pressure (kg/cm^2)
- P = internal design pressure (kg/cm^2)
- St = allowable stress at test temperature (kg/cm^2)
- S = allowable stress at design temperature (kg/cm^2) (see Table I, Appendix A, ASME 31.3)
- When St and S are equal, test pressure is 1.5 * P.

Where test pressure as defined above would produce a stress in excess of specified minimum yield strength at test temperature, test pressure shall be reduced to a pressure at which stress shall not exceed specified minimum yield strength at test temperature.

Maximum test pressure at which stress produced shall not exceed specified minimum yield strength may be calculated by the following equation:

$$Pm = 2*S*E*t$$

where
  - Pm = maximum test pressure (kg/cm^2)
  - S = specified minimum yield strength at test temperature (kg/cm^2)
  - t = specified pipe wall thickness minus mill tolerance (cm)
  - D = outside diameter (cm)
  - E = quality factor (see ASME B 31.3 Table A-1 B)

13.25.9 HYDROSTATIC TESTING OF PIPING DESIGNED FOR EXTERNAL PRESSURE (JACKETED LINES)

Lines meant for external pressure service shall be subjected to an internal test pressure of 1½ times the external differential design pressure but not less than a gauge pressure of 1.055 kg/cm^2 (15 psi).

In jacketed lines, internal line shall be pressure tested on the basis of internal or external design pressure, whichever is critical. This test shall be performed prior to completion of jacket.

Jacket shall be pressure tested on the basis of jacket design conditions.

13.25.10 PNEUMATIC TESTING

Pneumatic testing of a piping system is not permissible unless prior approval is obtained from the stakeholders. Many times, clearance for such testing is given based on stored energy calculations furnished by EPC contractor considering various risks involved.

Pneumatic test shall be performed generally in accordance with engineering specifications and codes referred therein (ASME B 31.3), at a test pressure of 1.1 times the design pressure in place of 1.5 times used in hydrostatic testing.

Dry nitrogen or clean and dry oil free air can be used as test medium.

Piping spread area (covered by test) and much beyond shall be cordoned off and access into cordoned-off area to be limited as safety precaution until pressure inside the line is brought down to atmospheric pressure.

Minimum metal temperature during pneumatic test of piping shall not be less than the temperature required by engineering design.

Piping system to be tested shall be protected by a relief valve, with set pressure equal to test pressure plus the lesser of 70 kPa or 7% of the test pressure.

Weak soapy water shall be applied on each and every weld joint covered by test pack and to be observed for leaks.

13.25.11 TEST DURATION

Codes and standards are often silent on duration of tests, either hydrostatic or pneumatic. However, as a minimum, test pressure needs to be maintained till inspection of all weld joints covered by test is completed. Therefore, for all practical purposes, duration ranging from ½ to 1 hour is considered ideal for the purpose, and most of the consultants through their specification insist for a duration of 1 hour for testing.

13.25.12 PREPARATION FOR TESTING

- All weld joints (butt and socket) covered by a test pack shall be accessible during test and shall not be painted, insulated, backfilled or otherwise covered until satisfactory completion of testing as per approved procedure. This procedure for hydrostatic testing is prepared requirements of applicable code, client specifications and other requirements if any included in the contract. Since surface preparation of painting of spools is comparatively easy to carry out in shops, in many projects contractors decide to apply painting of spools after completion of welding, NDT and PWHT as applicable. In such cases, hydrostatic testing of spools also needs to be completed prior to surface preparation and painting. This methodology, however, requires yet another hydrostatic testing after completion of all field welds, but before application of coating on field welds. Upon satisfactory completion of field welds and its NDT, PWHT, etc., field welds can also be coated.
- During field hydrostatic testing, all vents and other connections (which can serve as vents) shall be open during filling so that all entrapped air is vented prior to applying test pressure to the system. Test vents shall be installed at highest points to avoid air entrapment in piping.
- Equipment which are not to be subjected to pressure test shall be either disconnected from piping or bypassed using temporary piping during test.
- Safety valves and control valves shall not be included in site pressure testing.
- Temporary spades and blanks installed for testing purposes shall be designed to withstand the test pressure without distortion. Presence of spades shall be clearly visible during testing.
- All control valves shall be removed or replaced with temporary spools or blinded off during pressure testing.
- Check valves shall have the flap or piston removed for testing, where pressure cannot be located on the upstream side of the valve. Locking device of the flap pivot pin shall be reinstated together with the flap and a new cover gasket shall be installed after completion of test.
- Spring supports shall be restrained as recommended by vendor or removed and expansion bellows removed during hydrostatic testing.
- Drain points for test medium and disposal methodology shall be in accordance with pre-approved procedure.
- Care shall be taken to avoid overloading any part of the supporting structures during hydrostatic testing.
- Piping which is spring or counterweight supported shall be blocked up temporarily to a degree sufficient to sustain the weight of test medium. Holding pins shall not be removed from spring supports until testing is completed and system is drained.
- Pressure in the system shall be introduced gradually until the pressure is lesser than one-half of test pressure or 170 kPa gauge. Maintain that pressure for 10 minutes and then gradually increase pressure in steps of one-tenth of test pressure until specified test pressure is attained.

13.25.13   Blinds for Pressure Test

Plain test blanks shall be used with 3 mm (1/16″) flat non-asbestos gaskets for blanking flat face, and raised face flanges. Provide full face blanks and gaskets to the extent possible. However, where permanent operational blinds are installed, they may be used for pressure testing. Plate material, extra length bolts and gaskets, which are made at construction site may be used temporarily for testing. Temporary spades and blanks installed for testing purposes shall be designed to withstand test pressure without distortion. Temporary test spades and blanks shall be readily identifiable by painting a red colour on the handle. For ring joint flanges, spare ring joint gasket shall be required as they are to be used only once.

13.25.14   Test Equipment

Equipment used in pressure testing comprises of pumps, safety devices, pressure gauges and recorder, temperature gauges and recorder. Pumps used shall be capable of developing required positive pressure so the measuring devices should also have required measuring range. Apart from that, all pressure gauges and recorders, temperature gauges and recorders shall be calibrated within 60 days prior to testing and shall have valid calibration certificates on the date of testing. The gauges shall be of a minimum face size of 150 mm in diameter and ranged to approximately twice the test pressure. However, ASME B 31.3 permits pressure gauge range from 1.5 to 4 times the test pressure.

A minimum of two gauges shall be provided for each test system. One gauge shall be located at the highest point and the other soon after the pump or at grade, but sufficiently away from pump outlet to avoid damage of gauge due to pressure jerks from pump.

13.25.15   Completion of Testing

Pressure test shall be considered complete only when following aspects are completed:

- All defective welds, defective materials, flange leaks, valve gland leaks or other such defects have been corrected and accepted by all parties (stakeholders) witnessing the test.
- All documentation and "test pack" information is complete and accepted by all parties. In this regard, test pack documentation (except final test reports) is prepared in advance and is reviewed by all stakeholders well in advance to avoid delays in the process.
- All temporary test blinds or spades and strainers have been removed, new gaskets installed and piping system reinstated.
- Sealing materials shall not be used to correct leaks at joints. Valve glands shall not be tightened to the extent that valve cannot be operated. If necessary, valve glands shall be repacked.

After hydrostatic testing of system is complete and approved by all concerned, all lines and equipment shall be completely drained of the test fluid. Piping system vents shall be opened while draining to avoid creation of vacuum inside piping. Care shall be taken when draining the test fluid to avoid damage to other items of equipment stored in near vicinity. Usually, drain water needs to be hosed out to a pre-determined spot for water disposal or, if scarce, to be stored for reuse. Special attention shall be given to points where water may be trapped, such as in valve bodies or low points and to drain them completely.

13.25.16 Test Records

Records for piping which require that pressure be held for a specified period of time shall include any corrections of test pressure due to temperature variations between start and finish of the test.

Records of all tests carried out and approved by all stakeholders shall be retained in construction records and shall form a part of the project record book.

13.26 CLEANING AND REINSTATEMENT

13.26.1 Cleaning

Upon satisfactory completion of hydrostatic test of test pack, piping falling under test pack shall be drained off completely. Thereafter, line may have to be flushed out to remove all foreign matter and dried up by blowing dry air through the piping.

For flushing, clean water used for hydrostatic testing itself can be reused, provided it is not getting contaminated in the process. Water to be used for flushing and cleaning austenitic stainless steel shall contain less than 1 ppm chlorides. While flushing lines, it is recommended that inline instruments are not flushed through.

While above being the common requirements, some more additional cleaning is required in the case of compressor and turbine inlets, lube oil piping, etc., where cleanliness of lines is of paramount importance in operation of connected equipment. In such instances, a separate cleaning procedure is required and shall be approved by all concerned, prior to start of hydrostatic testing itself.

Since a considerable time period often exists between hydrostatic testing and start of commissioning activities, it is always desirable to dry the piping and all connected equipment to a reasonable level and to maintain the same till commissioning. This requirement is very important in case of process plants to be erected in humid areas.

13.26.2 Reinstatement

After successful completion of pressure testing, the lines need to be reinstated. This means that all temporary piping components shall be replaced with original items required in respective isometrics. Further, all dropped items like valves, bellows, etc., need to be reinstated. After reinstatement of lines according to drawings, extent of piping system covered by test pack concerned can be considered ready for pre-commissioning/commissioning activities.

13.27   INSPECTION OF VALVE INSTALLATION

Valves are major components of a piping system and require careful attention during selection, manufacturing and installation phases. Selecting a type of valve is based upon required function such as to block flow, throttle flow or prevent flow reversal. There are numerous types of valves with widely varying advantages and disadvantages, and this makes selection of type of valve a bit tricky. Most commonly used type of valve in process industry is the gate valve, which approximates to about 75% of requirements.

In addition to basic minimum requirements spelled out to applicable standard, client specifications often ask for additional valve design requirements. Once a valve is selected, its flange rating class must be specified based on its design pressure/temperature and maximum available operating pressure (MAOP) of the class. Accordingly, valves shall be inspected during its manufacture and they are often individually tested.

With regard to installation of valves, the inspector shall verify the following:

- Ensure that right type of valve is placed in piping as per drawing
- Connecting surfaces have similar preparations as per specification
- Flow direction (if applicable) is as envisaged in drawing

13.28   PIPE SUPPORTS

- Install all pipe supports, anchors, guides and other support attachments at specified locations in accordance with details given in drawings. Pipe supports can also be fabricated in accordance with the drawings whenever required.
- Welding preparations for pipe-to-pipe support welding shall be in accordance with drawings and shall be in accordance with specified WPS, which is prequalified. Welding shall be done by qualified welders.
- All pipe supports shall be individually identified by number and this number shall be marked on piping layout plans.
- All pipe supports and attachment welds to pipe shall be welded in accordance with same welding procedure used for that piping class.
- Temporary supports shall be used during pipework installation to prevent overstressing of pipe work. These temporary supports are to be removed from site after completion of installation.

13.28.1   GENERAL GUIDELINES FOR SUPPORTING

Piping systems shall be routed in such a way that it uses minimum supports and restraints, using minimum foundation and structural requirements to minimize cost of construction. Following guidelines shall be considered while finalizing piping:

- Piping system shall support itself to the extent possible to minimize amount of additional structural steel that is required to provide support.

- Piping with excessive flexibility may require additional restraints to minimize excessive movement and/or vibration that may be caused by fluid flow, wind or earthquake. Therefore, piping systems shall be designed with only just enough flexibility that is needed to accommodate expected thermal movement without causing excessive pipe stresses or end reaction loads. This means that piping systems shall not be overly flexible.
- Piping that is prone to vibration, such as reciprocating compressor suction and discharge systems, shall be supported independently from other piping systems. This independent support keeps the effects of vibration-prone system confined to that system and contained within directly associated structures. The effects are not transmitted to other systems.
- Piping that is located in structures shall be routed beneath platforms and near major structural members, at points that permit added loading. Routing beneath platforms avoids access interference problems. Routing near major structural members minimizes the need to increase size of structural members or to provide additional local reinforcement, due to increased bending moment.
- When possible, piping shall be routed near existing structural members to minimize the need for additional structure and foundations.
- Layout of piping system shall consider the safety of personnel who may be near the pipe. Major pieces of equipment, particularly heat exchangers, vessels and tanks, shall be accessible to fire-fighting equipment. Pipe racks or ways shall be routed to provide this access. There shall be adequate space under pipe-ways for people to walk and work and pipelines including pipe supports and restraints.

A piping system needs supports and restraints to accommodate the following:

- Permit piping system to function under normal operating conditions without failure of pipe or connected equipment.
- Support piping system weight loads.
- Keep sustained longitudinal pipe stress within allowable limits.
- Limit pipe sag to avoid process flow problems.
- Limit loads on connected equipment.
- Control or direct thermal movement of the pipe.
- Keep pipe thermal expansion stresses within allowable limits.
- Absorb other loads imposed on piping system.

Selection of a particular type of support or restraint depends on the following factors:

- Weight load
- Restraint load
- Clearance available for attachment to pipe
- Availability of nearby existing structural steel
- Direction of loads to be absorbed or movement to be restrained
- Design temperature
- Allowance required for thermal movement of pipe

13.28.2 Inspection of Supports and Restraints

Pipe supports are used to support the weight of piping system and contents of the system. The supports keep the pipe elevated at a desired height above the ground. The drawings issued contain the position of each pipe support and also indicate the type of support to be used at each location. The location and type of support are very important, and hence during installation, it is ensured the following need to be taken care:

- To keep pipe stresses caused by dead and live weight within allowable limits.
- To prevent excessive pipe sag.
- To prevent excessive reaction loads at equipment connections.

When practical, new piping shall use existing supports to minimize costs. As much as possible, new piping shall be located in existing pipe-ways. Intersecting pipe-ways are located at different elevations to facilitate access and future piping installation. Standard piping drawing specifies minimum spacing of lines that are supported on sleepers or pipe racks.

Restraints control or limit movement of pipe in one or more directions. Such restraint may be required to reduce thermal expansion reaction loads at equipment connections, or to limit pipe vibration. Some restraints keep the pipe from moving vertically or laterally but allow the pipe to move longitudinally.

Other restraints do not allow the pipe to move in any direction. A support is a specialized type of restraint that prevents pipe movement under vertical weight loading.

13.29 TYPES OF SUPPORTS

This is to give an overview of various types of supports and restraints used in piping systems.

The two general classes of supports are as follows:

- Rigid
- Flexible or resilient

It is the duty of an inspector to verify that right type of pipe support is placed at the right location, at required spacing and that alignment of pipe support is as indicated in applicable drawing. In case spacing has to be altered due to practical feasibility, prior approval shall be obtained from field engineering team either through revised drawing or through a DCN.

Inspector shall confirm that supports provide for a minimum of 300 mm clearance between bottom of pipe to finished grade or as specifically indicated in project documents. In addition, a minimum clearance of 50 mm shall be provided for inspection and freedom of pipe movement between above-ground piping crossing with any structure (including pipe support structure). This clearance is also required for above-ground piping crossing with another pipe.

13.29.1   Rigid Supports

Rigid supports are more common of the two support types. Engineers use rigid supports when weight support is needed and no provision to permit vertical thermal expansion is required. A rigid support does the following:

- Allows lateral movement and rotation.
- May or may not prevent movement up.
- Prevents movement down.

Figure 13.3 illustrates some rigid support types that are commonly used in piping. Selection of a particular type of rigid support depends primarily on the following:

- Amount of load to be carried
- Distance to solid attachment (structure, grade, etc.)
- Point of attachment to pipe (horizontal or vertical run, elbow, etc.)

Inspector verifying supports shall ensure that a 6 mm weep hole is drilled for all dummy supports. The weep hole shall be located near base plate for all vertical dummy supports, and near run pipe at 6 o'clock position for all horizontal dummy supports.

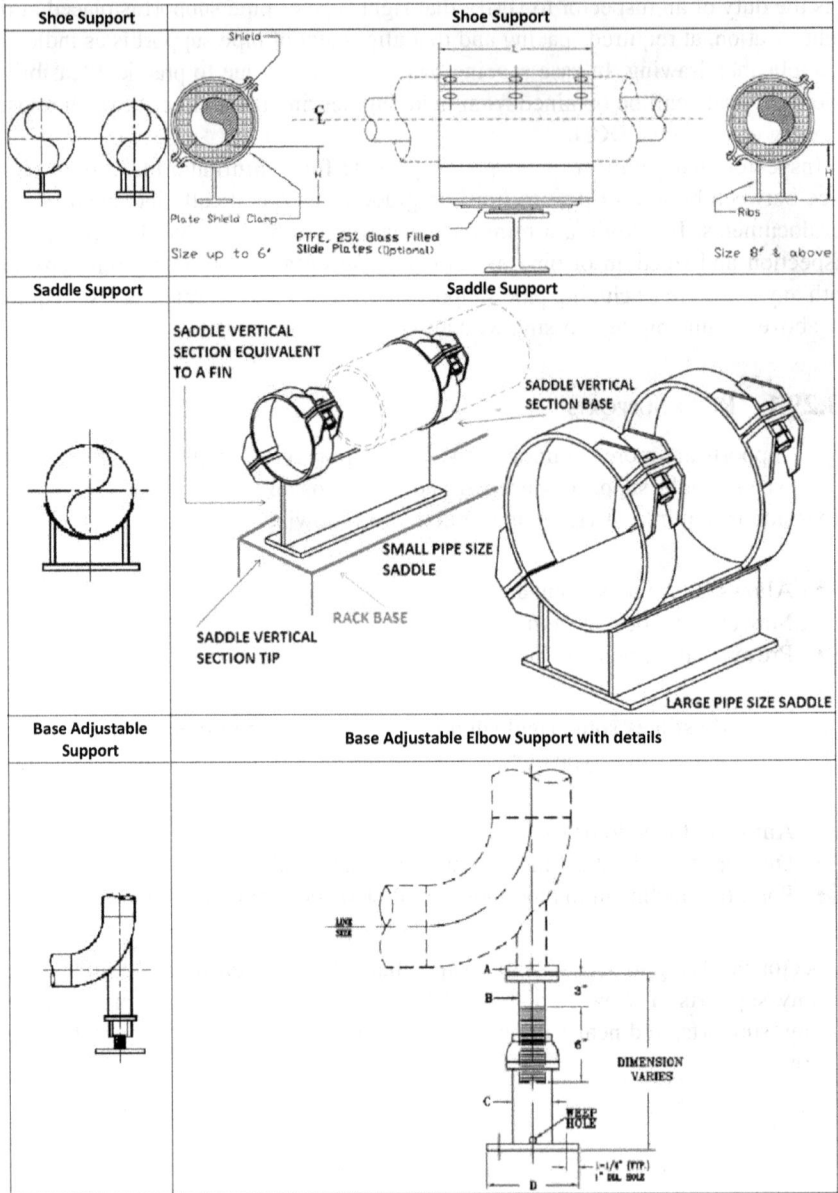

FIGURE 13.3 Pipe supports (Piping Technology & Products, Inc., 2022a; Terenzi and Marcolini, 2017; Piping Technology & Products, Inc., 2022b; and Bentley Communities, 2013).

(Continued)

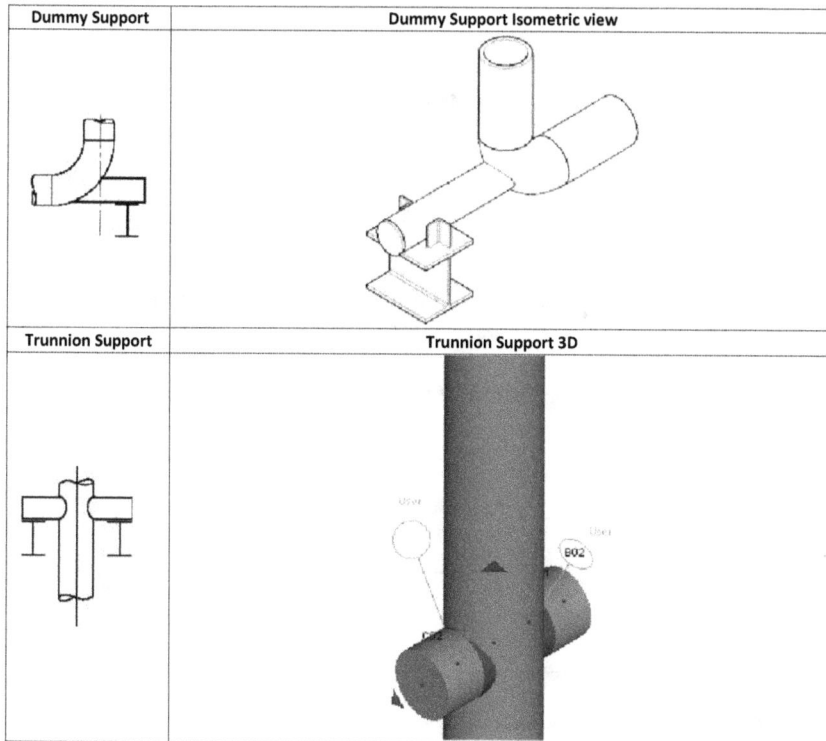

FIGURE 13.3 (*Continued*) Pipe supports (Piping Technology & Products, Inc., 2022a; Terenzi and Marcolini, 2017; Piping Technology & Products, Inc., 2022b; and Bentley Communities, 2013).

13.29.2 PIPE HANGERS

Pipe hangers are also a form of rigid support. Pipe hangers support pipe from structural steel or other facilities that are located above the pipe and carry pipe weight (dead plus live) in tension. A pipe hanger rod moves freely parallel and perpendicular to pipe axis; therefore, thermal expansion is not restricted longitudinally or laterally. The rod does restrict vertical thermal expansion and shall be long enough so that it does not restrict pipe lateral or longitudinal movement.

The following restrictions on use of rod hangers need to be verified by the inspector:

- Rod hangers are not generally used for lines NPS 300 (12″) and larger in liquid service or for lines designed for multiphase flow.
- All hangers shall be provided with means for vertical adjustment.
- Suitable locking devices provision shall be available at all threaded connections of the hanger assembly (use of double nuts).
- Rod hangers shall be subjected to tensile loading only.
- Practicality for replacing them with rigid pipe supports shall be evaluated and implemented, during construction in consultation with field engineering team.

Figure 13.4 shows some examples of pipe hangers.

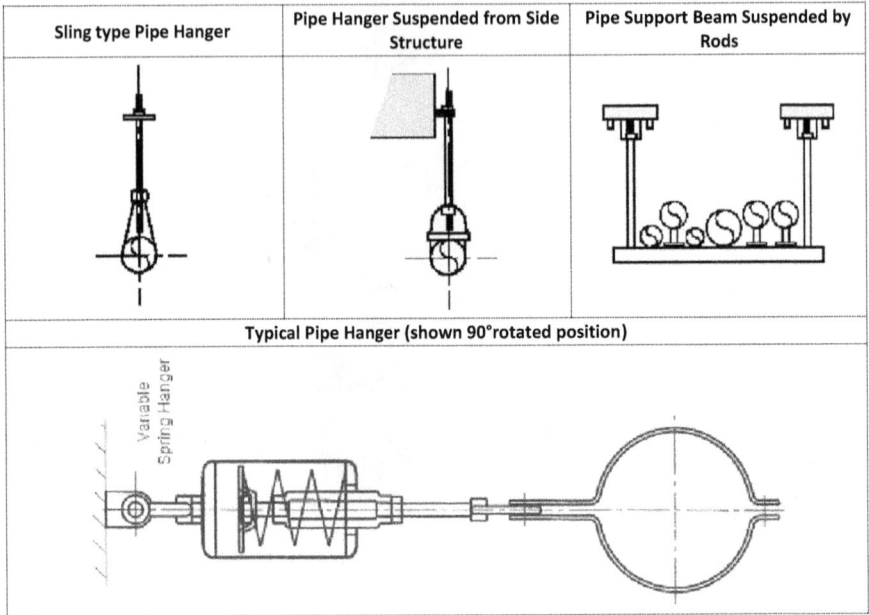

Sling type Pipe Hanger	Pipe Hanger Suspended from Side Structure	Pipe Support Beam Suspended by Rods
Typical Pipe Hanger (shown 90°rotated position)		

FIGURE 13.4   Pipe hangers (Sölken, n.d.).

13.29.3 FLEXIBLE OR RESILIENT SUPPORTS

Flexible or resilient type supports carry weight of piping and allow piping system to move in all three directions. A coil spring with right stiffness and pre-compression supports weight of piping. Spring being resilient, it permits vertical movement while carrying weight of piping. Ability to move vertically allows support to carry weight while permitting pipe to expand and contract as needed due to thermal expansion. Two basic types of flexible supports are as follows:

- Variable load
- Constant load

Type of flexible support selected from standard available models is based on the following factors:

- Design load
- Required movement
- Installation geometry
- Standard models available

13.29.3.1   Variable Load Flexible Support

Variable load flexible support is more common of the two types of flexible support. With variable load supports, pipe movement stretches or compresses the spring, changing the load that spring exerts on pipe. The spring is selected to provide required amount of support load to pipe throughout the movement range (Figure 13.5).

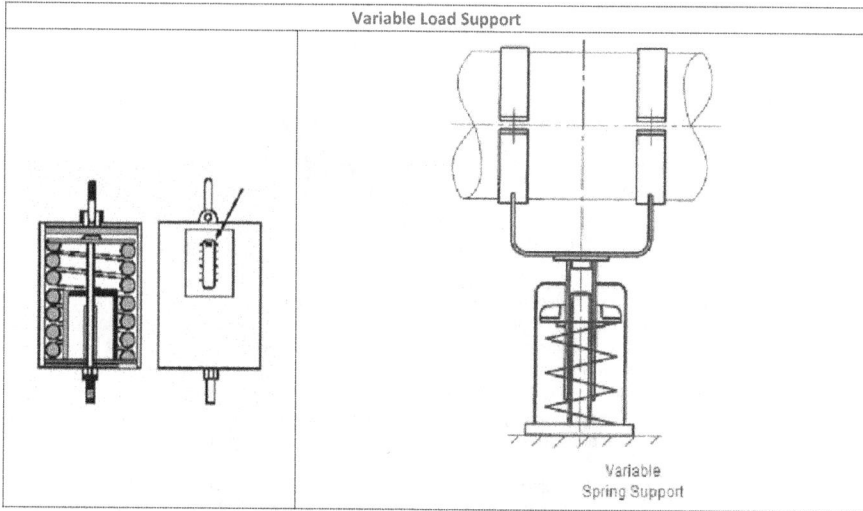

FIGURE 13.5   Variable load supports (Sölken, n.d.).

13.29.3.2   Constant Load Flexible Support

With constant load flexible supports, load that is exerted by support on pipe remains constant throughout movement range. Use of a variable-length internal-moment arm mechanism accomplishes this constant load. This type of support is required when load variation caused by vertical thermal movement in a variable-load-type spring is too large to be accommodated by piping system, or when thermal movement is greater than approximately 3″ (75 mm).

Figure 13.6 shows an example of a constant load support.

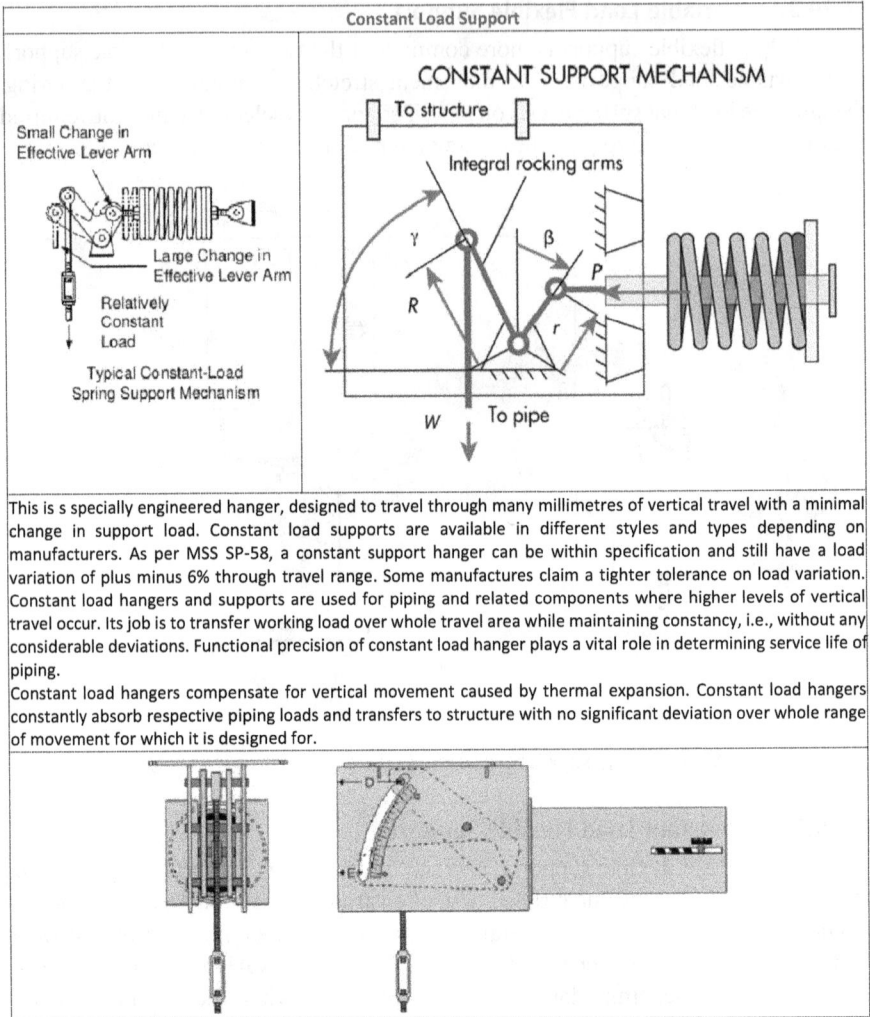

Constant Load Support

CONSTANT SUPPORT MECHANISM

This is s specially engineered hanger, designed to travel through many millimetres of vertical travel with a minimal change in support load. Constant load supports are available in different styles and types depending on manufacturers. As per MSS SP-58, a constant support hanger can be within specification and still have a load variation of plus minus 6% through travel range. Some manufactures claim a tighter tolerance on load variation. Constant load hangers and supports are used for piping and related components where higher levels of vertical travel occur. Its job is to transfer working load over whole travel area while maintaining constancy, i.e., without any considerable deviations. Functional precision of constant load hanger plays a vital role in determining service life of piping.

Constant load hangers compensate for vertical movement caused by thermal expansion. Constant load hangers constantly absorb respective piping loads and transfers to structure with no significant deviation over whole range of movement for which it is designed for.

FIGURE 13.6 Constant load support (Google Images, n.d.; Piping Technology & Products, Inc., 2017; and Machine Design, n.d.).

13.30 RESTRAINTS

Restraints have following two primary purposes in a piping system:

- They control unrestricted thermal movement of pipe by directing or limiting it. Generally, a piping system is totally restrained at its end connections to equipment.
- They control, limit or redirect thermal movement to reduce thermal stress in pipe or loads exerted due to thermal movement on equipment connections.
- They absorb loads imposed on pipe by other conditions. Examples of these other conditions are as follows:
  - Wind
  - Earthquake

- Slug flow
- Water hammer
- Flow-induced vibration

Several types of restraints are used in piping depending on requirements. Selection of type of restraint and its specific design details depends primarily on the following:

- Direction of pipe movement to be restrained.
- Location of restraint point.
- Magnitude of load to be absorbed.

One or more types of restraints or support may be combined at one location, depending on piping system design needs. Three types of restraints are as follows: Refer Figures 13.7 and 13.8

- Stops
- Guides
- Anchors

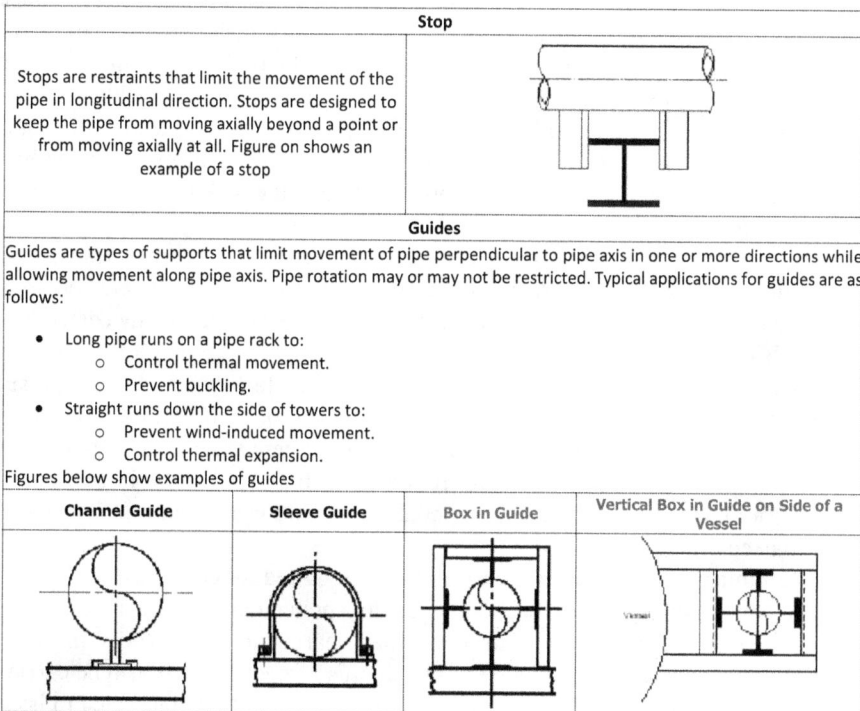

Stop	
Stops are restraints that limit the movement of the pipe in longitudinal direction. Stops are designed to keep the pipe from moving axially beyond a point or from moving axially at all. Figure on shows an example of a stop	

Guides			

Guides are types of supports that limit movement of pipe perpendicular to pipe axis in one or more directions while allowing movement along pipe axis. Pipe rotation may or may not be restricted. Typical applications for guides are as follows:
 	
• Long pipe runs on a pipe rack to:
  ○ Control thermal movement.
  ○ Prevent buckling.
• Straight runs down the side of towers to:
  ○ Prevent wind-induced movement.
  ○ Control thermal expansion.
Figures below show examples of guides			

Channel Guide	Sleeve Guide	Box in Guide	Vertical Box in Guide on Side of a Vessel

FIGURE 13.7 Restraints.

Anchors
Anchors stop pipe movement in all three translational directions. Engineers use anchors to isolate one section of a piping system from another section in terms of loading and deflection. A total anchor that eliminates all translation and rotation at one location is not as common as one or more restraints that act at a single location. It is difficult to design effective rotational anchors or restraints. Plant piping more commonly uses directional anchors that restrain pipes only in their translational directions.
Figures below show examples of anchor types that are typically used in aboveground plant piping systems.

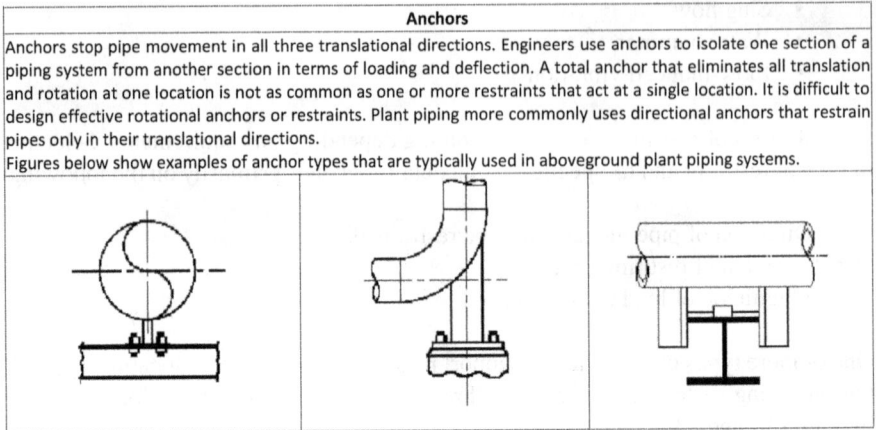

FIGURE 13.8   Anchors.

13.31   SURFACE PREPARATION AND COATING OF FIELD WELDS AND REPAIR OF COATING DAMAGES

All field welds shall have to be surface prepared and coated as required in contract after field hydrostatic testing of the line. Depending on feasibility on site, suitable methodology for surface preparation shall be selected based on surface texture requirements recommended by primer/paint manufacturers. Similarly, procedure for application of coating also shall be decided based on feasibility on site. Methodology adopted for surface preparation and painting of field welds could be slightly different from those followed for surface preparation and painting of spools in workshop. Therefore, this shall be clearly addressed in surface preparation and painting procedure (if issued as a single procedure) dealing with surface preparation and painting.

Similarly, spools already coated also may be damaged during storage, transportation to site or installation. Therefore, procedure shall also address the methodology proposed to rectify coating damages expected during erection of already coated piping spools on site.

Apart from the ITP for piping spool fabrication, installation and testing (usually in two parts: one for spool fabrication and other for installation, field welding and testing) covering major mechanical works, there shall be a separate ITP for surface preparation and painting. If required this also can be in two parts: one for spool and the other addressing coating of field welds and repair of damaged coating.

Since some more mechanical works would be required for completing the punch list (to be discussed in the next section) items, it is preferable to conduct final visual inspection for damages during construction after clearing all punch works. In case it is not possible to wait till completion of punch, this inspection shall also be carried out along with punch list and later punched area alone shall be re-inspected to issue mechanical completion.

13.32   SURFACE PREPARATION AND PAINTING OF PIPING AFTER INSTALLATION

As it is presumed that spools have undergone full painting at shops after NDT, PWHT (as applicable) and hydrostatic testing, painting required on site after installation essentially contains the following.

- Surface preparation and painting of all field welded joints.
- Surface preparation and painting of coating damages caused during transportation, handling or installation.

Both works mentioned above are carried out according to established procedures duly approved by all concerned and inspected essentially during surface preparation and coating according to established ITP. ITP for surface preparation and painting can be made either as a single document or two separate documents as desired. Surface preparation and painting is considered complete when all piping systems are coated as required and inspected and cleared according to inspection and test plan.

As this activity is being carried out predominantly on pipe racks, often rotary power brush cleaning is considered as the practical solution for both field welds and coating damages caused during installation of spools.

13.33   PUNCH LISTING

A punch list is a document prepared nearly at fag end of construction of process plant piping project. It is a list of works observed to be deviating from contract specifications that EPC contractor is expected to complete prior to final handing over of the plant.

In most of the EPC contracts for process plants, the general conditions to contract (GCC) require the contractor to declare the construction to have reached "substantial completion", when they consider so, and request a "Punch Listing Inspection".

Punch listing is extremely important as it is the method of logging deviations of process piping, inline equipment, supports, structural steel work and so on of the entire project. In other words, it is the method of capturing and recording:

- Damaged, defective, missing and incomplete equipment and piping.
- Non-conforming to specification items used in construction.
- Components, piping elements and workmanship which deviate from contractual specifications.

Punch list can also be considered as "works to-do list" and can be utilized at various stages of construction as required and the two main stages are:

- Handover to commissioning from construction.
- Handover from commissioning to operations.

If not handled professionally, this can be a very messy process and may become a real problem if there is no proper system for managing and controlling the "punch list" and remedial actions to resolve discrepancies. Based on the severity of discrepancies, items in punch list are categorized into three, namely "A", "B" and "C".

13.33.1 CATEGORY A PUNCH ITEMS

Items that are related to safety or items that prevent commissioning or production operations are placed under this category such as:

- Missing items
- Damaged equipment
- Incorrectly fitted or connected equipment, loose wires, loose bolts in flanges, etc.

Above punch items shall be completed by construction before handover to commissioning team, as they may prevent pre-commissioning and commissioning activities.

13.33.2 CATEGORY B PUNCH ITEMS

Items that can be finished during commissioning phase of contract, but prior to handover of plant to client are placed under this category, such as:

- Long bolts in flanges
- Missing piping labels
- Missing or temporary signs

Ideal situation is that all three punch items be completed before handing over the plant for commissioning. However, in reality, this is not possible and longest stretch permissible to complete these items shall be up to completion of commissioning. If this is not done even by that time, it may affect handing over process and thereby the handing over date, which will have a bearing on defect liability period as well. With taking over of plant by operations team of the client, contractor's construction and commissioning phase is over and hence no clients will be ready to take over any plant without completing all punch works from all categories.

13.33.3 CATEGORY C PUNCH ITEMS

Items that do not prevent plant or equipment from working and are purely cosmetic and do not prevent plant or equipment startup are placed under this category. These items can be remedied during the full-fledged operation of the plant on as-agreed basis, such as:

- Damages/scratches on painted surfaces of equipment/piping or any other related structures
- Missing non-safety-related signs

These items can be completed even after the equipment/plant/installation started running but still needs to be remedied and completed.

13.33.4   THE PUNCH LISTING PROCEDURE

There shall be only be one official punch list often called Master Punch List (MPL). However, it is quite usual that each discipline within construction team maintain discipline wise punch list for easy tackling of the same.

As official punch list is generated right from inspections at the inspection and test record (ITR) stage, punch list shall be relevant to ITR and those items listed on applicable checklist, which in itself shall prevent duplication of punch list items. However, it is the responsibility of punch list coordinator to identify and remove duplicated punch items.

When recording punch list items, listing shall be specific as possible and general comments are not to be included which are often difficult to remedy. Punch items shall have a unique serial number, followed by descriptions such as name, line number, cable number, fitting number and exact location of the defect and kind of defect so that it can be located and remedied easily. Even though it is better that the person who highlighted the defect shall ensure that the particular item is cleared, this may not be possible due to rotations of manpower during operations or due to any other circumstance where the original person is not available. However, it would be better if the punch list originator's name is clearly defined and entered into the database rather than the name of data entry clerk often seen included in punch list, for probable clearance of punch item by the originator himself.

Generic listing such as "bolts missing in flange" is not specific enough if the specific flange is not identified in punch list. In such cases, the location shall be marked on either a P&ID or applicable isometric and attached along with punch list for locating it easily for remedial actions.

If a punch list clearance team follows punch listing team, it is possible for the accompanying team to remedy many of the punch points on the spot, such as loose bolts so that the punch list will have only major works which cannot be remedied on the spot.

13.34   RELEASE FOR COMMISSIONING

When all mechanical works are practically complete and when all inline equipment and valves are connected with supports and hangers in position, line can be released for pre-commissioning activities like flushing, pickling, etc. as applicable. Though it is mandatory to clear all category "A" punch items prior to handing over for commissioning, category "B" punch items can spill over to commissioning phase as well.

For lines connecting turbines and compressors, often steam blowing is required to clear off all debris from the line apart from clearing all punch items, which are often considered as a pre-commissioning activity.

Flushing, gas purging, inertization of piping systems are covered by yet another procedure addressing all safeguards needed to protect connected instrumentation in that process.

14 Construction Documentation for Plant Piping (Mechanical)

14.1 GENERAL

This chapter describes various documentations to be included in the construction record book (CRB) for plant piping, which shall be the authentic construction document pertaining to plant piping, containing all salient construction details, including material test certificates, material traceability records, welding, non-destructive testing (NDT), inspection and tests, surface preparation and painting, for future reference during service life of the process plant. As this document does not have any immediate use, often its due importance is not recognized well. But as and when some modification or rectification is needed during service, many technical details would be required to carry out design for proposed modification, alteration or repair of the piping. In such instances, properly prepared CRB would be of immense help to the technical services department of client to gather it from a single source, the CRB.

With that intention, it would be appropriate to think about the contents of CRB prior to start of piping itself and compilation of the same shall start along with piping construction. If this is deferred till completion of actual construction work, compilation of an error-free CRB would be impractical due to the abundance of documents to be segregated, correlated and compiled with proper link and logic.

In this regard, it would be better to start compilation of CRB in a systematic manner along with construction work in a progressive manner so that as and when work is completed, documentation shall be available immediately for review or approvals as needed. The first major review in this regard would take place at the time of hydrostatic testing of test packs. At this point, the proposed test pack is reviewed for all associated documentation up to that point pertaining to each of the packages.

14.2 PROJECT RECORD BOOK

For an EPC project, the total documentation pertaining to a project is often termed as project record book (PRB). For projects involving multidisciplinary scope of work, PRB is usually structured in the following pattern (Figure 14.1).

- **Part A Guidelines to Retrieve Information and PRB Index**
  - This section essentially contains the detailed index to PRB. Being a voluminous document, retrieving a specific document from this is a herculean task, if not aware of the way the records are organized. In this regard, this section also contains guidelines about how the records are

DOI: 10.1201/9781003328124-14

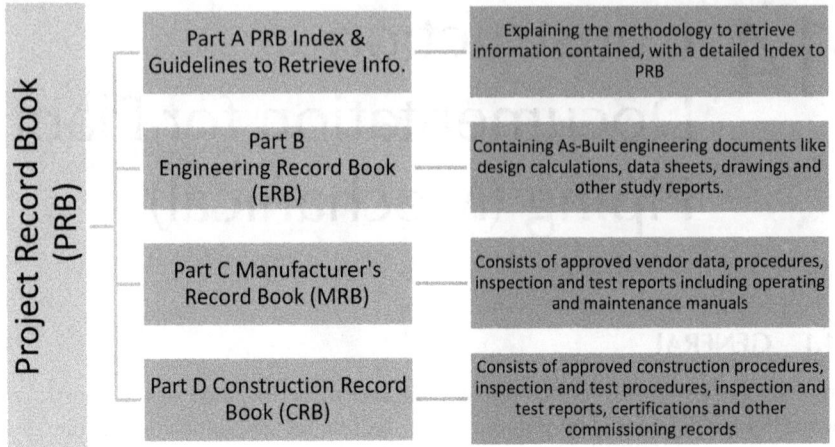

FIGURE 14.1   Project record book layout.

organized and explaining the methodology to retrieve a specific docu-
ment from the massive documentation.
- **Part B Engineering Record Book (ERB)**
  - This section contains all engineering documentation including the AFC
    documentation and as-built documentation of those engineering docu-
    ments. Documents like design calculations, data sheets, drawings and
    other study reports are often included under this section.
- **Part C Manufacturer's Record Book (MRB)**
  - This section essentially deals with the record books furnished by
    respective vendors like vendors for pressure vessels, heat exchanger,
    storage tanks, pumps, turbines and other package equipments, consist-
    ing of approved vendor data, procedures, inspection and test reports
    including operating and maintenance manuals as applicable.
- **Part D Construction Record Book**
  - This section essentially consists of documentation pertaining to construc-
    tion activities happening on site, of which a major chunk is from plant piping.
    It consists of all piping-related approved construction drawings, approved
    construction procedures, inspection and test procedures, inspection and test
    reports, certifications and other commissioning records as required.

Each of the above books shall be designed as a standalone document, with proper index
to all the four books in order to locate the volume of interest easily. It may be noted that
each of the books will run into many volumes depending on the size and intricacy of the
project and also the specific requirements for PRB from client. Further, most of the client specifications propose the sizes for each type of documentation from handling point

of view. The general practice is to standardize documents to A4 size for all descriptive documents and A3 size for drawings like piping key plans and isometric drawings and A1 size for static and rotating equipment general arrangement and details.

14.3   DOCUMENTATION MODE FOR PRB

In the earlier days, PRBs were in print and bound volumes and often around ten copies were required as part of final documentation. The copies are shared by various stakeholding section of the client as per their division of responsibility for maintenance.

With the massive digitalization taking place across the world, the hard copy requirements are paving way for soft copy documentation on account of storage problems and difficulty in search associated with hard copy documents.

While dealing with soft copy documents, type of documentation plays a vital role, such as PDF. While as-built documentations are furnished in PDF format, editable versions of drawings are often required in AutoCAD format and that of data sheets in Word or Excel formats. While this requirement may vary from client to client depending on the enterprise resource planning (ERP) systems followed by the client, the PDF, AutoCAD, Word and Excel are universally acceptable to most of the ERP systems.

Sl. No.	Description of Document	Format Suggested	Version
1	Process Design Basis	MS Word	A4 (Portrait)
2	Process and Instrumentation Diagrams	AutoCAD	A3/A4 (Landscape)
3	Process Flow Diagrams	AutoCAD	A3/A4 (Landscape)
4	Material Selection Diagrams	AutoCAD	A3/A4 (Landscape)
5	Piping Key Plans	AutoCAD	A3/A4 (Landscape)
6	Piping Isometric Drawings	AutoCAD	A3/A4 (Landscape)
7	Procedures	MS Word	A4 (Portrait)
8	Data Sheets (Process and Mechanical)	MS Excel	A4 (Portrait)
9	Rotating Equipment Data Sheets (Process and Mechanical)	MS Excel	A4 (Portrait)
10	Electrical and Instrumentation Data Sheets	MS Excel	A4 (Portrait)
11	Static Equipment Drawings (GA and Details) (Pressure Vessels, Heat Exchangers), including Storage Tanks	AutoCAD	A1/A3 as required
12	Rotating Equipment Drawings (GA and Details) (Pumps, Compressors, Turbines, etc.)	AutoCAD	A1/A3 as required
13	Piping Drawings (GA and Details)	AutoCAD	A1/A3 as required
14	Steel Structural drawings for equipment and piping support and other critical constructions associated with plant	AutoCAD	A1/A3 as required

14.4 PRB PREPARATION AND REVIEW RESPONSIBILITY

The flow chart in Figure 14.2 indicates the process involved in preparation of PRB including collection of data, compilation, review and submission to client as final approved PRB. Colour legends are used to represent the agency responsible for each of the activities. As seen in the flow chart, various documents are compiled by various agencies like vendors, sub-contractor or the main contractor.

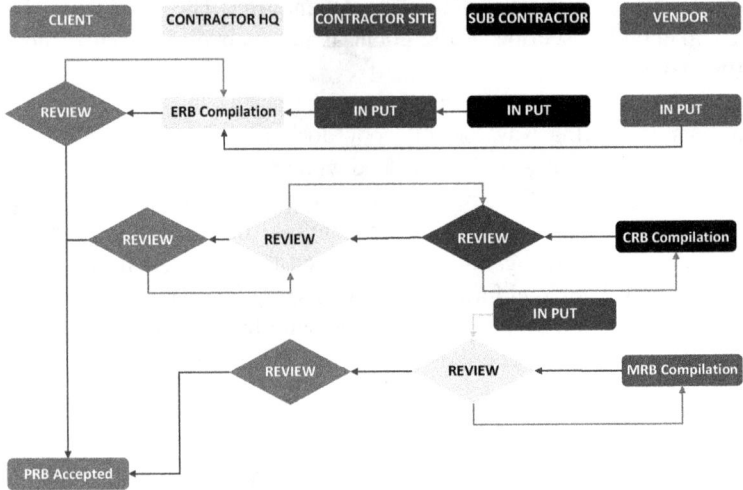

FIGURE 14.2   Project record book preparation flow chart.

14.5 CONSTRUCTION RECORD BOOK

While the above being the general overall requirement for the preparation of PRB for a large project, this section is only expected to give an overview of the probable or desirable contents of CRB, since one of its main constituents is the piping-related documentation.

In sections below, a comprehensive list of documents to be included in CRB is provided. Depending upon specific requirement and applications, additional documentation might be required. However, for all general purposes, the documents discussed below shall suffice. It may be noted that documents listed hereunder shall cover mechanical items pertaining and relevant to piping only. In order to limit the size of the documentation and repetition in PRB of which CRB is only a part, the following documents are often omitted from the scope of CRB:

- Documentations pertaining to static and rotating equipment
- Documents pertaining to valves and other similar intervening inline equipment in piping

It may further be noted that the list provided is not comprehensive. The list pertains only to piping-related documentation and construction documents pertaining to other disciplines of work are totally excluded in the table below.

Sl. No.	Description of Document	Status	
		AFC	AB
1	**Piping Engineering**		
1.1	Process Design Basis	X	X
1.2	Process Flow Diagram (PFD) with Heat and Material Balance	X	X
1.3	Material Selection Diagram (MSD) and Material Selection Guidelines (MSG) separate or together	X	X
1.4	Process and Instrumentation Diagram (P&ID)	X	X
1.5	Piping Key Plan	X	X
1.6	Piping Isometric Drawings (Isometrics)	X	X
1.7	Piping Material Specification (PMS)	X	X
1.8	Commodity Specification (part of PMS or separate)	X	X
1.9	Material Take-Off (MTO)	X	X
1.10	Material Requisition (MR)	X	X
1.11	Vendor Data Requirements (part of MR)	X	X
1.12	Technical Bid Evaluation (TBE)	X	X
1.13	Purchase Order (Piping Bulk Items)	X	X
1.14	Vendor Data Requirements (part of Purchase Order)	X	X
1.15	Packing/Delivery Instructions (part of Purchase Order)	X	
1.16	Hydrostatic Test Packages	X	
2	**Construction Documents**		
2.1	Pipe Spool Manufacturing Procedure	X	X
2.2	Pipe Spool Identification and Storage Procedure	X	X
2.3	Pipe Spool Transportation Procedure		
2.4	Pipe Spool Installation and Alignment Procedure	X	X
2.5	Site Welding, Inspection and Testing Procedure	X	X
2.6	Static and Rotary Equipment Installation Procedure	X	X
2.7	Inline Instruments and Equipment Installation Procedure	X	X
2.8	Alignment, Levelling and Grouting Procedure	X	X
2.9	Punch Listing Procedure	X	X
2.10	Scaffolding Procedure	X	X
2.11	Overall Construction Quality Plan	X	X
3	**QA/QC Documents**		
3.1	Piping Quality Plan	X	X
3.2	Inspection and Test Plan (Piping Spool Fabrication)	X	X
3.3	Inspection and Test Plan (Piping Spool Installation and Welding)	X	X
3.4	Welding Procedure Specification Summary with Coverage	X	X
3.5	Welding Procedure Qualification Records	X	X
3.6	Welder/Welding Operator Qualification Test Records	X	X
3.7	Welding Consumables Control Procedure	X	X
3.8	Welding Inspection Reports	X	X
3.9	NDT and Other Test Procedures for Pipe Spools and Field Welds	X	X
3.10	NDT Summary and NDT Reports	X	X
3.11	Post-Weld Heat Treatment (PWHT) of Pipe Spool and Field Welds (as Applicable)	X	X

(Continued)

Sl. No.	Description of Document	Status	
		AFC	AB
3.12	Hydrostatic Test Procedure for Piping Spools and Piping	X	X
3.13	Pneumatic Test Procedure for Reinforcement Pads (as Applicable)	X	X
3.14	Hardness Test Procedure	X	X
3.15	Punch List	X	X
3.16	Inspection and Test Plan (Mechanical Erection of Equipment)	X	X
3.17	Leak Testing Procedure	X	X
3.18	Procedure for Calibration of Welding Equipment	X	X
3.19	Procedure for Calibration of Electrode Oven	X	X
3.20	Surface Preparation and Coating Procedure (for Pipe Spools and Erected Piping)	X	X
3.21	Inspection and Test Plan (Surface Preparation and Painting/Coating)	X	X
	NDT Procedures		
3.22	Liquid Penetrant Test Procedure (Shop and Field Welds)		X
3.23	Magnetic Particle Test Procedure (Shop and Field Welds)		X
3.24	Ultrasonic Test Procedure (Shop and Field Welds)		X
3.25	Radiographic Test Procedure (Shop and Field Welds)		X
3.26	Visual Examination Procedure (Shop and Field Welds)		X
4	**Construction Records**		
4.1	Material Summary for Piping Materials (Cross-Reference between PO and MTC)		X
4.2	Material Test Certificate	X	
4.2.1	Pipes	X	
4.2.2	Fittings	X	
4.2.3	Flanges	X	
4.2.4	Fasteners	X	
4.2.5	Gaskets	X	
4.2.6	Valves	X	
4.2.7	Strainers and Other Small Equipment	X	
4.3	Weld Joint Fit-Up/Welding/Visual Inspection Report		X
4.4	Welding Summary		X
4.5	Weld Joints NDT Reports		X
4.6	Weld Joints NDT Summary		X
4.7	Weld Joints PWHT Reports		X
4.8	Weld Joints PWHT Summary		X
4.9	Pad Air Test Reports (if Applicable)		X
4.10	Spool Hydrostatic Test Reports		X
4.11	Spools Release Report		X
4.12	Pipe Spool Erection Report (Field)		X
4.13	Field Joint Fit-Up/Welding/Visual Inspection Reports		X
4.14	Field NDT Reports		X

(Continued)

Sl. No.	Description of Document	Status	
		AFC	**AB**
4.15	Field Pad Air Test Reports		X
4.16	Bill of Materials (BOM) Check Report (Test Pack Wise)		X
4.17	Hydrostatic Test Report (Test Pack Wise)		X
4.18	Test Pack Documentation		X
4.19	Signed-Off ITP for Pipe Spools/Test Packs		X
4.20	Surface Preparation and Painting Reports		X
4.21	Adhesion/Other Coating Inspection Test Reports		X
4.22	Signed-Off ITP for Surface Preparation, Painting		X
4.23	Punch List		X
4.24	Handing Over Report		X

Documents marked "X" are required in CRB. In the table above, the documents under Piping Engineering that are highlighted are not essential, as these documents are to be included in the ERB by default. However, for the sake of completeness of CRB, some clients require these documents also to be included under CRB (though duplication) to make CRB as a single source of reference.

14.6   DOCUMENTS SPECIFIC TO PIPING

While describing the contents of CRB, some of the documents are developed for easiness in tracing the construction process and are not at all required in any of the applicable standards. However, these documents are specified in list above, to corroborate each and every document included in CRB to respective component or weld joint shown in piping isometric drawings, including traceability of materials used for construction.

Moreover, from the list of documents mentioned, one may have a doubt regarding the requirement of both approved for construction (AFC) and as-built copy of the same document. Both the documents are required to understand the slight modifications or material substitution made on site under approval by field engineering team to tide over some issues that came up during construction. Though this requirement apparently looks like a repetition, it is meant for revealing the construction history, which will not be known when the construction team withdraws from site and operational crew takes over.

Similarly some more documents like weld maps, weld and NDT summary are also of paramount importance in ensuring smooth corroboration of documents pertaining to each and every part that constitute a component of the storage tank and ensures completeness of coverage (especially that of NDT and other tests) as envisaged in contract. While describing the relevance of each would be impractical, some are explained below since these are considered as the key in correlating documents to applicable drawings and to part numbers indicated therein.

14.6.1 Material Summary

Material summary is nothing but a tabulation of list of part numbers (as per isometric drawing) to material test certificates. In other words, a summary serves as a guide to locate applicable material test certificate against each piping component shown in isometric drawing. Therefore, this table is prepared in relation to each of the isometrics used in piping. With the help of this document, at one glance, the actual material used during construction for a particular piping component in specific isometric can be traced back. At a minimum, material test certificates pertaining to all pressure-containing parts of piping need to be compiled systematically with the help of a summary as mentioned above.

14.6.2 Weld Summary

Weld summary table is also prepared in line with the unique weld joint identifications provided in referred isometric drawing. Weld summary table contains details such as welding process used in welding, welding procedure specification (WPS) employed, welder(s) deployed for welding against each of the pressure-containing weld joints identified in isometric. In addition to the above, non-pressure-containing welds, such as that used to attach critical supports directly to piping (if identified in isometric) are also considered important and included in weld summary. Also, weld summary table for other structural members used in piping construction is not commonly prepared as it increases the volume of CRB. However, it shall be the prerogative of the client to ask for even such details as part of CRB.

14.6.3 NDE Reports

- For all non-destructive examinations performed on site, original reports shall be provided with due endorsement from contractor/consultant or client as applicable.
- Reports shall contain details of components or welds radiographed with unique identifications provided to each of weld seams/components.
- The personnel deputed to carry out NDT and prepare its report shall be qualified adequately for the envisaged works and shall be evidenced through proper qualifications.
- Details of examination techniques, equipment, consumables and extent of testing shall be presented in detail in the report. When technique sheets have been separately approved, reports need only contain a reference to these sheets.
- As in other reports, both performance date and report date shall be clearly indicated.
- When records consist of more than one page, all pages shall be numbered as "page x of y", followed by report number on all pages to ensure inclusion of complete report.
- Report shall clearly indicate stage at which NDT was carried out. For example, when heat treatment or forming is required in manufacturing, NDT reports shall indicate whether required examination was carried out before or after the heat treatment operation.

- Acceptance or rejection criteria adopted in evaluation also shall be indicated in report.
- Minimum achieved sensitivity levels or calibration standards also shall be reported together with acceptance criteria in report.

14.6.4 WELD JOINT NDT SUMMARY

NDT summary table is also prepared in line with the unique weld joint identification provided in referred isometric drawing. It includes the details of each of the NDT carried out on each weld with a reference to concerned NDT reports so that the completeness of NDT as envisaged can be ensured easily by reviewing the referred NDT reports. As in the case of weld summary, NDT summary also shall essentially cover all pressure-containing welds and optionally some critical non-pressure-containing welds as identified in isometric drawing.

14.6.5 HEAT TREATMENT RECORDS AND CHARTS

- When heat treatment of materials or components is performed to satisfy code/specification requirements, it shall be sufficient for the manufacturer to declare that requisite heat treatment was carried out, as in the case of normalizing, wherein temperatures need not be declared. However, for other heat treatments like tempering and simulated PWHT, heating and cooling rates, soaking temperature and time need to be specified in accompanying certificate.
- When heat treatment is carried out on materials to meet NACE MR0175/ISO 15156 requirements, sufficient details shall be reported to verify conformance. Hardness testing on heat-treated components also shall be required subsequent to heat treatment and results of the same shall be in accordance with NACE MR 0175/ISO 15156.
- When heat treatment is carried out on fabricated components like cleanout doors, heat treatment certificate and chart of multipoint temperature recorder shall be provided as record, with due endorsement from contractor and client/consultant.
- Records and reports shall always have direct reference to heat-treated components like part no., equipment tag number, etc.
- Actual chart speed, time of start, heating, soaking and cooling zones shall be clearly identified on chart along with direction of time progression.
- Location and identification of thermocouples shall be indicated in a sketch provided with report.
- When heat treatment is performed against a written procedure, reference to this procedure also needs to be reflected in report.
- As in the case of other inspection reports, both performance date and report date shall be clearly indicated in heat treatment record as well.
- When records consist of more than one page, all pages shall be numbered as "page x of y", followed by report number on all pages to ensure inclusion of complete report.

- When records are allowed in lieu of recorder charts, actual holding temperature and time shall be reported. For post-weld or stress-relieving treatments, maximum actual heating and cooling rates shall also be recorded.

14.6.6   WELD JOINT PWHT SUMMARY

As in weld and NDT summary, PWHT summary table is also prepared in line with unique weld joint identifications provided in referred isometric drawing, of course with reference to relevant PWHT reports. This document also serves as an easy checklist to verify completeness of PWHT and shall contain details pertaining to all types of weld joints requiring PWHT as per code or as per client specification.

14.6.7   PAD AIR TEST SUMMARY REPORTS

Branch weld joints identified in piping isometric with reinforcement pads need to undergo pad air test. If there are large numbers of reinforcement pads used in piping, all pads shall undergo pad air test. Since many reinforcement pads can be tested in a day, and the test pressure is the same, it is possible that all such pad tests can be covered by a single report if completed in a single day. The reports shall specifically identify the pads using the joint numbers provided in isometric drawing.

14.6.8   BOM CHECK REPORT (TEST PACK WISE)

Isometric drawing usually contains the BOM as well wherein all components to be fitted shall be identified. Therefore, BOM check on completed piping would reveal any missing components or welds, except for the site welds. Shortfalls if any identified during BOM check shall be reported and if required to be included in punch list-based criticality, so that all such issues will be addressed prior to handover.

14.6.9   TEST PACK DOCUMENTATION

The contents of test pack documents are described in detail in Chapter 13. As piping loop covered by test pack approaches final testing, test pack documentation is compiled in advance and submitted for review to all concerned inspections teams so that the shortfalls shall be known in advance and remedial actions can be taken not to affect the testing schedule. Since test pack documentations are voluminous in nature, its thorough review is time consuming and if done after completion of all piping works, and that is the reason why this document is submitted for review with some known shortfalls so that on the day of test the shortfalls alone need to be verified again to ensure completeness.

14.6.10   PUNCH LIST

In spite of efforts as above, some works may not be complete at the time of hydrostatic test of the test pack system and these are captured in the punch list, which are categorized as A, B and C punches as described in Chapter 13. All A punch items are completed prior to hydrostatic testing of piping loop. B and C items are carried out in due course and completion of those punch items may go up to handing over.

14.6.11 Mechanical Completion and Handing Over Report

As mentioned above, all punch items (A, B and C) shall be completed prior to handing over. When all works are completed, punch lists are closed and based on that, mechanical completion and handing over report is prepared. Thereafter, the commissioning activities and performance tests start. Upon completion of performance tests, the plant is ready for handover to client for regular operations and thereafter the responsibility to maintain plant shall be with the client.

14.7 GENERAL REQUIREMENT FOR CONTENTS OF CRB

- Documents forming part of CRB, requiring client approval shall have adequate evidence to prove so, especially in the case of engineering documents, which requires prior approval of client/consultant.
- The contractor/vendor shall perform checks by its engineering and QA teams prior to submitting documents for client certification. Corrections are not usually permitted in test certificates issued by respective manufacturers during this CRB review; however, inclusion of additional information if considered important may be endorsed on documents.
- One hard copy of the document shall be submitted in loose leaf binders, with proper indexing available in each volume for cross-reference. The recommended paper sizes for hard copy are also specified in table under Section 14.3.
- Hard copy shall be accompanied by a soft copy in PDF format, which shall be converted electronically into PDF format to maintain requisite quality. Soft copy shall be in searchable PDF format. In addition to above, all drawings shall be submitted in AutoCAD and data sheets in Excel format as well for implementing additions or deletions in future during service life of the tank.
- CRB shall be compiled on an ongoing basis during manufacture.
- Duplication of documents within a dossier shall be avoided to the extent possible.
- All pages within the CRB shall be clearly marked with proper captions, volume/page numbers as required.
- All documents shall be prepared in English language. In case some of the certifications are issued in languages other than English, an English translation of the same shall be enclosed with original certification with due endorsement of contractor as true translation.
- CRB shall be compiled according to a proposed CRB index duly approved by client right at the beginning. After compilation single hard copy of the same shall be submitted for review of client to ensure proper compilation and completeness of the document. Document thus approved by client alone shall be submitted as CRB in required number of hard and soft copies.

14.8   RECORDS AND REPORTS OF INSPECTIONS, TESTS AND CALIBRATIONS

- When specifying documents to be included in CRB, attention shall be paid to distinction between certificates, reports and records. Reports and records typically include useful data and definitive statements, whereas certificates can consist of unsupported statements certifying that tests, inspections or calibrations have been carried out with satisfactory results. Such statements from vendors or sub-contractors have no value to client, especially after expiry of the guarantee.
- Records and reports shall explicitly indicate the concerned item, material or equipment through proper identification system like item number, serial number or tag number, etc. as required.
- When inspections or tests are performed on sampling basis, this shall be made clear in the inspection certificate issued for the lot represented by the sample.
- When inspections or tests are performed against a written procedure, reference to this procedure shall be present in the test/inspection report.
- When results of inspection/tests are compared and accepted based on standards or specifications, reference of this document also shall be made on inspection/test report.
- Inspection and test report shall contain details like performance date and report date, along with a description of the type of test/inspection carried out.
- When reports consist of more than one page, all pages shall have the report number reflected in, along with "Page x of y" at appropriate place to ensure completeness of the report.
- Similarly, attachments if any to a report also shall be clearly indicated in report including information as to the number of pages of each attachment, in report.
- Any corrections, alterations or additions made to reports after endorsement by client inspectors shall be clearly traceable and dated.

14.9   CERTIFICATION FOR MATERIALS

- Material traceability is of paramount importance and hence as far as possible original certifications shall be maintained in CRB.
- When original material test certificate is issued for large quantity (of which only a small portion is required for the tank, like pipes, fittings, flanges, etc.), the original certificates shall be available with supplier or trader. Such availability of original certification shall be ensured through endorsements made by inspection engineers of the contractor/consultant/client at manufacturer's or trader's premises in their respective inspection reports for materials inspected and released from stockists.
- Materials without any of the above certifications shall not be entertained even for very small quantities.
- It is mandatory that certificates for wrought butt weld fittings shall be supported by certification for the parent plate or pipe as applicable. However, it

is not essential that such supporting certification need be original, as long as product specification requirements are established through sample testing after forming of the component.

- Corrections to certificates shall not be acceptable under any circumstances. When errors are found in certificates, it shall be reissued at source or else the material shall be rejected.
- When supplementary tests (tests not carried out my manufacturer) are carried out on materials with original certification, this shall be reported on separate sheets with cross-reference to original certification and to be endorsed by contractor/client inspectors as applicable.
- Transcribed data on material certificates shall normally be acceptable under the following circumstances:
  - i.   Heat analyses for wrought materials
  - ii. Certificates issued by stockists for bolting materials or screwed and socket weld fittings, which contain data and test results taken from manufacturer's certification and certified as having been accurately transcribed.
- When material is required to satisfy a carbon equivalent limit as determined by the long formula $CE=C+Mn/6+(Cr+Mo+V)/5+(Ni+Cu)/15$, then all component elements of this formula shall be determined and reported. It shall not be acceptable to assume zero content for any unreported elements.
- For non-ferrous alloys, where applicable standard gives a minimum value for predominant element, it is not acceptable for this element to be certified as "remainder" as this does not take into account the levels of impurities, which may be present.

14.9.1   Components Requiring "Material Certification" in Piping

- Pipes
- Fittings
- Flanges
- Gaskets
- Fasteners
- Valves
- Piping specials
- Supports
- Hangers
- Anchors
- Welding consumables including consumables for overlay cladding if required.

14.9.2   Contents of "Certification Dossier" for Bought Out Items (as Applicable)

- Index of the dossier
- Code Data Form or Certificate of Conformity
- Third Party Final Report or Certificate of Inspection
- Inspection and Test Plans (ITP) signed by all inspection authorities

- List of materials or layout sketch, showing position of component, cross-referenced to page numbers of material certificates and any supplementary reports
- Material test certificates including hardness testing as required
- Material repair records
- Weld key map/seam identification sketch (When manufacturer opts to maintain a written record of work performed by each coded welder in lieu of stamping welder's identification against his welds, this record shall be included in certification dossier. This record need not extend beyond "Code" welds.)
- Welding procedures and qualification records
- Production test records
- NDE reports and records. When repairs have been carried out, reports of original examinations shall be included
- Reports of any required special tests (Hydrogen Induced Cracking (HIC) and Sulfide Stress Cracking (SSC) etc.)
- Heat treatment charts/records as applicable
- Pressure and leak test reports
- Coating or lining application records and examination reports
- Nameplate rubbing (facsimile)
- As-built General Arrangement Drawing and isometric drawings for piping
- Records of critical dimensions
- All design calculations (including relief valves supplied with packaged vessels)
- Function test records for actuated valves
- Valve pressure/temperature ratings
- Special certificates (fire safe, etc.)
- Bend manufacturing procedures and qualification tests
- Pump performance test records, vibration records and function test records for instruments and controls
- Pipeline welds traceability, pressure testing charts and gauging, cleaning and drying records
- Dimensional records

15 Formats for Plant Piping Documentation

15.1 GENERAL

For maintaining traceability of materials and to ensure compliance of all inspection and tests are carried out in accordance with contract/code requirements, it is essential that duly certified documents are available for verification by any of the agencies authorized to do so.

Apart from material traceability, stage wise inspections and tests carried out also need to be recorded and countersigned by all concerned. In order not to miss out on any of the basic requirements, typical formats are often filled up and signed off after every inspection. This chapter provides a set of documents developed for piping construction, if filled up properly without omission shall form a sufficiently descriptive and logic documentation with regard to piping. It may be noted that information required to fill in the format is sufficient enough to meet code (ASME B 31.3) requirements, and thus also most of the client specifications as well.

Purpose of these formats is to provide uniformity in its structure, usefulness and completeness of information required, so that these reports would be helpful while carrying out maintenance, repair and modifications that are often required during the service life of any process plant.

15.2 FORMATS FOR SPOOL FABRICATION AND INSTALLATION

Sl. No.	Description	Document Code	Remarks
1	Request for Inspection	RFI	
2	Material Inspection Report	MIR	
3	Consumable Inspection Report	CIR	
4	Welding Consumable Control Log	CCL	
5	Material Traceability Report	MTR	
6	Fit-up Inspection Report	FIR	
7	Welding Visual Inspection Report	VIR	
8	Weld Summary Report	WSR	
9	Radiographic Test Report	RTR	
10	Ultrasonic Test Report	UTR	
11	Magnetic Particle Test Report	MPR	
12	Dye Penetrant Test Report	DPR	
13	NDT Summary	NDS	

(Continued)

DOI: 10.1201/9781003328124-15

Sl. No.	Description	Document Code	Remarks
14	Pneumatic Test Report (Reinforcement Pads)	PTR	
15	Dimensional Inspection Report (Spool)	DIR	
16	Dimensional Inspection Report (Piping Installation)	DIR	
17	PWHT Report	HTR	
18	Hardness Survey Report	HSR	
19	Spool Inspection Release	SIR	
20	Spool Installation Completion Report	SIC	
21	PWHT Summary Report	HTS	
22	Construction Summary Report	CSR	
23	Attachment Inspection Report	AIR	
24	Punch List	PL	
25	Test Pack Verification Check List	TVC	
26	Line Pneumatic Test (Line Leak Test) Test Pack	LLR	
27	Line Hydrostatic Test Report Test Pack	LHR	
28	Mechanical Completion Certificate	MCC	

Above listed formats cover documentation up to mechanical completion, leaving apart surface preparation and painting of piping. Though it is known that mechanical completion includes surface preparation and painting of lines as well, surface preparation and painting by itself being a specialized topic, it was considered appropriate not to include that in this book. However, an inspection and test plan activities for surface preparation and painting is described in the Appendix A to the book just to provide an overview of scope of work included under surface preparation and coating.

Since many works carried out during spool fabrication are repeated after installation of piping on site, many formats listed above may have to be made for works completed after installation and welding/bolting of spools.

As mentioned earlier, the following works essentially take place after spool fabrication at construction site.

- Installation of spools and field welding
- Inspection and testing of field welds
- Installation of valves and other inline equipment
- Installation of piping supports, anchors, hangers, etc.
- Final hydrostatic testing of piping as test packs as described earlier
- Pneumatic Leak Testing if applicable
- Surface preparation and painting of field weld joints and repair of coating damaged locations
- Punch listing
- Handing over for commissioning
- Punch clearance (B&C) items
- Piping completion

In case a different construction sequence is used at site, all the formats listed in the table above and below may not be applicable and hence left to judicious selection by the readers.

The 28 formats listed above and enclosed herewith are developed with an intention to have 100% compliance to inspection and test requirements and to generate a positive verifiable documentation at any point of time during construction or during service life of process plant. It may be noted that formats are developed based on sequence of construction described in previous chapters. Therefore, these formats may require some modifications here and there when used at piping construction sites following an entirely different sequence based on constraints at site. Above all, careful reading through code and client specification is also recommended prior to developing formats for piping construction works so that none of the essential requirements shall be missed out.

Logo		Request for Inspection		Ref: XXXX-RFI-001 Page : 1 of 1

Project Name	
Purchase Order No.	
Delivery Reference	
TPI Certificate Reference	

Delivery Type	First		Partial		Final		Replacement		Others	

From		To		
QA/QC Mgr. (Sub-Con.)		Client		Thru' proper Channel
Signature				

Notification Date		Inspection Date	
Notification Time		Inspection Time	

	Pipe		Fittings		Flange		Valves		Bolts/Nuts		Gaskets

	Supports		Inline Equipment		Others (Specify)-----------------------------

Location	

ITP Clause Reference		Specification Reference	
Activity Reference		No. of Attachments	

Inspect the following: ☐ WITNESS ☐ HOLD

Sl. No	Description of Items

Remarks (Sub-Contractor)	Remarks (Contractor)	Remarks (Client)

Note:
1. To be issued for all "W" and "H" points in ITP with required notification time as per contract

Sub Contractor		Contractor		Client	
Name		Name		Name	
Signature		Signature		Signature	
Date		Date		Date	

Logo	**Material Inspection Report**	Ref: XXXX-MIR-001
		Page : XX of YY

Project Name								
Purchase Order No.								
Delivery Reference								
TPI Certificate Reference								
RFI Reference								
Delivery Type	First		Partial		Final		Replacement		Others	

Notification Date		Inspection Date	
Notification Time		Inspection Time	

	Pipe		Fittings		Flange		Valves		Bolts/Nuts		Gaskets

	Supports		Inline Equipment		Others (Specify)-------------------------

Location	

ITP Clause Reference		Specification Reference	
Activity Reference		No. of Attachments	

ITP Clause Reference		Specification Reference	
Activity Reference		No. of Attachments	

Inspected the following: | | WITNESS | | HOLD

Sl. No	Description of Items	Remarks
1		
2		
3		
4		
5		
6		

Remarks (Sub-Contractor)	Remarks (Contractor)	Remarks (Client)

Note:

1. To be prepared for all incoming raw materials and shall be referred against each part number in applicable drawing.

Sub Contractor		Contractor		Client	
Name		Name		Name	
Signature		Signature		Signature	
Date		Date		Date	

Logo	Consumable Inspection Report	Ref: XXXX-CIR-001
		Page : XX of YY

Project Name	
Purchase Order No.	
Delivery Reference	
TPI Certificate Reference	

Delivery Type	First		Partial		Final		Replacement		Others	

From		To		
QA/QC Mgr. (Sub-Con.)		Client		Thru' proper Channel
Signature				

Notification Date		Inspection Date	
Notification Time		Inspection Time	

Location	

ITP Clause Reference		Specification Reference	
Activity Reference		No. of Attachments	

Inspect the following: ☐ WITNESS   ☐   HOLD

Sl. No.	Product Details	Brand	Size	Qty.	Lot No.	Batch Test Report	Remarks

Note:
1. To be prepared for all types of welding consumables like electrodes, filler wires, fluxes etc. Shielding / trailing gases may be exempted.

Sub Contractor		Contractor		Client	
Name		Name		Name	
Signature		Signature		Signature	
Date		Date		Date	

Logo	Welding Consumable Control Log	Ref: XXXX-CCL-001 Page : XX of YY

Project Name	
Purchase Order No.	
Delivery Reference	
TPI Certificate Reference	

Sl. No.	Date	Electrode /Flux	Size	Qty.	Lot / Batch No.	Baking Temp.	Baking Period	Remarks

Note:
1.   To be prepared for all types of welding consumables like electrodes, filler wires & fluxes

Sub Contractor		Contractor		Client	
Name		Name		Name	
Signature		Signature		Signature	
Date		Date		Date	

Logo	Material Traceability Report or Material Summary	Ref: XXXX-MTR-001
		Page : XX of YY

Project Name	
Purchase Order No.	
Delivery Reference	
TPI Certificate Reference	
P&ID Reference	

Line Number		Isometric No.	

Sl. No.	Part No.	Material Specification	Heat No.	MTC No.	MIR Reference	Remarks

Note:

1. To be prepared for all pressure containing parts against each part number of each Isometric.

Sub Contractor		Contractor		Client	
Name		Name		Name	
Signature		Signature		Signature	
Date		Date		Date	

Logo	Fit-up Inspection Report	Ref: XXXX-FIR-001 Page : XX of YY

Project Name	
P&ID Reference	

Line Number		Isometric No.	

Sl. No.	Isometric Drawing No.	Joint No.	Joint Type	Material Spec	WPS No.	Remarks

Note:

1. To be prepared for all Joints identified in Isometric drawing including welding BW, SW, BRW and other structural welding to piping components

Sub Contractor		Contractor		Client	
Name		Name		Name	
Signature		Signature		Signature	
Date		Date		Date	

Logo	Weld Visual Inspection Report				Ref: XXXX-VIR-001 Page : XX of YY

Project Name	
P&ID Reference	

Line Number		Isometric No.	

Sl. No.	Isometric Drawing No.	Joint No.	Joint Type	Welding Process	WPS No.	Remarks

Notes:
1. To be prepared for all Joints identified as pressure containing welds including that of reinforcing pads.
2. For structural attachment welds, Check list as in format WCC-001 would suffice

Sub Contractor		Contractor		Client	
Name		Name		Name	
Signature		Signature		Signature	
Date		Date		Date	

Logo	Weld Summary Report	Ref: XXXX-WSR-001 Page : XX of YY

Project Name	
P&ID Reference	

Line Number		Isometric No.	

Sl. No.	Weld No.	Description	WPS No(s)	Process	Welder	Visual Inspection Report Reference	Remarks
1							
2							
3							
4							
5							
6							
7							
8							
9							
10							
11							
12							
13							
14							
15							
16							
17							
18							
19							
20							
21							
22							
23							
24							
25							
26							
27							
28							
29							
30							
31							
32							

Note:
1. To be prepared for each isometric
2. Description means type of joint such as BW, SW, BRW etc.

Sub Contractor		Contractor		Client	
Name		Name		Name	
Signature		Signature		Signature	
Date		Date		Date	

Logo	Radiographic Test Report	Ref: XXXX-RTR-001
		Page : XX of YY

Project Name	
P&ID Reference	

Line Number		Isometric No.	

Radiographic Details

Source		Film Type		SFD	
Source Strength		Film Brand		Density	
Source Size		IQI		Sensitivity	
Screen		Technique		☐ SWSI ☐ DWSI ☐ DWDI	
Material		Weld Condition		☐ As Welded ☐ After PWHT	

Description of Welds Radiographed

Sl. No.	Joint No.	Segment Identification	Thickness	Welder No.	Interpretation	Location	Evaluation	Remarks

Film Size	100X240 mm	100X400mm	100X 200 mm	Others	

Radiographer		Processor		Interpreter	

Abbreviations for Evaluation	Acc.	Acceptable	Rep.	Repair	RT	Re-take

Interpretation Codes		EUC	External Under Cut	RC	Root Concavity
NSD	No Significant Defects	ICP	Incomplete Penetration	IP	Isolated Porosity
ISI	Isolated Slag Inclusion	BT	Burn Through	CP	Cluster Porosity
ESI	Elongated Slag Inclusion	IF	Incomplete Fusion	HB	Hollow Bead Porosity
IRP	Inadequate Root Penetration	TI	Tungsten Inclusion	LF	Lack of Fusion
IUC	Internal Under Cut	EP	Excess Penetration	CR	Crack

Note:
1. Every weld and segment radiographed shall be covered in this report

Sub Contractor		Contractor		Client	
Name		Name		Name	
Signature		Signature		Signature	
Date		Date		Date	

Logo	Ultrasonic Test Report	Ref: XXXX-UTR-001 Page : XX of YY

Project Name	
P&ID Reference	

Line Number		Isometric No.	

Ultrasonic Test Details

Equipment Type		Area Scanned		Time Base Range	
Model / Sl. No.		Surface Condition		Defect Report Level	
Certificate No.		Couplant		Defect Reject Level	
Calibration Block		Transfer Correction		Stage	Before/After PWHT
Material & Thickness		Sensitivity		Reference dB	
				Scanning dB	

Details of Probes

Angle	Frequency	Size	Type	Angle	Frequency	Size	Type
(1)				(2)			
(3)				(4)			
(5)				(6)			

Description of Welds Tested

Sl. No.	Joint No.	Location	Probe < Indication	Defect Type	Defect Size (mm) Length	Depth	Reference Height (+dB)	Evaluation

Test Results & Comments(if any)	

Technician		Witnessed by		Interpreter	

Notes:
1. Every joint or segment ultrasonically tested shall be clearly covered by this report.
2. Additional sketches (if required) also shall be attached to describe methodology.

Sub Contractor		Contractor		Client	
Name		Name		Name	
Signature		Signature		Signature	
Date		Date		Date	

Logo	**Magnetic Particle Test Report**	Ref: XXXX-MPR-001
		Page : XX of YY

Project Name	
P&ID Reference	

Line Number		Isometric No.	

Magnetic Particle Test Details

Equipment Type		Magnetic Particle	
Model / Sl. No.		Material	
Certificate No& Validity		Surface Condition	
Contrast Paint		Stage	Before /After PWHT
Method ☐ Wet ☐ Dry ☐ Florescent		Lighting ☐ Natural Violet ☐ Artificial ☐ Ultra	

Description of Welds Tested

Sl. No.	Joint No.	Location	Indication	Defect Size	Evaluation

Test Results & Comments(if any)	

Technician		Witnessed by		Interpreter	

Notes:
1. Every segment tested shall be clearly covered by this report.
2. Additional sketches (if required) also shall be attached to describe methodology.

Sub Contractor		Contractor		Client	
Name		Name		Name	
Signature		Signature		Signature	
Date		Date		Date	

Logo	Dye Penetrant Test Report	Ref: XXXX-DPR-001 Page : XX of YY

Project Name	
P&ID Reference	

Line Number		Isometric No.	

Dye Penetrant Test Details

Equipment Type		Cleaner Type	
Material		Cleaner Application	
Surface Condition		Developer Type	
Penetrant Type		Development Time	
Dwell Time		Stage		Before /After PWHT
Method ☐ Color dye ☐ Florescent dye	Lighting ☐ Natural ☐ Artificial ☐ Ultra Violet		

Description of Welds Tested

Sl. No.	Joint No.	Location	Indication	Defect Size	Evaluation

Test Results & Comments(if any)	

Technician		Witnessed by		Interpreter	

Note:
1. Every joint /segment of weld tested shall be clearly covered by this report.

Sub Contractor		Contractor		Client	
Name		Name		Name	
Signature		Signature		Signature	
Date		Date		Date	

Logo	NDT Summary Report	Ref: XXXX-NDS-001
		Page : XX of YY

Project Name	
P&ID Reference	

Line Number		Isometric No.	

Sl. No.	Weld No.	Description	NDT Report Reference				Remarks
			RT	UT	MPT	LPT	
1							
2							
3							
4							
5							
6							
7							
8							
9							
10							
11							
12							
13							
14							
15							
16							
17							
18							
19							
20							
21							
22							
23							
24							
25							
26							
27							
28							
29							
30							
31							
32							

Note:
1. To be prepared for each isometric
2. Description means type of joint such as BW, SW, BRW etc.
3. One or more reports may have to be indicated against each NDT for ensuring completeness

Sub Contractor		Contractor		Client	
Name		Name		Name	
Signature		Signature		Signature	
Date		Date		Date	

Logo	**Pneumatic Test Report (Reinforcement Pads)**	Ref: XXXX-PTR-001 Page : XX of YY

Project Name	
P&ID Reference	

Line Number		Isometric No.	

Pressure Gauge Details

Identification		Certificate No.	
Make		Validity	
Range		Bubble Solution	
Date of Calibration		Metal Temperature	

Test Pressure	

Sl. No.	Branch Connection Identification	Joint No.	Welder No.	Remarks

Note:
1.   Nozzle identifications of all nozzles shall be correlated to respective drawings and weld maps

Sub Contractor		Contractor		Client	
Name		Name		Name	
Signature		Signature		Signature	
Date		Date		Date	

Logo	Dimensional Inspection Report (Spool)	Ref: XXXX-DIR-001 Page : XX of YY

Project Name	
P&ID Reference	

Line Number		Isometric No.	

Spool Number (For individual Spools)	

Sl. No.	Description of Dimension	Identification (if any)	Dimension		Remarks
			Required	Actual	

Note:
1. All salient dimensions shown in isometric drawings shall be reported in this format

Sub Contractor		Contractor		Client	
Name		Name		Name	
Signature		Signature		Signature	
Date		Date		Date	

Logo	Dimensional Inspection Report (Piping Installation)	Ref: XXXX-DIR-001 Page : XX of YY

Project Name	
P&ID Reference	

Line Number	

Isometric drawing number	

Sl. No.	Description of Dimension	Identification (if any)	Dimension		Remarks
			Required	Actual	

Note:
1. All salient dimensions calculated based on respective isometric drawings shall be reported in this format

Sub Contractor		Contractor		Client	
Name		Name		Name	
Signature		Signature		Signature	
Date		Date		Date	

Logo	PWHT Report (Shop & Field Welds)	Ref: XXXX-HTR-001 Page : XX of YY

Project Name	
P&ID Reference	

Heat Source			
Type	Furnace / Local	Heat Source	Flame / Electrical/Induction
Thermocouple used		Chromel Alumel	
Recorder		Single point / Multipoint (XX Points)	
Chart Speed		(XX mm per hours)	

Recorder Connection Details											
1	2	3	4	5	6	7	8	9	10	11	12

PWHT Cycle			
Heating Rate(°C/hr)	Soaking Temp. (°C)	Soaking Time (Hrs)	Cooling Rate(°C/hr)

Recording above (°C)		Still air Cooling below (°C)	

Sl. No.	Line Number	Isometric No.	Weld Identification	Remarks

Notes:
1. Proposed to be prepared against each isometric when 100% PWHT is required
2. Type of heat treatment shall be indicated as Furnace/ Electrical/ Induction etc.
3. Used thermocouple is to be specified in place of Chromel Alumel indicated, which is common
4. Recorder connection details indicates the location where the thermocouples (12 points) are connected
5. Chart speed selected shall be clearly recorded

Sub Contractor		Contractor		Client	
Name		Name		Name	
Signature		Signature		Signature	
Date		Date		Date	

Logo	**Hardness Survey Report**	Ref: XXXX-HSR-001
		Page : XX of YY

Project Name	
P&ID Reference	

Line Number		Isometric No.	

Hardness Test Details

Equipment Type		
Model / Sl. No.		
Certificate No& Validity		
Material		
Stage	Before /After PWHT	

Description of Welds Tested

Sl. No.	Joint No.	Location & Hardness Values (BHN)					Remarks
		PM(A)	HAZ(B)	Weld (C)	HAZ(D)	PM(E)	

Test Results & Comments(if any)	

Technician		Witnessed by		Interpreter	

Notes:
1. Every segment tested shall be clearly covered by this report.
2. Additional sketches (if required) also shall be attached to describe methodology.

Sub Contractor		Contractor		Client	
Name		Name		Name	
Signature		Signature		Signature	
Date		Date		Date	

Logo	Spool Inspection Release	Ref: XXXX-SIR-001
		Page : XX of YY

Project Name	
P&ID Reference	

Line Number	

Sl. No.	Isometric Drawing No.	Spool No.	Remarks

Welding Completion	Yes/No	PWHT Completion	Yes/No/N A
NDT Completion	Yes/No/N A	Hardness Survey	Yes/No/N A
Branch weld air test	Yes/No/N A		

Notes:
1. To be prepared for all identified pipe spools in piping isometrics
2. N A (Not Applicable) shall be used wherever the attribute is not required to be applied
3. Incase the answer is "No", it requires further explanation

Sub Contractor		Contractor		Client	
Name		Name		Name	
Signature		Signature		Signature	
Date		Date		Date	

Logo	**Spool Installation Completion Report**	Ref: XXXX-SIC-001 Page : XX of YY

Project Name	
P&ID Reference	

Line Numbers	

Sl. No.	Line No.	Isometric Number	Remarks

Welding Completion	Yes/No	PWHT Completion	Yes/No/N A
NDT Completion	Yes/No/N A	Hardness Survey	Yes/No/N A
Gasketing & Bolting	Complete/Incomplete	Remarks	
Installation of Valves	Complete/Incomplete	Remarks	
Installation of Others	Complete/Incomplete	Remarks	

Notes:
1. To be prepared for all identified test packs
2. N A (Not Applicable) shall be used wherever the attribute is not required to be applied
3. Incase the answer is "No", it requires further explanation

Sub Contractor		Contractor		Client	
Name		Name		Name	
Signature		Signature		Signature	
Date		Date		Date	

Logo	PWHT Summary Report (Shop & Field Welds)	Ref: XXXX-HTS-001
		Page : XX of YY

Project Name	
P&ID Reference	

Sl. No.	Line Number	Isometric Number	Weld Identification	PWHT Report Reference	Remarks

Notes:
1. Proposed to be prepared against each isometric when 100% PWHT is required

Sub Contractor		Contractor		Client	
Name		Name		Name	
Signature		Signature		Signature	
Date		Date		Date	

Logo	**Construction Summary Report**	Ref: XXXX-CSR-001
		Page : XX of YY

Project Name	
P&ID Reference	

Line Number		Isometric No.	

Sl. No.	Weld No.	Joint Type	Fit-up	WPS	Welder	Visual	RT/UT	MPT/ LPT	Pad Test	PWHT	Hardness Report	Remarks
(1)	(2)	(3)	(4)	(5)	(6)	(7)	(8)	(9)	(10)	(11)	(12)	

Explanation Notes (not to be part of format)	
(1)	Weld number as given in isometric drawing
(2)	BW, SW or BRW as applicable
(3)	Fit up inspection report reference
(4)	WPS Number
(5)	Welder Identification
(6)	Visual inspection release report reference
(7)	RT / UT Report reference (both tests may not be required together)
(8)	MPT / LPT Report reference (both tests may not be required together)
(9)	Reinforcement Pad Air Test Report Reference
(10)	PWHT Report Reference
(11)	Hardness Survey Report Reference
(12)	Any other observations or notes

Notes:
1. All pressure retaining weld joints identified in Isometric including that of reinforcing pads and welds of critical support & hangers direct to piping shall be included in this report.
2. Report reference numbers shall be provided against each test to ensure completeness.

Sub Contractor		Contractor		Client	
Name		Name		Name	
Signature		Signature		Signature	
Date		Date		Date	

Logo	Supports, Hanger, Anchor & other Attachment Inspection Report	Ref: XXXX-AIR-001
		Page : XX of YY

Project Name	
P&ID Reference	

Line Number		Isometric No.	

Sl. No.	Description of Item	Identification Number	Location & Number		Remarks
			Required	Provided	
1					
2					
3					
4					
5					
6					
7					
8					
9					

Notes:
1. To be prepared for each isometric and all types of supports anchors etc connected to piping need to be included
2. List out any other attachments connected to piping other than those specified in drawing and to write specific remarks for the ratification of engineering team

Sub Contractor		Contractor		Client	
Name		Name		Name	
Signature		Signature		Signature	
Date		Date		Date	

Logo	Punch List	Ref: XXXX-PL-001 Page : XX of YY

Project Name	
P&ID Reference	

Line Number (s)

Sl. No.	Isometric Number	Description of O/S Work	Category of Punch			Remarks
			A	B	C	

Notes:
1. To be prepared against each line covering all isometrics in test pack
2. Punch items need to be categorized as A, B and C depending on severity of each
3. Category "A" is for items that are to be carried out before Hydrostatic Testing
4. Category "B" is for items that can be carried even after Hydrostatic Testing, but before   surface preparation & painting
5. Category "C" is for items that can be carried even after painting but before commissioning, without affecting any of the completed works, like replacement of a valve or a fitting etc.

Sub Contractor		Contractor		Client	
Name		Name		Name	
Signature		Signature		Signature	
Date		Date		Date	

| Logo | Test Pack Verification Check List | Ref: XXXX-TVC-001 |
| | | Page : XX of YY |

| Project Name | |
| P&ID Reference | |

| Test Pack Number | |

Test Pack Envelope

Sl. No.	Line No.	Isometric Number	Remarks

Verification	Status	Remarks
Material Traceability Reports	Complete/Incomplete	
Weld Completion	Complete/Incomplete	
NDT Completion	Complete/Incomplete	
Pad Air Test Reports	Complete/Incomplete	
PWHT Reports	Complete/Incomplete	
Hardness Survey Reports	Complete/Incomplete	
Flange Assembly	Complete/Incomplete	
Installation of Others	Complete/Incomplete	

Notes:
1. To be prepared for all identified test packs
2. Incase the answer is "Incomplete", it requires further explanation

Sub Contractor		Contractor		Client	
Name		Name		Name	
Signature		Signature		Signature	
Date		Date		Date	

Logo	Pneumatic Test Report (Line Leak Test)	Ref: XXXX-LLR-001 Page : XX of YY

Project Name	
P&ID Reference	

Test Pack Number	

Pressure Gauge Details

Identification		Calibration Certificate No.	
Make		Date of Calibration	
Range		Validity	

Details of Test Medium

Bubble Solution	
Method of application	

Test Pressure	

Test Envelope

Sl. No.	Line Number	Isometric Number	Remarks

Note:
1. To be prepared against each test pack covering all line numbers & isometrics covered by test pack.

Sub Contractor		Contractor		Client	
Name		Name		Name	
Signature		Signature		Signature	
Date		Date		Date	

Logo	Line Hydrostatic Test Report	Ref: XXXX-LHR-001 Page : XX of YY

Project Name	
P&ID Reference	

Test Pack Number	

Pressure Gauge Details

Identification		Calibration Certificate No.	
Make		Date of Calibration	
Range		Validity	

Pressure Recorder Details (Optional)

Identification		Calibration Certificate No.	
Make		Date of Calibration	

Test Pressure		Metal Temperature (°C)	

Details of Test Medium

Test Medium	
Inhibitors (if any)	
Dosing Rate	

Test Envelope

Sl. No.	Line Number	Isometric Number	Remarks

Notes:
1. To be prepared for each test pack covering all lines and isometrics within that test pack.
2. Calibration Reports for Gauges shall be attached with this report
3. Test report for medium used also shall be enclosed
4. Dosing details with its MSDS for chemicals also shall be enclosed.

Sub Contractor		Contractor		Client	
Name		Name		Name	
Signature		Signature		Signature	
Date		Date		Date	

Logo	**Piping Mechanical Completion Certificate**	Ref: XXXX-MCC-001 Page : XX of YY

Project Name	
Client Name	
Contract No.	
Scope of Work	Engineering, Procurement, Construction Inspection & Testing of Plant Piping
Status of Work	Completion of all Mechanical Works

Reference Documents

Description	Document Nunber	Remarks
Process & Instrumentation Diagrams	Provided as Attachment	
Line List	Provided as Attachment	

Details of Piping Works

Description	Status	Remarks
Installation of Spools	Complete/Incomplete	
Installation of Valves	Complete/Incomplete	
Installation of in line equipment	Complete/Incomplete	
Installation of supports hangers etc.	Complete/Incomplete	
Non Destructive Testing	Complete/Incomplete	
Post Weld Heat Treatment	Complete/Incomplete	
Other Inspections	Complete/Incomplete	
Hydrostatic Testing of Lines	Complete/Incomplete	
Line Leak Testing	Complete/Incomplete	
Surface preparation & Painting	Complete/Incomplete	

This is to certify that the inter connecting Plant Piping of -------------(Project Name) ---------------------------

set up at --(location) --
is designed, constructed, inspected and tested generally in accordance ASME B 31.3 Edition---(year)----
and also meeting the client required specified in above mentioned contract and is ready in all respects
to initiate commissioning activities.

Piping Contractor (Name &Address)		
Authorized Representative	(Name)	Seal
Signature		
Date		

Client/Commissioning Team (Name &Address)		
Authorized Representative	(Name)	Seal
Signature		
Date		

Notes:
1. List of applicable P&ID's are provided as Attachment (1)
2. Line list enclosed as Attachment (2)

Appendix A
Inspection and Test Plan (ITP) for Piping

A.1 GENERAL

Inspection and test plan (ITP) for piping is basically a consolidation of the inspection and tests to be carried out at various stages of fabrication and installation of piping. It gives an overview of various inspections required at each and every crucial stage of fabrication which has a bearing on the overall quality of piping. It involves the assembly of so many small piping components of varying diameters and configurations along with installation of all online valves and equipment. Apart from the above, it also provides an overview of quantum inspections to be carried out and witnessed. In order not to miss any of the salient activities during inspection, as required in codes and specifications, it is always better to have an ITP covering all these activities described fairly well, so that every activity shall be inspected and recorded.

As all are aware, piping involves works related to other disciplines as well and hence inclusion of all those activities under ITP for piping is practically impossible. Since the book is dealing with mechanical construction aspects of piping alone, typical ITP described in ensuing section covers only aspects related to mechanical works included under piping.

Rather than adopting the typical ITP blindly, readers are encouraged to modify the ITP document provided hereunder to suit specific project requirements, scope of work and manufacturing sequence proposed on site.

A.2 TYPICAL INSPECTION AND TEST PLAN FOR PROCESS PLANT PIPING

Typical Inspection & Test Plan for Process Plant Piping ❶

No.	Inspection/Test Activity	Reference Document ❸	Acceptance Criteria ❸	Verifying Document	Contractor	Consultant	Client	Remarks ❷
						Activity By		
A. Prior to Start of Work on Site								
1	Pre-Inspection Meeting (PIM)	Specification (Spec.)	Specification (Spec.)	MOM	H	H	H	
Documentation								
2	Piping Spool Manufacturing Procedure	ASME B 31.3, Spec.	ASME B 31.3, Spec.	Acceptance Stamp	H	R/A	R/A	
3	Welding Procedure Specifications (WPS)	ASME Sec IX. Spec.	ASME Sec IX, Spec.		H	R/A	R/A	
4	Welding Procedure Qualifications Records (PQR)				H	R/A	R/A	
5	Welder/Operator Qualifications Records (WQT)				H	R/A	R/A	
6	Weld Repair Procedure				H	R/A	R/A	
7	Inspection and Test Plan	ASME B 31.3, Spec.	ASME B 31.3, Spec.		H	R/A	R/A	
8	NDT Procedures	ASME Sec. V, Spec.	ASME Sec. V, Spec.		H	R/A	R/A	
(a)	Radiographic Test Procedure (RT)				H	R/A	R/A	
(b)	Ultrasonic Test Procedure (UT)				H	R/A	R/A	
(c)	Magnetic Particle Test Procedure (MPT)				H	R/A	R/A	
(d)	Liquid Penetrant Test Procedure (LPT)				H	R/A	R/A	

(Continued)

Typical Inspection & Test Plan for Process Plant Piping ❶

No.	Inspection/Test Activity	Reference Document ❸	Acceptance Criteria ❷	Verifying Document	Contractor	Consultant	Client	Remarks
						Activity By		
9	Post-Weld Heat Treatment (PWHT) Procedure	ASME B 31.3, Spec.	ASME B 31.3, Spec.		H	R/A	R/A	
10	Hardness Survey Procedure	NACE MR 0175, Spec.	NACE MR 0175, Spec.		H	R/A	R/A	
11	Hydrostatic Test Procedure (Spools)	ASME B 31.3, Spec.	ASME B 31.3, Spec.		H	R/A	R/A	
12	Surface Preparation and Painting (Spools)	Specification	Specification		H	R/A	R/A	
13	Pipe Spool Identification and Storage Procedure	Good Engineering Practice (GEP) / Specification	GEP/Specification		H	R/A	—	
14	Pipe Spool Transportation Procedure	Specification			H	R/A	—	
15	Pipe Spool Installation and Alignment Procedure				H	R/A	—	
16	Site Welding, Inspection and Testing Procedure	ASME Sec. IX, Spec.	ASME B 31.3, Spec.		H	R/A	R/A	
17	Static and Rotary Equipment Installation Procedure	GEP / Specification	GEP / Specification		H	R/A	—	
18	Inline Instruments and Equipment Installation Procedure				H	R/A	—	
19	Alignment, Levelling and Grouting Procedure				H	R/A	—	
20	Punch Listing Procedure	ASME B 31.3, Spec.	ASME B 31.3, Spec.		H	R/A	R/A	
21	Scaffolding Procedure	GEP/Specification	GEP/Specification		H	R/A	—	

(Continued)

Typical Inspection & Test Plan for Process Plant Piping ❶

No.	Inspection/Test Activity	Reference Document ❷	Acceptance Criteria ❸	Verifying Document	Activity By			Remarks
					Contractor	Consultant	Client	
B. Materials								
1	Material Receipt Inspection-Pipes	ASME Sec II	Codes and	Inspection Reports	H	W	RW1	
2	Material Receipt Inspection – Fittings and Flanges	ASME Sec II, B 16.5	Specifications		H	W	RW1	
3	Material Receipt Inspection – Fasteners and Gaskets, etc.	Codes			H	W	RW1	
4	Material Receipt Inspection – Valves and Inline Equipment				H	W	RW1	
5	Material Receipt Inspection – Welding Consumables	ASME Sec II			H	W	RW1	
C. Piping Spool Manufacturing (Shop)								
1	Marking, Cutting and Bevelling of Pipes and Material Traceability	Isometrics, std. drg	ASME B 31.3	Traceability Reports	H	SW	–	
2	Bevel Preparation of Fittings, Flanges and Material Traceability	Isometrics, std. drg	ASME B 31.3		H	SW	–	–
3	Fit-up of Spools	WPS/Spec./std. drg	ASME B 31.3	Fit-Up Report	H	RW2	–	–
4	Monitoring during Welding and Visual Inspection on Completion	WPS/ASME Sec. IX/ Spec.	WPS/Procedure/ ASME B 31.3	Welding Report	H	RW2	–	–
5	NDT of Welds	ASME Sec. V/Spec.	ASME B 31.3/ASME Sec. VIII Div1	NDT Reports	H	R	R	
6	PWHT (if Any)	WPS/Spec.	ASME B 31.3	PWHT Report	H	W	RW1	–
7	Hardness Survey	ISO 15156/Spec.	Procedure/ISO 15156/Spec.	Hardness Report	H	RW1	–	
8	Final Inspection of Spools	Isometrics/Spec.	Isometrics/Spec./ ASME B 31.3	Reports	H	H	RW1	–
9	Identification of Spools and Control	Specification	Procedure	Reports	H	W	–	

(Continued)

Typical Inspection & Test Plan for Process Plant Piping ❶

No.	Inspection/Test Activity	Reference Document ❷	Acceptance Criteria ❸	Verifying Document	Activity By			Remarks
					Contractor	Consultant	Client	
D. Surface Preparation/Painting of Spools								
1	Surface Preparation	SSPC, Specification	SSPC/Procedure	Surface Preparation/Painting Reports	H	W	RW1	–
2	Application of Primer				H	RW1	–	
3	Dry Film Thickness (DFT) Measurement				H	RW1	—R	–
4	Application of Intermediate Coat				H	RW1	–	
5	DFT Measurement				H	RW1	–	
6	Application of Final Coat				H	RW1	–	
7	Final DFT Measurement				H	W	RW1	
8	Identification and Tagging of Spools	Specification	Spool Identification Procedure	Spool Report	H	W	–	
E. Field Installation of Spools and Assembly								
1	Internal Cleanliness Prior to Installation	Specification	Installation Procedure	Installation Report	H	RW1	–	–
2	Assembly/Fit-up of Field Joints (Welds or Flange)	WPS/Spec.	Isometric/ASME B 31.3/Spec.	Assembly/Fit-Up Report	H	W	W	–
3	Verification for Induced Stresses on Installed Line	Specification	ASME B 31.3 Installation Procedure	Installation Report	H	W	–	–
4	Verification or Slopes and Other Key Dimensions (Surveying)	Specification	Isometrics/ASME B 31.3, Spec.	Surveying Report	H	W	RW1	–
5	Monitoring during Welding and Visual Inspection on Completion	WPS/ASME Sec. IX/Spec.	WPS/Procedure/ASME B 31.3	Welding Report	H	W	–	–
6	NDT of Welds	ASME Sec. V/Spec.	ASME B 31.3/ASME Sec. VIII Div1	NDT Request	H	R	R	–

(Continued)

Typical Inspection & Test Plan for Process Plant Piping ❶

No.	Inspection/Test Activity	Reference Document ❷	Acceptance Criteria ❸	Verifying Document	Activity By			Remarks
					Contractor	Consultant	Client	
7	PWHT (if Any)	WPS/Spec.	ASME B 31.3	PWHT Report	H	W/R	R	–
8	Hardness Survey	ISO 15156/Spec.	Procedure/ISO 15156/Spec.	Hardness Report	H	RW1	–	
9	Installation of Pipe Support and Hangers, Anchors, etc.	Isometrics/P&ID	Isometrics/P&ID	Pipe Support Report	H	W	RW1	–
10	Installation of Valves and Other Inline Equipment	Isometrics/P&ID	Isometrics/P&ID	Pipe Support Report	H	W	RW1	
11	Line Check (Punch Listing)	Isometrics/P&ID	Isometrics/P&ID	Punch List Report	H	W	W	–
12	Test Package Compilation	Isometrics/P&ID/Spec.	Supporting Documentation	Clearance for Line Test	H	R/A	R	–
13	Line Pressure Test	ASME B 31.3/Spec.	ASME B 31.3/Spec./Hydrostatic Test Procedure	Pressure Test Report	H	H	H	–
14	Valve Installation (Uninstalled during Test)	Isometrics/Spec.	Isometrics/Spec.	Reinstallation Report	H	W	W	–
15	Final Inspection (Reinstatement)	Isometrics/P&ID/Spec.	Isometrics/P&ID/Spec.	Reinstatement Certificate	H	H	W	–
F. Surface Preparation and Coating of Field Welds and Repair of Coatings Damaged during Installation on Spools								
1	Surface Preparation of Fields Welds and Adjacent Areas	SSPC, Specification	SSPC/Procedure	Surface Preparation/Painting Reports	H	W	RW1	
2	Application of Primer				H	RW1	–	
3	DFT Measurement				H	RW1	–	
4	Application of Intermediate Coat				H	RW1	–	
5	DFT Measurement				H	RW1	–	

(Continued)

Typical Inspection & Test Plan for Process Plant Piping ❶

No.	Inspection/Test Activity	Reference Document ❸	Acceptance Criteria ❸	Verifying Document	Activity By			Remarks
					Contractor	Consultant	Client	
6	Application of Final Coat				H	RW1	—	
7	Final DFT Measurement				H	W	RW1	
8	Preparation of Coating Damaged Locations				H	W	—	
9	Application of Coatings				H	RW1	—	
10	Final DFT Measurement at Repair Locations				H	W	—	
11	Surface Preparation for Wrapping Under Ground (UG) Piping	SSPC, Specification	SSPC/Procedure	Wrapping Report	H	W	RW1	—
12	Wrapping for UG Piping				H	RW1	R	
13	Holiday Test				H	H	W	—
G. Punch List Clearance and Handing Over for Commissioning								
1	Clearance of Punch Items, B and C				H	W	W	
2	Handing Over Certificate (for Commissioning)				H	H	H	

Notes and Legends

❶ ITP is based on the production sequence in which pipe spool fabrication is completed in all aspects including spool hydrostatic testing, followed by surface preparation and painting of spool leaving an uncoated end with a width of about 1″ at field joints.

❷ PIM to be scheduled after completion of one round of review of all inspection/test-related documentation pertaining to project.

❸ Documents mentioned under the columns are only indicative and not comprehensive. Try to include specific documents under each.

Abbreviations used

(Continued)

Typical Inspection & Test Plan for Process Plant Piping ❶

No.	Inspection/Test Activity	Reference Document ❷	Acceptance Criteria ❸	Verifying Document	Activity By			Remarks
					Contractor	Consultant	Client	
MOM	Minutes of Meeting							
H	Hold Point, Hold on the production till the stakeholders attend or perform inspection.							
W	Witness Point, Contractor to notify all stakeholders but need not hold on the production. Client can waive this inspection based on their discretion.							
R	Document Review, which includes engineering documents, procedures such as WPS, PQR, NDT Procedures, etc.							
A	Approval can be with or without any comments.							
RW1	Random Witness (10%), Witness randomly approximately at 10% of instances. Being random, all stakeholders need to be notified of the stage. For example, one random visit for whole UT tests or one or two visits during whole surface preparation works for painting.							
RW2	Same as above at 20%							
P&ID	Process & Instrumentation Diagram							
Isometrics	Isometric drawing							
Drg.	Drawing							
Spec	Specification							

Notes.

(1) Wherever codes are mentioned under "reference document" and "acceptance criteria" columns, users are encouraged to specify the exact codes they are expected to follow.

(2) Similarly, wherever procedures are mentioned, its reference number is also to be mentioned to be more specific.

Appendix B
Welding Procedure Specification (WPS), Procedure Qualification Record (PQR) and Welder Qualification Record (WQR)

B.1 INTRODUCTION

Though a large number of welding techniques are in use, manual, semi-automatic or automatic electric arc welding processes are used extensively in piping fabrication. Predominantly used welding processes in plant piping are Manual Metal Arc Welding (MMAW), Gas Tungsten Arc Welding (GTAW), Submerged Arc Welding (SAW) and Gas Metal Arc Welding (GMAW)/Flux Cored Arc Welding (FCAW).

Any organization certified to ISO 9001 is expected to carry out validation of processes they use in the production of goods and services. As required in ISO, organizations shall validate any processes for production and service provision where resulting output cannot be verified by subsequent monitoring or measurement and therefore, deficiencies become apparent only after the product is in use or the service has been delivered. Validation shall demonstrate the ability of these processes to achieve planned results. In fabrication environment like piping, this process is accomplished through preparation of WPS, qualification of procedure through PQR and qualification of welders through WQT.

Apart from the above, applicable Process Plant Piping Code, ASME B 31.3, also requires preparation of WPS, conduct of PQR and qualification of welders prior to start of production welding in piping. Therefore, development of a Welding Procedure Specification (WPS) and generation of PQR to validate the WPS are carried out by every contractor, broadly in accordance with requirements of ISO and specifically in compliance to applicable code of construction (for process plant piping, ASME B 31.3) and based on special requirements from the client.

Since ISO requires only validation of the WPS, by default, all specific requirements for welding in process plant piping come from ASME B 31.3, which require welding to be carried out according to a written-down procedure. ASME B 31.3 insists on following ASME Section IX code for preparation and establishment of welding procedure, whereas that for pipeline constructions as per ASME B 31.4 and 31.8, the code referred is API 1104.

As mentioned earlier, the piping contractor is responsible for all welding performed by their organization. Hence, the contractor is responsible to develop their own WPSs, in accordance with applicable codes and standards and also based on restrictions (over and above applicable code) provided in the contract, for use at piping prefabrication shop or at plant site itself. Welding in piping is permitted only after satisfactory qualification of applicable WPSs, followed by qualification of welders or welding operators. No welding shall be permitted in piping without a properly qualified WPS and welding shall be done only by a qualified welder for that WPS, that too within permissible range specified in ASME Section IX.

Commonly followed standards for different types of welding is provided in table below:

Type of Work	Applicable Standard
Structural Steel Welds	AWS D1.1
Piping and Pressure Vessel Welds	ASME SEC.IX
Pipeline Welds	API 1104

B.2   WELDING PROCEDURE SPECIFICATION

WPS is nothing but instructions on how to complete a particular type of weld and is prepared to provide clear direction and instructions for making production welds. It lists out various welding parameters, which is claimed by contractor to produce weld joints with properties satisfying design requirements. As this is only a claim, it needs to be established through a procedure qualification test, followed by destructive and non-destructive tests spelled out in applicable codes, and documentation pertaining to this test is known as Procedure Qualification Record (PQR).

Prior to start of production welding, it is absolutely essential that a suitable WPS is developed for type of welding required. This may consist of one or more welding process, the suitability of which broadly depends on the following considerations:

- Suitability for material to be welded
- Availability of welding equipment
- Material specification of piping components to be welded
- Diameter and thickness ranges and construction schedules
- Practical feasibility
- Productivity
- Quality of welds produced
- Finish and aesthetics of weld beads
- Ease of operation
- Schedule and other economic considerations
- Skill levels of available welders

Because of the above factors, experience of contractor also plays a vital role in selecting most suitable welding procedure capable of delivering welds with requisite quality and good productivity.

In piping, completion of piping isometrics is considered the starting point for preparation of WPS. Upon preparation of piping isometrics and material take-offs for the project, contractor shall be in a position to summarize various types of piping materials required in the project with specifics such as material specifications, diameter range, thickness ranges, impact test requirements, PWHT requirements, etc. With the above information and also based on limitations/restrictions imposed by codes and client specifications, welding engineers shall be in a position to estimate the number of WPSs required to cover the entire project work.

Even though it is possible to prepare a consolidated list of WPSs required for the project based on above information, the same needs to be juxtaposed with probable welding process that is proven and feasible for the project with reasonable productivity.

The flowchart below shows progression for preparation of WPS, generation of PQR, followed by qualification of required number of welders and conditions thereto. QW 482 of ASME Section IX provides a recommended format for WPS.

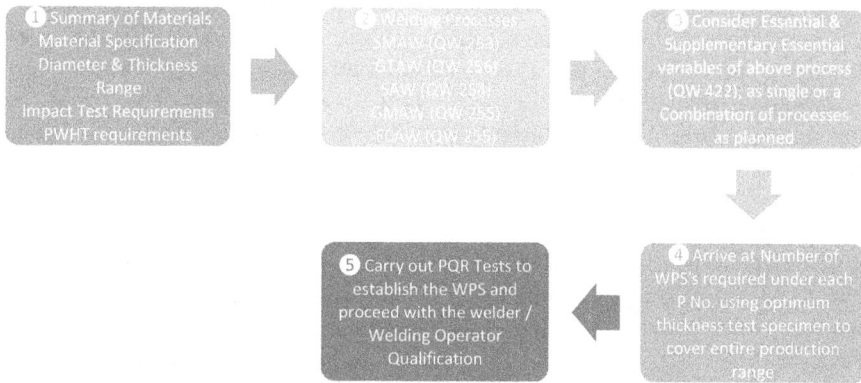

In this regard, physical parameters consolidated in Block (1) of flow chart above are considered against essential and supplementary essential variables (QW 253–QW 269 of ASME Section IX) for each welding process in Block (2) of flowchart. Tables QW 253–QW 269 of ASME Section IX provides list of essential, supplementary essential and non-essential variables (including those for special processes like weld overlay) for eighteen (18) different welding processes covered by ASME Section IX. With this consideration, WPSs required under each process can be quantified, usually prepared as a table. In this process, it is not essential that a single welding process alone shall be considered appropriate for welding a joint. Code allows use of combination welding processes as well, subject to qualifications limits specified for each process.

Outcome of activities mentioned in Block (1) eventually produces a consolidated table as below with information mentioned therein.

Material Specification	P No.	Thickness and Diameter	Impact Test	PWHT

Considering economic feasibility based on material specification, diameter and wall thickness, welding processes or combinations are decided upon, and table to be provided with another column to reflect applicable welding process (process selection in Block 2).

Now considering the parameters finalized till now, the number of test specimens required (with different thicknesses to cover the thickness range to be welded) is arrived and consolidated in yet another column added to the above table.

Material Specification	P No.	Thickness and Diameter	Impact Test	PWHT	Number and Thickness of Specimens

When impact test is made mandatory for welds based on process requirements, it applies further restriction on the range of weld metal thickness qualified by a test coupon of specified thickness. This condition may call for more number of specimens with different thicknesses to cover desired thickness range to be welded in that material. In addition, in such instances, all supplementary variables shall become essential, any change in which also would require yet another WPS. Similarly, addition or deletion of PWHT to a weld is a change in essential variable requiring separate qualification.

As mentioned earlier, WPSs so prepared are just preliminary documents, which need to be validated through welding a PQR test specimen, to be followed by NDT and the various mechanical tests on specimens prepared out of the test coupon plate welded in PQR test. Consequently, the original (or preliminary) WPS prepared cannot carry the reference number of the supporting PQR as it is not established at the time of preparation of WPS. This is exactly the reason for addressing the preliminary WPS as a claim, based on past experience of the contractor. Subsequent to documentation of PQR, WPS shall be revised to incorporate PQR reference number in it and because of this reason, the document revision number for an approved or established WPS shall be R1 at a minimum, assuming that R0 was assigned to preliminary WPS, which is unsupported.

B.3   PROCEDURE QUALIFICATION RECORD

As mentioned earlier, PQR test is the validation of proposed WPS, without which a WPS shall not be considered for any work. This validation is done through conducting a PQR test, using similar welding parameter as in proposed WPS and carried out by a proven welder or welding operator as the case may be. Test specimens (colloquially called coupon) so produced are later subjected to Non-Destructive and Destructive Test (NDT & DT) as required in codes (ASME Section IX and ASME B 31.3) and also in accordance with project specifications and are evaluated against acceptance norms as specified in those documents. When all test results are in acceptable range as specified in code and project specifications, all parties involved sign off the PQR document. The steps involved in establishing a PQR document is shown in following flow diagram.

| Conduct PQR Test | Radiography of Test Specimen | Mark up Specimens as per QW 463 | NDT & Mechanical Tests | PQR Documentation |

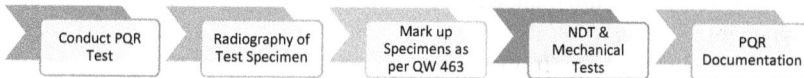

Upon consolidating PQR specimen requirement based on final table described under WPS, the next step shall be to procure materials to carry out PQR test. If project schedule permits, it can wait till receipt of bulk piping material on site. Since preparation of test specimens and carrying out tests take a considerable amount of time, it is better to procure materials (except special exotic materials which are scarce in the open market) required for carrying out PQR test from the local market with proper material certification.

For carrying out PQR test, specimen sizes to be welded shall be of sufficient dimensions to include all test specimens required by ASME Section IX (QW 453). In case of plate specimens, this is possible by increasing the length of specimen, whereas in pipe specimens, multiple specimens may be required to produce required number of test specimens, especially in the case of small diameter specimens.

PQR is the document which records all actual welding parameters used during welding of PQR Test Specimens, as also the test results.

QW 483 provides recommended format for PQR. As a general practice, all actual welding parameters used while carrying out PQR test are recorded meticulously against each weld pass used in filling groove. It is desirable to fill in all applicable parameters in that format, being authentic record of test, which can be used in the future. This includes details such as but not limited to electrode size, current, voltage, gas flow rates, pre-heat/interpass temperatures, etc. as applicable.

While welding parameters form only a part of format in QW 483, it can be completed only after satisfactory completion of mechanical tests.

When mechanical tests conducted on PQR specimens meet code-/project-specific requirement, associated WPS can be considered as "Qualified". When WPS is qualified through PQR, welder or welding operator who produced the PQR test specimen is automatically eligible to be qualified within range specified by code based on thickness of specimen and impact test requirements. Often, testing of specimens (such as tensile, bend, impact, radiography or macro etching hardness) is not permitted at own labs of the contractor and is required to be conducted at an independent laboratory.

B.4   DOCUMENTATION OF PQR

In principle, the filled-in format (QW 483), signed by all concerned, is good enough as minimum documentation. The standard template given may not be sufficient to hold all information mentioned above and hence needs to be expanded to record all salient parameters used during test. In addition to that, being a documentation with indefinite validity, the present industry practice is to have the following supporting evidences well enclosed with the document in the order of preference below:

1. Duly filled-in format as per QW 483
2. Mechanical Test Reports (Tensile, Bend and Notch Toughness from the lab duly attested by all present during test)

3. Micro/Macro Examination Report (from laboratory conducting test) as applicable
4. Weld Metal Chemistry (from laboratory conducting test) if required
5. Radiography Report
6. Material Test Certificate for base pipe material
7. Material Test Certificate for Filler Rod or Electrodes
8. PWHT Report (if applicable)
9. Preliminary WPS

As the PQR is considered a permanent record, which can be used for many more years to come, utmost care shall be taken to make it error free, meaning even free of typographical errors.

B.5   MECHANICAL TESTS FOR PQR

Depending on requirements in client specification applicable, all or a few selected properties below of weld metal may have to be verified to consider the WPS as qualified. They are based on amount of deformation which metal can withstand under different modes of force application listed below:

- *Malleability*: Ability of a material to withstand deformation under static compressive loading without rupture
- *Ductility*: Ability of a material to undergo plastic deformation under static tensile loading without rupture. Measurable elongation and reduction in cross section area
- *Toughness*: Ability of a material to withstand bending or the application of shear stresses by impact loading without fracture
- *Hardness*: Measurement of a material's surface resistance to indentation from another material by static load
- *Tensile strength*: Measurement of the maximum force required to fracture a material bar of unit cross-sectional area in tension

The above properties as applicable shall be established through the following standard mechanical tests:

Tests for Required Properties	Tests for Weld Quality
• Tensile test (Transverse, All Weld)	• Macro testing
• Toughness testing (Charpy, Izod)	• Fillet weld fracture testing
• Hardness tests (Brinell, Rockwell, Vickers)	• Butt weld nick break testing
• Bend testing	• NDE (VT, PT, MT, RT, UT)
• CTOD (Crack Tip Opening Displacement)	
• Corrosion tests, HIC, SOHIC, IGC tests	

a. **Reduced Section Tension Test**
   Tensile testing, also known as tension testing, is a fundamental materials science test in which a sample is subjected to a controlled tension until

failure. Test involves straining a test specimen in tension, generally to fracture, for determining one or more of mechanical properties and is usually carried out at ambient temperature unless otherwise specified.

b. **Bending Test (Transverse Specimen)**

Bend testing determines ductility or strength of a material by bending the material over a given radius. After bending, the specimen is inspected for cracks on outer bent surface. Bend testing provides insight into modulus of elasticity and bending strength of a material, here that of the weld.

c. **Nick Break Test**

Nick break test has been devised to determine whether weld metal of a welded butt joint has any internal defects, such as slag inclusions, gas pockets, poor fusion, and/or oxidized or burnt metal.

d. **Charpy Impact Test (Weld Joint Section)**

Test consists of breaking a V-notched specimen by one blow from a swinging pendulum, under defined conditions, with V-notch at the middle and specimen resting on two supports. Parameter measured here is the energy absorbed for breaking specimen in impact. Many times, this test is required to be conducted at temperatures even as low as $-193°C$.

e. **Vickers Hardness Number (HV)**

A diamond indenter in the form of a right pyramid with a square base and with a specified angle between opposite faces at the vertex is forced into the surface of a test specimen followed by measurement of diagonal length of the indentation left on metal surface after removal of test load. Test load often used is 5 kgf.

f. **Macro Examination and Photo**

Macro-examinations are also performed on a polished and etched cross section of a welded metal. During examination, a number of features can be determined including weld run sequence, important for weld procedure qualifications tests. Apart from this, any defects on specimen at polished cross section can be assessed for compliance with relevant specifications. Slag, porosity, lack of weld penetration, lack of sidewall fusion and poor weld profile are among the features revealed in such examinations. It is normal to look for such defects either by standard visual examination or at magnifications of up to 50×. It is also routine to photograph the section to provide a permanent record, known as a photomacrograph.

B.6 TYPES AND PURPOSES OF TESTS SPECIFIED IN ASME SECTION IX FOR THE TYPES OF WELDS REQUIRED IN PIPING

a. **Mechanical Tests**

Test	Purpose
Tension Test	Ultimate tensile strength (UTS) of groove welds (QW 150)
Guided Bend Test	Degree of soundness and ductility of groove welds (QW 160)
Fillet Weld Test	To determine size, contour and degree of soundness of fillet welds (QW 180)
Notch Toughness Test	Notch toughness of weldments (QW 171 and 172)

b. Special Examinations for Welders

Radiographic or ultrasonic examination	As a substitute for mechanical testing of QW-141 for groove-weld performance qualification, to prove the ability of welders to make sound welds (QW 191)

c. Visual Examination

Visual examination	To determine that final weld surfaces meet specified quality standards (QW-194)

B.7   ROLE AND RESPONSIBILITY MATRIX DURING PQR TESTING

Task	Responsibility				
	(1)	(2)	(3)	(4)	(5)
Review of Preliminary WPS	P	V	V		
Preparation of PQR Test specimen	P	I	-	-	-
Inspection of Materials for PQR Specimen	I	I	I	-	I
Verification of MTCs for Specimens, Electrodes, Filler, etc.	V	V	R, V	-	R, V
Inspection of PQR Specimen for Dimension and Fit-up	I	I	I	-	-
PQR Qualification					
• Record current/voltage during each pass	D	D	V	-	V
• Record travel speed in each pass	D	D	V	-	V
• Gas flow rate for each pass (shielding/ trailing)	D	D	V	-	V
• Verify the root penetration	A	A	A	A	A
• Verify calibration of all measuring devices	V	V	V	-	V
Review of radiograph of PQR specimen	A	A	A	A	A
Inspection of marking of test specimens and stamping	I	I	I	-	-
Preparation of samples and transfer of identifications	I	I	I	-	-
Mechanical Tests					
All Weld Tensile	W	W	W	W	W
Transverse Tensile	W	W	W	W	W
Root and Face Bend	W	W	W	W	W
Side Bend Tests	W	W	W	W	W
Nick Break Test	W	W	W	W	W

(*Continued*)

Task	Responsibility				
	(1)	**(2)**	**(3)**	**(4)**	**(5)**
Charpy Impact Test	W	W	W	W	W
Hardness Test	W	W	W	W	W
Special Tests like IGC, HIC, SOHIC, etc.	W	W	W	W	W
Macro Etching and Examination	W	W	W	W	W
Micro examination and photo	V, A	V, A	V, A	-	V, A
Preparation of PQR with all attachments	P	V	V	-	V
Revision of WPS (incorporation of PQR No)	P	V	V	-	V
Approval of WPS and signing of PQR	A	A	A	A	A

Notes / Legends

(1)	Sub-contractor
(2)	Main Contractor
(3)	Consultant
(4)	Owner
(5)	Third-Party Inspection
P	Preparation
I	Inspection
W	Witness
V	Verify
R	Random
A	Approval
D	Documentation

B.8   MECHANICAL TESTING FOR WELD TEST SPECIMENS (OF PQT AND WQT)

As mentioned previously, the test coupons welded for establishing WPS (PQR Test Coupon) qualification are subjected to the same cycle of PWHT (if specified) and NDE as required for production welds, prior to preparation of specimens for mechanical testing. Specimens shall be taken outside the region of probable gross imperfections (e.g. porosity, slag inclusions, etc.) revealed by NDE and also shall be clear of the discard areas specified in ASME Section IX at the beginning and end of specimen. Specimens for mechanical testing shall be removed from test coupon by mechanical methods in accordance with figures shown in QW-463 of ASME Section IX. Number of specimens required for mechanical tests and acceptance criteria as in ASME Section IX goes as below.

Transverse Tensile Tests	Number of Specimens	2

Acceptance Criteria

- Specified minimum tensile strength of base metal; or
- Specified minimum tensile strength of weaker of two base metals; or
- If specimen breaks in base metal outside weld or fusion line, test shall be accepted as meeting Code requirements, provided strength is not more than 5% below specified minimum tensile strength of base metal.

Guided Bend Tests		Number of Specimens	4
Mandrel Diameter	As per QW 466.1, ASME IX	Bending Angle	180°

Other Requirements
- For $T < \frac{3}{4}''$ (19 mm): 2 face bends + 2 root bends or 4 side bends
- For $T \geq \frac{3}{4}''$ (19 mm): 4 side bends where T = base metal or specimen thickness

Acceptance Criteria
- No open defects > 1/8″ (3 mm) in weld or heat-affected zone (HAZ) in any direction on the convex surface.

NB. As a guide, the length of the bend test specimens should be at least $1.5 \times$ mandrel OD with a minimum of 150 mm and a maximum of 400 mm.

Notch Toughness Tests	Charpy V-Notch	Number of Specimens	As per Spec
Requirements		To be done when required by other sections.	

Acceptance Criteria
- The acceptance criteria shall be in accordance with that section specifying impact requirements

Notch Toughness Tests	Drop Weight	Number of Specimens	As per Spec
Requirements		To be done when required by other sections.	

Acceptance Criteria
- The acceptance criteria shall be in accordance with that section requiring drop weight tests.

Macro Examination
Requirements   When required by other sections

A transverse section of weld is removed from any suitable position from test coupon by mechanical means. Later, this is prepared for metallographic examination by conventional grinding, polishing and viewed under low magnification (up to ×50) after etching.

Acceptance Criteria
- Visual examination of the cross section of the weld metal and HAZ shall show complete fusion and freedom from cracks, except that linear indications at the root not exceeding 1/32″ (0.8 mm) shall be acceptable.
- Weld shall not have a concavity or convexity greater than 1/16″ (1.5 mm).
- There shall be not more than 1/8″ (3 mm) difference in the lengths of the legs of the fillet.

Hardness Examination

Requirements   When required by other sections

Hardness measurements when carried out shall be according to QW 462.12, ASME IX, on base material, HAZ and weld metal on the transverse section taken for macrographic examination. For PQRs carried out for sour service environment, hardness survey is carried out as per Clause 7.3.3.3 of NACE MR 0175/ISO 15156-1. Hardness test results so obtained also shall be included as supporting documentation of PQR.

Acceptance Criteria

* Hardness values so obtained shall not exceed guidance values given in applicable code.

B.9   WELDER/WELDING OPERATOR QUALIFICATION TESTS AND RECORDS

Upon satisfactory qualification of WPS as described in sections above, next step is to qualify welders or welding operators (in the case of automatic welding). Following are the predominant factors that influence the number of welders required under each WPS:

* Process selected and its productivity
* Quantum of weld
* Project schedule
* Work locations

Block diagram depicting process of welder qualifications is provided below.

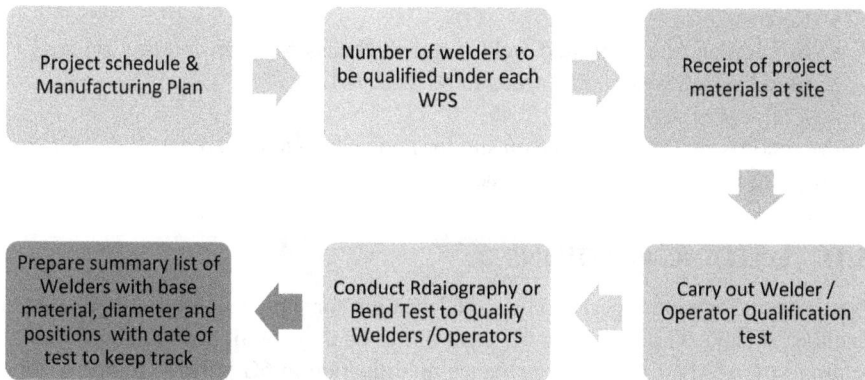

Upon finalization of the number of welders/welding operators required under each WPS, starting point for welder qualification is the receipt of project materials (pipe/fittings) on site.

Main objective in carrying out welder/welding operator qualification test is to validate the ability of a welder to follow verbal or written instructions (WPS) and to verify that the weld produced meets required quality standards, which is ensured through radiography or mechanical testing.

Welder qualification test is much simpler compared to procedure qualification test. Inspectors witnessing WQT shall ensure that welding parameters are used by each welder in accordance with approved WPS (within permitted tolerance provided by code), such as current, voltage and travel speed, gas flow rates, deposited weld metal thickness, etc. Though it is not mandatory to include this information in format (QW 484) for WQT, contractor is expected to have this information at least in their in-house records.

B.10   GENERAL REQUIREMENTS

As mentioned earlier, qualification of required number of welders/welding operators to meet project schedule is the primary responsibility of the piping contractor.

Performance test records of welders issued from previous contracts in lieu of welder qualification tests are not generally accepted.

Piping contractor is responsible to maintain records of qualifications of all their welders and tracking record shall essentially consist of the following:

- WPS used for qualification
- Base metal and thickness of specimen
- Qualification range, base metal and thickness range, pipe diameters
- Qualified positions
- Renewal date
- In addition, the deposited weld metal thickness for each welding process used by each welder also needs to be recorded.

Above records shall be available on site for review by authorized personnel of client at any point in time during the contract.

B.11   WELDING POSITIONS

Test coupon for welder performance qualification for pipe welding shall be welded in either 2G and 5G positions or 6G positions to qualify for all positional welding as required by ASME IX. For performance qualification in 6G position, range of pipe diameters, thicknesses, joint geometry and steel grades requiring qualifications must be according to job requirements estimated by engineering.

B.12   VISUAL EXAMINATION

Visual examination of performance test coupons shall show complete joint penetration and fusion between base metal and weld metal over entire circumference, both inside and outside.

B.13   MECHANICAL TESTING FOR WQT

Type and number of guided bend test specimens required for welder qualification is governed by pipe thickness and position in which it was welded. QW 463 provides guidance for removal of test specimens. Similarly, acceptance criteria for guided bend tests are given in QW-163, ASME IX. Actual testing operations such as guided bend testing and radiography shall be performed by independent laboratory on test specimens removed from the qualification test coupon under the direction of the inspector.

B.14   RADIOGRAPHIC EXAMINATION

Performance test coupon produced for welder qualifications on site is examined by radiographic method in lieu of mechanical testing. The results of radiographic examination shall be assessed in accordance with QW-191, ASME IX.

B.15   QUICK REFERENCE TO ASME B 31.3 ON WPS, PQT AND PRODUCTION WELDS

Various Clauses of ASME B 31.3 applicable to Welding Procedure Specification, Procedure Qualification Test, Welder Qualification Test and Production Welds are summarized in below table.

Sl. No.	Applicable Clause(s)	Description of Requirement or Restriction
1	Chapter II Part 4 – Design Clause 311	Weld joints permitted in any material for which it is possible to qualify welding procedures, welders and welding operators in conformance with the requirements in Chapter V
2	311.2.5 Socket Welds (a)	Recommended to avoid socket welded joints in any service where crevice corrosion or severe erosion is possible
3	311.2.5 Socket Welds (c)	Socket welds larger than DN 50 (NPS 2) are not permitted under severe cyclic conditions
4	311.2.7 Seal Welds	Seal welds (para. 328.5.3) are permitted to prevent leakage of threaded joints and are not considered as strength welds
5	Chapter III Figure 323.2A and Table 323.3.1, Table K323.3.1	Impact test requirements on weld specimens provided
6	Chapter V Clause 328.1	Each employer is required to have their own WPS/PQR and WQT
7	Chapter V Clause 328.1	WPS/PQR and WQT shall be carried out as per BPV Code, Section IX except as modified in Chapter V
8	Chapter V Clauses 328.2.1(b) to (f)	Exemptions to BPV Section IX requirements
9	Chapter V Clause 328.2.2	Permits procedures qualified by others
10	Chapter V Clauses 328.2.2(a) (1) and (2)(b) to (i)	Specifies conditions to accept procedures qualified by others (processes permitted are SMAW and GTAW and their combinations only)

(Continued)

Sl. No.	Applicable Clause(s)	Description of Requirement or Restriction
11	Chapter V Clause 328.2.3	Permits WQT by others and requirements specified
12	Chapter V Clause 328.2.4	Specifies requirements on records to be maintained
13	Chapter V Clause 328.3	Specifies electrodes/filler metal requirements in Clauses 328.3 1(a) to (e)
14	Chapter V Clause 328.3.2	Requirements for Weld Backing Materials (including non-ferrous and non-metallic) are provided in Clauses 328.3.2(a) to (c)
15	Chapter V Clause 328.3.3	Requirements for Consumable Inserts are provided
16	Chapter V Clause 328.4	Preparations for welding – 328.4.1 Cleaning, 328.4.2 End Preparation, 328.4.3 Alignment
17	Chapter V Clause 328.5	Welding Requirements – 328.5.1 General, 328.5.2 Fillet and Socket Welds, 328.5.3 Seal Welds, 328.5.4 Welded Branch Connections, 328.5.5 Fabricated Laps, 328.5.6 Welding for Severe Cyclic Conditions
18	Chapter V Clause 328.6	Weld Repair
19	Chapter V Clause 330	Pre-heating – 330.1 General, 330.1.1 Requirements and recommendations, 330.1.2 Unlisted Materials, 330.1.3 Temperature Verification, 330.1.4 Pre-heat Zone, 330.2 Specific Requirements, 330.2.3 Dissimilar Materials, 330.2.4 Interrupted Welding
20	Chapter V Clause 331	Heat Treatment – 331.1 General, 331.1.1 Heat Treatment requirements, 331.1.3 Governing Thickness, 331.1.4 Heating and Cooling, 331.1.6 Temperature Verification, 331.1.7 Hardness Test, 331.2 Specific Requirements, 331.2.1 Alternative Heat Treatment, 331.2.2 Exceptions to Basic Requirements, 331.2.3 Dissimilar Materials, 331.2.4 Delayed Heat Treatment, 331.2.5 Partial Heat Treatment, 331.2.6 Local Heat Treatment

Apart from above the clauses similarly numbered as above under various categories with prefixes A, M, MA, K, U, UM shall also be applicable when piping falls under any of those categories of services identified in ASME B 31.3.

B.16   SPECIAL QUALIFICATIONS FOR WELDERS/ WELDING OPERATORS

Welding in spaces with limited access or other similar conditions may attract special qualification of welders. Such special welder qualifications (if any expected in project) shall be identified in project specification, that is, welding on a liquid filled/hot and/or purged pipeline/branch welds with limited access etc. Welding on in-service pipelines is to be completed only by company trained and qualified welders under the supervision of operations/pipeline maintenance personnel.

B.17   WELD REPAIR PROCEDURE QUALIFICATION

Qualification of production welding procedures and repair welding procedures shall be performed concurrently. WPS qualified on groove welds are applicable for weld repairs to groove and fillet welds and for weld build-up. However, following thickness limits are applicable in case of groove welds.

- Thickness range for base metal and deposited weld metal for each welding process shall be in accordance with QW-451, with no upper limit on base metal thickness when qualification is done on base metal having a thickness of 38 mm (1½″) or more.

B.18   GENERAL INSTRUCTIONS TO WELDERS/ WELDING SUPERVISORS

1	Welders shall only work within the limit of their qualification range.
2	Welding supervisor/welders have the responsibility to ensure that right WPS is selected from the list of approved WPSs applicable to specific piping class and thickness.
3	Welders shall withdraw consumables only from designated welding stores and endorse initials on the "Welding Consumables Register" maintained by consumable store
4	Withdraw only sufficient quantities of welding consumable for a 4-hour period of work.
5	Low-hydrogen electrodes shall be conveyed to job site in heated "quivers" only.
6	All low-hydrogen welding electrodes shall be maintained in heated "portable quivers". The lid of the quiver shall be closed following withdrawal of each electrode.
7	During change of shifts, all "quivers" shall be returned to consumable store for verification. All issued and unused low-hydrogen electrodes shall be scrapped (reuse permitted only when manufacturer recommends so).
8	Fit-up geometry shall be in accordance with applicable engineering documents and approved WPS.
9	Welding parameters specified in WPS shall be always adhered to.
10	Pre-heat and interpass temperatures shall be monitored either using "Tempil Stiks" or calibrated (analogue or digital) pyrometers. Pre-heat temperature shall be as per WPS.
11	Pre-heat shall be applied through full thickness of joint and hence advisable to check from opposite side wherever possible.
12	For repair welds, pre-heating temperature shall be 50°C (122°F) above that used for original weld. Maximum pre-heat for repair welding is 150°C (302°F).
13	Weld joint fit-ups with Internal misalignment in excess of 1.5 mm (1/16″) shall not be welded.
14	Deposit the root pass and six successive passes or 1/3 of the weld volume prior to interruption (allowing to cool to ambient temperature).
15	Buttering (build-up) is permitted as follows: • Buttering shall not exceed the lesser of 10 mm or 1/3 base metal thickness. • If buttering to exceed 10 mm or 1/3 base metal thickness then this shall be welded with the approval of client and with additional NDT as considered appropriate.
16	Back welding is permissible for all applications, so long as same electrodes and process is used as for the fill pass.

(Continued)

17 The following points shall be observed when fabricating/welding stainless and non-ferrous materials:

- Weld edge preparation and filler materials shall be degreased using an appropriate solvent.
- Weld preparation surfaces shall be buffered using flapper wheels.
- Argon hoses shall be checked for any loose connections or leakage, etc.
- Bridge tacks to be used while fitting up joints. Avoid tacking directly on to the adjacent pipe wall. Bridge tack within fusion faces of joint wherever possible.
- Prior to welding, back purge shall be set up in accordance with approved WPS for a sufficient period of time to ensure an oxygen content of less than 0.5%. When available, "purge monitoring devices" shall be used.
- Back purge shall be maintained for a minimum of four passes. Pipe dams shall be left in place until completion of welding.
- The end of partly used filler wire shall be snipped off prior to restart.
- Maintain tip of filler rod within gas shroud during welding to avoid contamination.
- Interpass temperatures for these materials is critical. Welder shall check temperature prior to next pass. Austenitic stainless welds may require "Ferrite" content test. Cleanliness during fabrication and monitoring of interpass temperatures have a direct effect on test results.
- "Iron-free" cutting and grinding disks shall alone be used while working with stainless steels.
- Similarly, markers free of sulphur, chlorine alone shall be used in stainless steels.
- All material handling tools and devices shall be adequately protected/lined, to prevent possibility of carbon "pick-up" (i.e. lifting clamps, work bench, transport trucks, supports, etc.).
- Wherever possible, tools shall be colour-coded to prevent cross-use with other material types.

	ASME Section IX		Short description of Welding Parameter	Welding Variables			
	Clause	Difference		SMAW (QW 253)	SAW (QW 254)	GMAW/FCAW (QW 255)	GTAW (QW 256)
QW-402 Joints	0.1	φ	in groove design	N	N	N	N
	0.4	−	of backing in single sided weld	N	N	N	
	0.5	+	of backing and chemical composition				N
	0.10	φ	in root spacing	N	N	N	N
	0.11	±	nonfusing retainers	N	N	N	N
QW-403 Base Metal	0.5	φ	in group number QW-422	S	S	S	S
	0.6	T Limits	T Limits impact	S	S	S	S
	0.7	T/t	T/t Limits > 8 in. (203 mm)				
	0.8	φ	φ T Qualified	E	E	E	E
	0.9	t pass	t pass > 1/2 in. (13 mm)	E	E	E	
	0.10	T	T Limits Qualified (Short Circuit Arc)			E	
	0.11	φ	P-No. Qualified	E	E	E	E
	0.13	φ	P-No. 5/9/11				
QW-404 Filler Metals	0.3	φ	Size				N
	0.4	φ	F-No.	E	E	E	E
	0.5	φ	A-No.	E	E	E	E
	0.6	φ	Diameter	N	E	N	
	0.7	φ	Diameter > 1/4 in. (6 mm)	S			
	0.9	φ	Flux/Wire Classification		E		
	0.10	φ	Alloy flux		E		
	0.12	φ	AWS classification	S		S	S
	0.14	±	Filler			E	
	0.22	±	Consumable insert				N
	0.23	φ	Filler metal product form (Solid/metal or flux-cored)			E	E
	0.24	± φ	Supplemental Filler Metal		E	E	
	0.25	±	Supplemental Powder Filler metal				
	0.26	>	Supplemental Powder Filler metal				
	0.27	φ	Alloy elements		E	E	
	0.29	φ	Flux designation		N		
	0.30	φ	t	E	E	E	E
	0.32	t Limits	t Limits (Short Circuit Arc)			E	
	0.33	φ	AWS Classification	N	N	N	N
	0.34	φ	Flux Type		E		
	0.35	φ	Flux/Wire Classification		S N		
	0.36		Recrushed Slag		E		
	0.50	±	Flux				N

Group	No.	Symbol	Variable								
QW-405 Positions	0.1	+	Position		N		N		N		N
	0.2	φ	Position	S				S		S	
	0.3	φ↑↓	Vertical Welding		N				N		N
QW-406 Preheat	0.1	Decrease	Decrease > 100°F (56°C)	E		E		E		E	
	0.2	φ	Preheat maintenance		N		N		N		
	0.3	Increase	> 100°F (56°C) (IP)	S		S		S		S	
QW-407 PWHT	0.1	φ	PWHT	E		E		E		E	
	0.2	φ	PWHT (Time & Temperature Range)	S		S		S		S	
	0.4	T Limits	T Limits	E		E		E		E	
QW-408 Gas	0.1	± or φ	± Trailing or φ composition						N		N
	0.2	φ	Single, mixture or %					E		E	
	0.3	φ	Flow rate						N		N
	0.5	± or φ	± or φ Backing flow						N		N
	0.9	- or φ	- Backing or φ composition					E		E	
	0.10	φ	Shielding or trailing					E		E	
QW-409 Electrical Characteristics	0.1	>	Heat input	S		S		S		S	
	0.2	φ	Transfer Mode					E			
	0.3	±	Pulsing I								N
	0.4	φ	Current or Polarity	S	N	S	N	S	N	S	N
	0.8	φ	I & E Range		N		N		N		N
	0.12	φ	Tungsten electrode								N
QW-410 Technique	0.1	φ	String/Weave		N				N		N
	0.3	φ	Orifice cup, or nozzle size						N		N
	0.5	φ	Method of cleaning		N				N		N
	0.6	φ	Method of back gouge		N				N		N
	0.7	φ	Oscillation						N		N
	0.8	φ	Tube to Work distance						N		
	0.9	φ	Multi to single pass per side	S	N	S	N	S	N	S	N
	0.1	φ	Single to multi electrodes			S	N	S	N	S	N
	0.11	φ	Closed to out of chamber welding							E	
	0.15	φ	Electrode spacing						N		N
	0.25	φ	Manual or Automatic		N		N		N		N
	0.26	±	Peening		N		N		N		N
	0.64		use of thermal process	E		E		E		E	

LEGEND

+ Addition	> Increase/greater than	↑ Uphill	← Forehand	φ Change	- Deletion
< Decrease/less than	↓ Downhill			→ Backhand	

E	Essential Variables which must be indicated on both the WPS and recorded on the PQR. Any changes to these variables require requalification of WPS.
S	Supplementary Essential Variables must be indicated on the WPS and Recorded on the PQR when Impact Testing is required. Changes to these variables when Impact testing is performed require requalification of WPS.
N	Nonessential variables must be indicated on the WPS but when changed do not require requalification of WPS.
Note	NOTE: WPS's are to indicate all Essential, Nonessential and Supplementary Essential variables. Do not indicate variables which are not used as NA they are applicable and should be entered on the WPS as None or Not used.

QW-482 SUGGESTED FORMAT FOR WELDING PROCEDURE SPECIFICATION (WPS)
(See Qw 200.1, Section IX, ASME Boiler and Pressure Vessels Code)

Organization Name	(1.1)	By	(1.2)
Welding Procedure Specification No.	(1.3)	Date	(1.4)
Revision No.	(1.5)	Date	(1.6)
Supporting PQR No(s)	(1.7)	Date(s)	(1.8)
Welding Process(es)	(1.9)	Type(s)	(1.10)
	(Automatic, Manual, Machine or Semi Automatic)		
Client	(1.11)	Consultant	(1.12)
Project	(1.13)		

JOINTS (QW-402)

Joint Design	(2.1)				
Root Spacing	(2.2)			(2.5)	
Backing:	Yes (2.3)	No (2.3)			
Backing Material (Type)	(2.4)				

(Refer to both backing and retainers)

(2.6)	Metal	(2.6)	No fusing Metal
(2.6)	Nonmetallic	(2.6)	Other

Sketches, Production Drawings, Weld Symbols or Written Description should show the general arrangement of parts to be welded. Where applicable, the details of weld groove may be specified

Sketches may be attached to illustrate joint design, weld layers and bead sequence (eg., for notch toughness procedures, for multiple process procedures etc.)

* BASE METALS (QW- 403)

P-No.	(3.1)	Group No.	(3.2)	to	P-No.	(3.1)	Group No.	(3.2)	
		OR							

Specification and type/grade or UNS Number		(3.3)	
to Specification and type/grade or UNS Number		(3.3)	
	OR		
Chemical Analysis & Mechanical Properties		(3.4)	
to Chemical Analysis & Mechanical Properties		(3.4)	

Thickness Range:

Base Metal:	Groove	(3.5)	Fillet	(3.6)	
Maximum Pass Thickness ≤ 13mm (1/2")			Yes (3.7)	No	(3.7)
Other	(3.8)				

*FILLER METALS (QW 404)	1	2
Spec. No. (SFA):	(4.1)	
AWS No. (Class):	(4.2)	
F. No.:	(4.3)	
A. No.:	(4.4)	
Size of Filler Metals	(4.5)	
Filler Metal Product Form	(4.6)	
Electrode Brand Name	(4.7)	
Supplemental Filler Metal	(4.8)	
Weld Metal		
Deposited Thickness		
Groove	(4.9)	
Fillet	(4.10)	
Electrode Flux (Class)	(4.11)	
Flux Type	(4.12)	
Flux Trade Name	(4.13)	
Consumable Insert	(4.14)	
Other	(4.15)	

* Each base metal-filler metal combination should be recorded individually

POSITIONS (QW 405)

Position(s) of Groove				(5.1)	
Welding Progression:	Up	(5.2)	Down	(5.2)	
Position(s) of Fillet				(5.3)	
Other				(5.4)	

PREHEAT (QW 406)

Preheat Temperature, Minimum	(6.1)
Interpass Temperature, Maximum	(6.2)
Preheat Maintenance	(6.3)
Other	(6.4)

(Continuous or special heating, where applicable, should be recorded)

POST WELD HEAT TREATMENT (QW 407)

Temperature Range	(7.1)
Time Range	(7.2)
Other	(7.3)

GAS (QW 408)

		Percent Composition		
	Gas(es)	(Mixture)	Flow Rate	
Shielding Gas	(8.1)	(8.2)	(8.3)	
Trailing	(8.4)	(8.5)	(8.6)	
Backing	(8.7)	(8.8)	(8.9)	
Other	(8.10)	(8.11)	(8.12)	

ELECTRICAL CHARACTERISTICS (QW 409)

Weld Pass(es)	Process	Filler Metal		Current Type & Polarity	Amps (Range)	Wire Feed Speed (Range)	Energy or Power (Range)	Volts (Range)	Travel Speed (Range)
		Classification	Diameter						
(9.1)	(9.2)	(9.3)	(9.4)	(9.5)	(9.6)	(9.7)	(9.8)	(9.9)	(9.10)

Other (eg. Remarks, Comments, Hot Wire Addition, Technique, Torch Angle etc.)

(9.11)

Amps and volts or power or energy range, should be recorded for each electrode size, position, and thickness etc.

Pulsing Current	(9.12)	Heat Input (max.)	(9.13)
Tungsten Electrode Size and Type (Pure Tungsten, 2% Thoriated etc.)		(9.14)	
Mode of Metal Transfer for GMAW (FCAW) (Spray Arc, Short Circuiting Arc, etc.)		(9.15)	
Other	(9.16)		

TECHNIQUE (QW 410)

String or Weave Bead	(10.1)
Orifice, Nozzle or Gas Cup Size	(10.2)
Initial and Interpass Cleaning (Brushing, Grinding, etc.)	(10.3)
Method of Back Gouging	(10.4)
Oscillation	(10.5)
Contact Tube to Work Distance	(10.6)
Multiple or Single Pass (Per Side)	(10.7)
Multiple or Single Electrodes	(10.8)
Electrodes Spacing	(10.9)
Peening	(10.10)
Other	(10.11)

Prepared		Approved	
Name	(11.1)	Name	(11.1)
Designation	(11.2)	Designation	(11.2)
Signature	(11.3)	Signature	(11.3)
Date	(11.4)	Date	(11.4)

QW-483 SUGGESTED FORMAT FOR PROCEDURE QUALIFICATION RECORD(PQR)

(See QW 200.2, Section IX, ASME Boiler and Pressure Vessels Code)

Record Actual Variables Used to Weld Test Coupon

Organization Name	(1.1)		By	(1.2)
Procedure Qualification Record No.	(1.7)	Date		(1.8)
WPS No.	(1.3)	Date		(1.4)
Welding Process(es)		(1.9)		
Types (Manual, Automatic, Semi-Automatic)		(1.10)		
Client	(1.11)		Consultant	(1.12)
Project	(1.13)			

JOINTS (QW-402)

(2.5)

Groove Design of Test Coupons

(For combination Qualifications, the deposited weld metal thickness shall be recorded for each filler metal and process used)

* BASE METALS (QW- 403)

Material Specification		(3.9)						
Type/Grade or UNS Number		(3.10)						
P-No.	(3.1)	Group No.	(3.2)	to P-No.	(3.1)	Group No.	(3.2)	
Thickness of Test Coupon		(3.11)						
Diameter of Test Coupon		(3.12)						
Maximum Pass Thickness		(3.13)						
Other		(3.8)						

FILLER METALS (QW 404)

FILLER METALS (QW 404)	1	2
SFA Specification	(4.1)	
AWS Classification	(4.2)	
Filler Metal F-No.	(4.3)	
Weld Metal Analysis A-No.	(4.4)	
Size of Filler Metal	(4.5)	
Filler Metal Product Form	(4.6)	
Electrode Brand Name	(4.7)	
Supplemental Filler Metal	(4.8)	
Electrode Flux Classification	(4.12)	
Flux Type	(4.13)	
Flux Trade Name	(4.14)	
Weld Metal Thickness	(4.9)	
Other	(4.16)	

POSITIONS (QW 405)

Position of Groove	(5.1)	
Welding Progression (Uphill, Downhill)	(5.2)	
Other	(5.4)	

PREHEAT (QW 406)

Preheat Temperature	(6.1)
Interpass Temperature	(6.2)
Other	(6.4)

POST WELD HEAT TREATMENT (QW 407)

Temperature	(7.1)
Time	(7.2)
Other	(7.3)

GAS (QW 408)

	Percent Composition			
	Gas(es)	(Mixture)	Flow Rate	
Shielding Gas	(8.1)	(8.2)	(8.3)	
Trailing	(8.1)	(8.2)	(8.3)	
Backing	(8.1)	(8.2)	(8.3)	
Other	(8.4)	(8.4)	(8.4)	

ELECTRICAL CHARACTERISTICS (QW 409)

Current	(9.5)		
Polarity	(9.5)		
Amps.	(9.6)	Volts	(9.9)
Tungsten Electrode Size and Type			(9.14)
Mode of Metal Transfer for GMAW (FCAW)			(9.15)
Heat Input	(9.8)		
Other	(9.16)		

TECHNIQUE (QW 410)

Travel Speed	(10.12)
String or Weave Bead	(10.1)
Orifice, Nozzle or Gas Cup Size	(10.2)
Oscillation	(10.5)
Initial and Interpass Cleaning (Brushing, Grinding, etc.)	(10.3)
Method of Back Gouging	(10.4)
Multiple or Single Pass (Per Side)	(10.7)
Multiple or Single Electrodes	(10.8)
Other	(10.11)

Tensile Test (QW-150)						
Specimen No.	Width	Thickness	Area	Ultimate Total Load	Ultimate Unit Stress (psi or Mpa)	Type of Failure and Location
(12.1)	(12.2)	(12.3)	(12.4)	(12.5)	(12.6)	(12.7)

Guided-Bend Tests (QW-160)	
Type and Figure No.	Result
(13.1)	(13.2)

Toughness Tests (QW-170)							
Specimen No	Notch Location	Specimen Size	Test Temperature	Impact Values			Drop Weight Break (Y/N)
				ft-lb or J	% Shear	Mils or mm	
(14.1)	(14.2)	(14.3)	(14.4)	(14.5)	(14.6)	(14.7)	(14.8)
Comments	(14.9)						

Fillet-Weld Test (QW-180)						
Result -Satisfactory:	Yes (15.1)	No (15.1)	Penetration into Parent Metal:		Yes (15.2)	No (15.2)
Macro-Results	(15.3)					

Other Tests	
Type of Test	(16.1)
Deposit Analysis	(16.2)
Other	(16.3)

Welder's Name	(17.1)		Clock No. (17.2)	Stamp No.	(17.3)
Tests Conducted by	(17.4)				

We certify that the statements in this record are correct and that the test welds were prepared, welded and tested in accordance with the requirements of Section IX of the ASME Boiler and Pressure Vessel Code

Organization	(17.5)	
Date (17.6)	Certified by	(17.7)

(Detail of record of tests are illustrative only and may be modified to confirm to the type and number of tests required by the Code)

QW-484 A SUGGESTED FORMAT FOR WELDER PERFORMANCE QUALIFICATION (WPQ)
(See QW 301, Section IX, ASME Boiler and Pressure Vessels Code)
Record Actual Variables Used to Weld Test Coupon

Welder's Name		Identification No.	

Test Description

Identification of WPS followed		Test Coupon		Production Weld	
Specification and type/grade or UNS Number of base metal(s)			Thickness		

Testing Variables and Qualification Limits

Welding Variables (QW-350)	Actual Values	Range Qualified
Welding Process(es)		
Type (ie, manual, semiautomatic) used		
Backing (with /without)		
Plate	Pipe (enter dia if pipe or tube)		
Base metal P No. To P No.		
Filler metal or eletrode specification(s) (SFA) (info only)		
Filler metal or electrode classification(s) (info only)		
Filler metal F Number(s)		
Consumable inserts (GTAW or PAW)		
Filler metal product form (GTAW or PAW)		
(solid/metal or flux cored/powder)		
Deposit thicknesss for each process		
Process 1	3 layers min.	Yes	No		
Process 2	3 layers min.	Yes	No		
Position qualified (2G,6G, 2F etc)		
Vertical Progression (uphill or downhill)		
Fype of fuel gas (OFW)		
Insert gas backing (GTAW, PAW, GMAW)		
Transfer mode		
(spray /globular or pulse to short circuit -GMAW)		
GTAW current type / polarity (AC, DCEP, DCEN)		

Results

Visual examination of completed weld (QW-302.4)		

Transverse face and root bends (QW-462.3(a))	Longitudinal bends (QW-462.3(b))
Side Bends (QW-462.2)	

Pipe bend specimen corrosion resistant weld metal overlay (QW-462.5(c))
Plate bend specimen corrosion resistant weld metal overlay (QW-462.5(d))
Pipe specimen, macro test for fusion (QW-462.5(b))
Plate specimen, macro test for fusion (QW-462.5(e))

Type	Result	Type	Result	Type	Result

Alternative volumetric examination results (QW-191)		RT	or	UT	(check one)
Field weld-fracture test (QW-181.2)		Length and percent of defects			
Fillet weld in plate (QW-462.4(b))		Fillet weld in plate (QW-462.4(c))			
Macro examination (QW-184)	Fillet size(in)		X	Concavity/convexity (in)	
Other tests					

Film or specimen evaluated by		Company	
Mechanical tests conducted by		Laboratory test no.	
Welding supervised by			

We certify that the statements in this record are correct and that the test coupons were prepared, welded and tested in accordance with the requirements of Section IX of the ASME Boiler & Pressure Vessel Code

Organization	
Date		Certified by	

QW-484 B SUGGESTED FORMAT B FOR WELDING OPERATOR PERFORMANCE QUALIFICATION (WOPQ)

(See Qw 301, Section IX, ASME Boiler and Pressure Vessels Code)

Record Actual Variables Used to Weld Test Coupon

Weldling operator's Name		Identification No.	

Test Description *(information only)*

Identification of WPS followed		Test Coupon		Production Weld	
Specification and type/grade or UNS Number of base metal(s)		**(2.4)**		Thickness	
Base metal P-Number		to	P-Number		Position (2G, 6G, 3F Etc)	
Plate		Pipe (enter dia if pipe or tube)			
Filler metal (SFA) specification		Filler metal electrode classification			

Testing Variables and Qualification Limits when using Automatic Welding Equipment

Welding Variables (QW-361.1)	Actual Values	Range Qualified
Type of welding (Machine)		
Welding process		
Filler metal used (Yes/No) (EBW or LBW)		
Type of laser for LBW (CO$_2$ to YAG etc.)		
Continuous drive or inertial welding (FW)		
Vacuum or out of vacuum (EBW)		

Testing Variables and Qualification Limits when using Machine Welding Equipment

Welding Variables (QW-361.2)	Actual Values	Range Qualified
Type of welding (Machine)		
Welding process		
Direct or remote visual control		
Automatic arc voltage control(GTAW)		
Automatic joint tracking		
Position qualified (2G,6G, 3F etc)		
Consumable inserts (GTAW or PAW)		
Backing (with / without)		
Single or multiple pass per side		

Results

Visual examination of completed weld (QW-302.4)			
Transverse face and root bends (QW-462.3(a))		Longitudinal bends (QW-462.3(b))	
	Side Bends (QW-462.2)		
Pipe bend specimen corrosion resistant weld metal overlay (QW-462.5(c))			
Plate bend specimen corrosion resistant weld metal overlay (QW-462.5(d))			
Pipe specimen, macro test for fusion (QW-462.5(b))			
Plate specimen, macro test for fusion (QW-462.5(e))			

Type	Result	Type	Result	Type	Result

Alternative volumetric examination results (QW-191)		RT	or	UT	(check one)
Field weld-fracture test (QW-181.2)		Length and percent of defects			
Fillet weld in plate (QW-462.4(b))		Fillet weld in plate (QW-462.4(c))			
Macro examination (QW-184)		Fillet size(in)	X	Concavity/convexity (in)	
Other tests					

Film or specimen evaluated by		Company	
Mechanical tests conducted by		Laboratory test no.	
Welding supervised by			

We certify that the statements in this record are correct and that the test coupons were prepared, welded and tested in accordance with the requirements of Section IX of the ASME Boiler & Pressure Vessel Code

Organization	
Date		Certified by	

Sl.No	QW 482	QW 483	Description
			General Information
1.1	X	X	Name of the organization, usually that of piping contractor / manufacturing work shop as applicable responsible to conduct prepare and qualify the WPS
1.2	X	X	Prepared by who in the organization with designation
1.3	X	X	Unique number to be assigned by the manufacturer to the WPS
1.4	X	X	Date of Preparation
1.5	X	X	Revision status such as Rev 0, (original or preliminary issue, followed by Rev 1, rev 2 etc.)
1.6	X	X	Date of revision
1.7	X	X	Reference number of Supporting PQR (s) . This shall be blank in original issue of WPS but Rev 1 onwards, it can be filled in with a PQR Number
1.8	X	X	Dates of referred PQR(s)
1.9	X	X	Welding processes envisaged in the WPS one or more can be mentioned
1.10	X	X	Type of processes proposed such as manual, semi auto, auto etc.
1.11	X	X	Client Name and location
1.12	X	X	Consultant Name (if there is one)
1.13		X	Project Name
			Joint Details
2.1	X		Type of weld joint, Groove, Fillet Etc and details like single V or double V etc.
2.2	X		Root gap
2.3	X		"X" mark "Yes" or "No" as appropriate
2.4	X		Type of backing material used. Apart from backing materials, retainers if any used also shall be indicated
2.5	X	X	Provide sketch as mentioned in notes below
2.6	X		"X" mark against appropriate
			Base Metal Details
3.1	X	X	P No. of Material to be joined on both sides
3.2	X	X	Gr. No. of Material to be joined on both sides
3.3	X		Specification/Type/grade of UNS number of material on both sides
3.4	X		Chemical analysis and Mechanical properties of material on both sides
3.5	X		Thickness range proposed for groove weld as required
3.6	X		Thickness range proposed for fillet weld as required
3.7	X		"X" mark for "Yes" or "NO" as applicable
3.8	X	X	Any other salient information to be indicated if of importance with regard to base metals
3.9		X	Material specification
3.10		X	Type/grade or UNS Number of base material
3.11		X	Specimen thickness used in test
3.12		X	Diameter of test specimen
3.13	X	X	Maximum thickness of passes deposited

Filler Metal Details

			(To be filled for different electrodes if used so)
4.1	X	X	SFA No(s) of filler metals used in weld
4.2	X	X	AWS No Class of filler metals used in weld
4.3	X	X	F No(s) of filler metals used in weld
4.4	X	X	A No(s) of filler metals used in weld
4.5	X	X	Size of filler metals used
4.6	X	X	Filler metal product form such as straight wire, coil etc.
4.7	X	X	Electrode brand name
4.8	X	X	Supplementary filling wires if any used
4.9	X	X	Deposited weld metal thickness-groove weld
4.10	X		Deposited weld metal thickness-fillet weld
4.11	X	X	Electrode flux class
4.12	X	X	Flux type
4.13	X	X	Flux trade name
4.14	X	X	Consumable inserts if any used
4.15	X	X	Any other salient information to be indicated if of importance with regard to filler metal

Position Details

5.1	X	X	Position of groove for test
5.2	X	X	"X" mark appropriate
5.3	X		Positions of fillet weld
5.4	X	X	Any other salient information to be indicated if of importance with regard to position of welding

Preheat Details

6.1	X	X	Minimum preheat temperature
6.2	X	X	Maximum interpass temperature
6.3	X		Preheat maintenance details
6.4	X	X	Any other salient information to be indicated if of importance with regard to preheat

Post Weld Heat Treatment Details

7.1	X	X	Temperature Range
7.2	X	X	Time range or soaking time
7.3	X	X	Any other salient information to be indicated if of importance with regard to PWHT

Gas Details

8.1	X	X	Gas used for shielding or trailing or backing purposes
8.2	X	X	If mixture of gases used for shielding or trailing or backing purposes, indicate % composition
8.3	X	X	Flow rate of gas used for shielding or trailing or backing purposes
8.4	X	X	Any other salient information pertaining to gas, its mixture and flow rate

Details of Electrical Charecteristics

9.1	X		Identification details for welding passes
9.2	X		Welding processes against each of welding passes identified in 9.1
9.3	X		Filler metal classification against each of welding passes identified in 9.1
9.4	X		Filler metal diameter against each of welding passes identified in 9.1
9.5	X	X	Type of current used and its polarity against each of welding passes identified in 9.1
9.6	X	X	Current (Amps) range against each of welding passes identified in 9.1
9.7	X		Wire feed range against each of welding passes identified in 9.1 if applicable
9.8	X	X	Energy or heat input against each of welding passes identified in 9.1
9.9	X	X	Voltage range against each of welding passes identified in 9.1
9.10	X		Travel speed range against each of welding passes identified in 9.1
9.11	X		Any other salient information pertaining to electrical characteristics under any of the items above
9.12	X		If pulsing current used, indicate "Yes"
9.13	X		Maximum heat input value during welding
9.14	X	X	Type of tungsten electrode, brief description as required in notes below
9.15	X	X	Mode of metal transfer for GMAW as required in note below
9.16	X	X	Any other salient information pertaining to electrical characteristics under any of the items 9.12 to 9.15 above

			Details of Technique
10.1	X	X	Indicate whether string or weave bead. If weave, indicate width of weave
10.2	X	X	Orifice, nozzle or gas cup size as appropriate
10.3	X	X	Initial and interpass cleaning such as brushing grinding or any other methodology
10.4	X	X	Method of back gouging
10.5	X	X	Oscillation if applicable
10.6	X		Contact tube to work distance
10.7	X	X	Multiple or single pass per side
10.8	X	X	Multiple or single electrode
10.9	X		Electrode spacing when multiple electrodes used
10.10	X		Peening if any done during or after welding
10.11	X	X	Any other salient information pertaining to technique
10.12		X	Travel speed (essential for automatic welding, preferable for manual welding)
			Preparation / Approval Details
11.1	X		Name of person who prepared/ approved the document
11.2	X		Designation of the person who prepared/ approved the document
11.3	X		Signature of the person who prepared/ approved the document
11.4	X		Date of preparation / Approval of the document
11.5	X		Name of person who prepared/ approved the document
			Tensile Test Details
12.1		X	Specimen identification number
12.2		X	Width of specimen
12.3		X	Thickness of specimen
12.4		X	Area of specimen
12.5		X	Ultimate total load shown in UTM
12.6		X	Ultimate unit stress in psi or Mpa
12.7		X	Type of failure such as ductile or brittle and location of failure
			Guided Bend Test Details
13.1		X	Type of test and figure number used
13.2		X	Test results such as satisfactory or unsatisfactory
			Toughness Test Details
14.1		X	Specimen identification number
14.2		X	Location of notch
14.3		X	Specimen size
14.4		X	Test temperature
14.5		X	Ultimate total load shown in UTM
14.6		X	Impact test value Jules
14.7		X	Impact test value % shear
14.8		X	Impact test value mils or mm
14.9		X	Drop weight break, indicate "Yes" or "No"
			Fillet Weld Test Details
15.1		X	Indicate test result as "Yes" or "No" against satisfactory as appropriate by placing "X"
15.2		X	Mention about penetration as "yes" or "No" by placing "X" appropriately
15.3		X	Observations in macro examination to be indicated here
			Other Test Details
16.1		X	Details of any other tests if done
16.2		X	Deposit analysis results if done
16.3		X	Any other salient information pertaining to such tests
			Authentication Details
17.1		X	Welders name
17.2		X	Welder (employee) number
17.3		X	Welder Code No.
17.4		X	Name & designation of person who conducted the test
17.5		X	Organization name and stamp
17.6		X	Date of certification
17.7		X	Certifying authority

Appendix C
Piping Progress Monitoring Based on Standard Point Factor

In piping construction sites, extensively seen progress monitoring methodology is the progress calculation based on the "inch dia" for piping and "inch meter" for pipelines. This methodology fails to consider the quantum involved in welds as it does not take into consideration the thickness of pipe joints. Therefore, to obtain a realistic progress status at any point of time of the project, it requires a comprehensive progress monitoring system with appropriate weightage factors based on actual work content and the appendix provides one such methodology as detailed in ensuing sections.

C.1 ENGINEERING PROGRESS

Engineering progress is calculated based on total number of isometrics generated and those already attained AFC (Approved for Construction) status as detailed in tabulation at the end. Different subsections that constitute the piping progress are listed out under Item (A) of the spread sheet, which is calculated as a percentage. To the percentage so obtained, for calculating the overall progress, the weightage factor of 0.3 (30%) is applied for piping engineering. In other words, if engineering progress is 60%, it will be reflected as $60 \times 0.3 = 18\%$ in overall progress. The same methodology (with different weightage factors proposed) shall be extended to all other attributes identified below.

C.2 PIPING FRONT AVAILABILITY FOR PIPE SPOOL FABRICATION

During the process of engineering and procurement action prior to start of piping spool manufacture at site workshop, piping front availability would be a good tool to monitor engineering and procurement, up to material receipt at site. In order to carry out this analysis, the following information from respective departments is necessary.

- Total number and list of isometric drawings (area wise or material category wise or as required by client) for entire piping (from piping engineering)
- Number and list of isometrics in Approved for Construction (AFC) status (engineering)
- Number and list of drawings for which all materials are received at site and accepted by QA/QC (procurement/planning)

For carrying out the above analysis, the receipt of materials, such as pipes, fittings and flanges, alone is considered. Materials like gaskets, bolts and nuts, valves, pipe supports, etc. are not considered essential for this purpose.

Priority segments also can be included under this section for better control of the project.

C.3   STANDARD POINT FACTOR

The common practice in measuring piping progress is "inch dia" and "inch meter" for offsite piping. This system does not give any consideration for the thickness of pipes welded. To state an example, while a butt weld joint in schedule 40 pipe can be completed very easily, that in schedule 160 pipe takes more time since the volume of welds to be deposited in the case of latter is more than double compared to the former. Moreover, when weld metal volume is high, chances of defect inclusion are also more and the time and effort required to repair a defect in thicker weld are also more. It is true that a 100% accurate piping progress monitoring is too complicated and hence not worth taking the pains in feeding in additional details, but a more realistic one based on standard point factor (SPF) is a more accurate, sensitive and easy methodology.

SPF is the factor based on thickness of pipe welded with reference to the thickness of schedule STD pipe in each diameter. As an example, the SPF for all commercially available wall thicknesses of NPS 100 (4″) diameter is provided below. It can be seen that SPF for schedule "STD" in all sizes shall be "1.0" and for thicknesses below "STD", SPF shall be less than "1.0" and for those above "STD" shall be above "1.0".

US	Metric	OD		Schedule Designations	Wall Thickness		Standard Point Factor
		(inches)	(mm)	(ANSI/ASME)	(inches)	(mm)	(SPF)
				5/5S	0.083	2.11	0.35
				10/10S	0.120	3.05	0.51
					0.156	3.96	0.66
					0.188	4.78	0.79
4	100	4.5	114.3	STD/40/40S	0.237	6.02	1.00
				XS/80/80S	0.337	8.56	1.42
				120	0.438	11.13	1.85
				160	0.531	13.49	2.24
				XX	0.674	17.12	2.84

In piping progress monitoring system based on SPF, the inch diameter and inch meter (commonly used progress measuring yard sticks) are multiplied by SPF to arrive at a thickness-sensitive progress figure. Therefore, progress measurement based on SPF is a closer approximation to actual progress. In addition, the ensuing sections provide the ways and means for including and considering all connected activities in piping spool manufacture, installation and testing in assessing the overall piping progress.

The SPF for pipe sizes from NPS 6 (1/8″) up to NPS 1,200 (48″) is provided in the table at the end of the appendix.

C.4 PIPING SPOOL FABRICATION PROGRESS (MEASURED USING SPF AS DESCRIBED)

In order to report realistic progress, it is essential that piping spool fabrication needs to be broken down into component activities and progress of those activities needs to be measured and consolidated to have overall progress.

The specific activities considered during fabrication of piping spools are

- Weld Joint Fit-Up
- Welding
- NDT of Welds
- PWHT of Welds
- Hardness Survey of Welds
- Hydrostatic Testing of Welds

a. **Weld Joint Fit-Up**
   i. Butt weld
      Fit-up release register is considered as the base document for this purpose. As and when the fit-up log is signed off by all stakeholders, the fit-up of a weld joint is considered complete. The progress in standard point system of weld joint in 10″ Sch 80 pipe is worked out as $10 \times SPF = 10 \times 1.63 = 16.3$ Inch Dia.
   ii. Socket welds
      Here also fit-up release register is considered as the base document. When all stakeholders sign off the joint fit-up log, it is considered that the stage is satisfactorily accomplished. Being fillet welds, it is not essential to consider the standard point system here as the variation is going to be very marginal, especially for the lower sizes 2″ and below, wherein socket weld joints are the predominant in numbers. The progress of socket weld henceforth is calculated as follows (for a 2″ Sch 80 socket weld pipe joint).

      SPF for socket welds is considered as 1 irrespective of size and schedule, resulting in simple inch dia itself for all socket welds. Moreover, socket weld sizes are often limited to 2″ and below. Progress is calculated as

$$Progress = 2 \times SPF = 2 \times 1 = 2.0 \text{ Inch Dia.}$$

b. **Welding**
   For welding stage as well, more or less the same methodology described under C 4 (a) (i) and (ii) applicable for categories butt and socket shall prevail. For claiming completion of this stage, signing off the weld visual inspection report or register is considered as the basis.
   i. Butt weld
      Progress of butt weld is calculated the same way as that proposed in Clause C 4 (a) (i) above for fit-up. For a 10″ Sch 80 pipe the butt weld progress in standard point system is worked out as $10 \times SPF = 10 \times 1.63 = 16.3$ Inch Dia.

ii. Socket welds

Socket welds in oil and gas industry are normally restricted to pie sizes 2″ and below and variation in pipe wall is considered somewhat significant here as well and SPF is applied here. Accordingly, for a 2″ Sch 80 pipe, the progress of socket weld joint is calculated as $= 2 \times SPF = 2 \times 1.42 = 2.84$ Inch Dia.

c. **Non-Destructive Testing of Welds**

Satisfactory completion of NDT is considered as a significant milestone in piping and hence this stage also needs to be given due consideration in piping progress. In case defects in welding were revealed in NDT, repair of the same is carried and progress for said milestone can be claimed only upon satisfactory completion of NDT of the repaired location as well.

Considering the variation in ease and expertise involved in different NDT techniques commonly deployed in piping, weightage factors are assigned to various NDT techniques. The methods commonly used include RT, UT, MPT and LPT. For precise progress monitoring, weightage factors assigned to NDT methods are provided below.

RT	1.0	UT	0.8	MPT	0.4	LPT	0.2

Since time taken by NDT is more (especially for RT) based on the thickness of weld (and also for repairs), the variation in thickness (SPF) is taken into consideration while reporting progress for RT.

In plant piping, various NDT methods are often deployed depending on criticality of the piping, physical feasibility and effectiveness of NDT method, considering configuration, material of construction, accessibility, etc.

i. NDT of butt welds

Progress of NDT of butt welds is calculated in the same way as in the progress of weld, which is claimed after successful completion of NDT (including re-NDT after repair if defect is revealed in NDT beyond acceptance criteria).

For a 10″ Sch 80 pipe, the progress claimed after NDT is worked out as $10 \times SPF = 10 \times 1.6 = 16.30$ Inch Dia.

While time taken for RT is very sensitive to wall thickness, other NDT methods like UT, MPT and LPT are not that sensitive to wall thickness and in true sense SPF need not be considered for these NDT methods. However, to avoid one more complication in progress reporting, for all NDT techniques, SPF correction is applied to UT, MPT and LPT as well, to make the methodology uniform and general for all NDT methods.

ii. NDT of socket welds

For socket welds also, which occurs often in sizes 2″ and below, progress of NDT is calculated similar to that of butt welds as $= 2 \times SPF = 2 \times 1.42 = 2.84$ Inch Dia. Please note that weightage factors proposed for NDT techniques shall be applicable here also.

d. **Post-Weld Heat Treatment of Welds**

Welds in piping may require Post-Weld Heat Treatment (PWHT) based on weld thickness as required in ASME B 31.3 or as service requirement is certain oil and gas services (such as low temperature or sour gas service). If the requirement of PWHT is based on thickness of weld as required in code, often socket welds would not fall under this. Whereas if the requirement is based on service requirement as mentioned earlier, both butt and socket welds would require PWHT. The methodology to calculate the progress of PWHT of both butt and socket welds is given below, which again is similar to NDT. The PWHT considered here is the one at around 650°C which is normally applicable for carbon steels. Other types of heat treatment required during production of piping spools if any need to be considered separately.

i. PWHT of butt welds

Progress of PWHT of butt welds is calculated in the same way as in progress of weld and is claimed after successful completion of PWHT.

For a 10″ Sch 80 pipe, progress claimed after PWHT is worked out as $10 \times SPF = 10 \times 1.63 = 16.30$ Inch Dia.

ii. PWHT of socket welds

For socket welds which occur often in sizes 2″ and below, normally PWHT shall not be applicable. However, as mentioned earlier, service (low temperature, sour gas, etc.) of the piping may require PWHT irrespective of thickness at weld joint. In such cases, thickness variation is not considered significant in time taken for PWHT and hence for calculating progress of PWHT in socket welds, SPF need not be taken into account. For a 2″ pipe, irrespective of schedule (meaning $SPF = 1$), progress is calculated as $2 \times SPF = 2$ Inch Dia for all schedules.

e. **Hardness Survey of Welds**

Welds in sour environment (often encountered in oil and gas industry) require hardness survey after completion of PWHT. If specified so, this service requirement is equally applicable to both butt and socket welds.

Progress of hardness survey of both butt and socket welds is calculated simply as the inch dia itself irrespective of pipe schedule. For a 10″ pipe, the hardness survey progress is 10 Inch Dia and that for 2″ pipe, it is 2 Inch Dia, irrespective of pipe schedules.

f. **Hydrostatic Test of Welds**

In plant piping, hydrostatic testing is carried out in two phases. In case of internally coated pipes, the spools are hydro tested before application of coating and the entire line is tested again after installation. Regarding calculation of progress, easy and effective methodology shall be to account progress when each and every weld undergoes satisfactory hydrostatic testing. As test pressure is related to wall thickness of pipe and as complication increases due to increase in diameter, progress of hydrostatic testing is calculated for both butt and socket welds considering SPF.

For a 10″ Sch 80 pipe weld, hydrostatic test progress is $10 \times SPF = 10 \times 1.63 = 16.30$ Inch Dia and that for a 2″ Sch 80 pipe weld is $2 \times SPF = 2 \times 1.42 = 2.84$ Inch Dia, for both butt and socket welds.

C.5 PIPING SPOOL INSTALLATION

a. **Piping Spool Installation**

Progress of piping spool installation is measured in terms of inch meter installed at site. The general practice is to ignore the pipe wall thickness for this calculation. When a wide range of thicknesses are involved, installation work varies widely. Especially for piping using high wall thickness pipes, installation and assembly of one spool to other consumes more time, may require heavier handling cranes and can pose more problems in aligning spools at site. Therefore, SPF is applied while calculating piping installation progress as a better approximation.

Once a piping isometric is completely installed, total inch meter that can be claimed can be pre-calculated (with SPF) using tools available in isometric generation software used for developing the isometric drawings.

For a 10″ Sch 80 pipe weld, piping spool installation progress per meter of pipe spool erected is $10 \times SPF \times 1 = 10 \times 1.63 \times 1 = 16.30$ Inch Meter and that for a 2″ Sch 80 pipe weld is $2 \times SPF \times 1 = 2 \times 1.42 \times 1 = 2.84$ Inch Meter, for both butt and socket welds.

Accordingly, for a piping spool of around 20 m in length (as per drawing) of 10″ Sch 80 pipe, when installed satisfactorily, the progress is calculated as $= 10 \times SPF \times 20 = 10 \times 1.63 \times 20 = 326$ Inch Meter.

When pipe spool involves branch connections of other sizes, they also shall be added to reported progress in the same way it is calculated for 10″ main pipe.

b. **Valves and Other Intervening Inline Equipment Installation**

The intervening inline items separately installed at site include components of widely varying weights and configurations, starting from small handheld valves up to very heavy motor operated/control valves with actuators. As this varies considerably, the progress monitoring can be based on the tonnage erected, including weight of control gear/actuators.

As and when purchase orders for valves are placed and acknowledged by vendors, vendors shall be in a position to indicate approximate tonnage of valves and this shall be taken for measuring the progress of installation of valves and other inline equipment.

c. **Field Weld Joint Fit-Up**
   i. Butt weld
      Calculated similar to that in Section C 4 (a) (i).
   ii. Socket welds
      Calculated similar to that in Section C 4 (a) (ii).

d. **Field Welds**
   i. Butt weld
      Calculated similar to that in Section C 4 (b) (i).
   ii. Socket welds
      Calculated similar to that in Section C 4 (b) (ii).

e. **Non-Destructive Testing of Field Welds**
   i. NDT of butt welds
      Calculated similar to that in Section C 4 (c) (i).

ii. NDT of socket welds
Calculated similar to that in Section C 4 (c) (ii).

f. **Post-Weld Heat Treatment of Field Welds**
   i. PWHT of butt welds
   Calculated similar to that in Section C 4 (d) (i).
   ii. PWHT of socket welds
   Calculated similar to that in Section C 4 (d) (ii).

g. **Hardness Survey of Field Welds**
   Calculated similar to that in Section C 4 (e).

h. **Installation Pipe Supports, Hangers, etc.**
   Progress calculated based on the tonnage erected as in the case of installation of valves as described in Clause C 5 (b).

i. **Hydrostatic Test of Field Welds/Golden Joints and Piping After Completion**
   Progress of such hydrostatic testing or leak testing using air is also measured similar to the piping installation progress as described in Section C 5.1 in Inch Meters. With regard to inline equipment, no specific contribution is considered in progress calculation under this head as these components are pre-tested and are included for sake of easiness in carrying out test.

j. **Punch Listing and Clearing of Punches**
   When all critical punch listing is completely rectified in a particular segment of piping, progress of that segment is also calculated, similar to that in Section C 5.1, Piping Installation, without taking into consideration the SPF.

   When all punch list of a piping segment consisting of length 150 m is made of 10″ Sch 80 pipe, the punch list clearing progress is calculated as 10″ × 150 m = 1,500 Inch Meter.

C.6 OVERALL PROGRESS USING WEIGHTAGE FACTORS

a. **Weightage Factors**
   The weightage factors assumed for calculating the overall progress on three major sections of work in piping are shown in the table below. When piping engineering is completed 100%, based on the weightage factors above, its reflection in overall progress shall be 30%. Similarly when piping spool fabrication is 100% complete, its reflection in overall progress shall be yet another 30% and that for piping installation 40%, all put together constituting 100%.

Sl. No.	Description of Activity	Weightage (%)
1	Piping Engineering Progress	30
2	Piping Spool Fabrication Progress	30
3	Piping Installation	40

Table below provides further split of above-mentioned weightage factor (for piping spool fabrication) for all the constituent attributes identified under it. The percentage listed under each item (1)–(6) in below table sums up to 30% assigned for Piping Spool Fabrication. For item (3), Non-Destructive

Testing, there are four sub-categories identified with different weightages in table below, constituting the 5% assigned for NDT.

Sl. No.	Description of Activity	Weightage
	Piping Spool Fabrication Progress	**30%**
1	Weld Joint Fit-Up	**4%**
2	Welding	**6%**
3	Non-Destructive Testing	**5%**
3 (a)	Radiography	40% of 5%
3 (b)	Ultrasonic Testing	30% of 5%
3 (c)	Magnetic Particle Testing	20% of 5%
3 (d)	Liquid Penetrant Testing	10% of 5%
4	Post-Weld Heat Treatment	**5%**
5	Hardness Survey of Welds	**2%**
6	Hydrostatic Testing of Pipe Spool	**8%**

All the bold percentages add up to 30% which is the 100% of Piping Spool Fabrication. In overall progress of Piping, 100% piping spool fabrication will only account for 30%.

Weightage factors of individual attributes identified under Piping Installation, summing up to 40% of progress, is provided in the table below.

Sl. No.	Description of Activity	Weightage (%)
	Piping Installation	**40**
1	Piping Spool Installation	10
2	Valve and Other Inline Equipment Installation	5
3	Field Weld Joint Fit-Up	2
4	Field Welding	3
5	NDT of Field Welds	2
6	PWHT of Field Welds	3
7	Hardness Survey of Field Welds	1
8	Installation of Pipe Supports, Hangers, etc.	4
9	Hydrostatic Test of Field Welds/Golden Joints and Piping After Completion	5
10	Punch List Clearance	5

Tables above provide a somewhat comprehensive list of attributes, many of which may not be available or required in many small piping projects. In such cases, redistribution of weightage factors within the three main head identified (engineering, spool fabrication and piping installation) shall be carried out judiciously.

Progress achieved on any day (daily or cumulative) is obtained in either inch dia or inch meter. In order to convert this into a percentage figure, the respective 100% figures against each attribute identified in tables above are required from day 1. Due to the extensive use of software for preparation of isometrics and material take-offs, obtaining these figures is quite easy upon freezing of P&IDs, location of equipment and pipe routing.

b. **Piping Progress Monitoring Spread Sheet**

The spread sheet provides the finer details of the progress calculation as described in above sections.

Section A deals with piping engineering and work front availability. Progress percentages may often be required under various heads based on scope of work and few salient, almost common ones are described therein.

Section B deals with piping spool fabrications progress under common milestones described therein. Though many of the figures worked out contribute only minutely to overall progress, these figures shall be very useful in identifying bottlenecks at various phases such as radiography, which is often possible only during night shifts.

Section C deals with installation progress. Weightage factors for various milestones identified are provided considering the work content is primarily based on inch meter rather than inch dia used in piping spool progress measurement.

	Piping Progress Monitoring (based on Stadard Point Factor)					
A	Piping Engineering & Font Availability	Number of Documents				Progress Percentage
1	Number of Isometrics for the plant (including CS, SS and non metallic Piping)	A=D+G+K				
2	Number of above Isometrics in AFC Status	B				B/A*100
3	Number of Isometrics with all materials in BOM at stores	C				C/A*100
4	Number of Isometrics in Carbon Steel Piping	D				
5	Number of above Isometrics in AFC Status (CS Piping)	E				E/D*100
6	Number of Isometrics with all materials in BOM at stores (CS Piping)	F				F/D*100
7	Number of Isometrics in Stainless & other Alloy Steel Piping	G				
8	Number of above Isometrics in AFC Status (SS & other AS Piping)	H				H/G*100
9	Number of Isometrics with all materials in BOM at stores (SS & other AS Piping)	J				J/G*100
10	Number of Isometrics in Non Metallic Piping	K				
11	Number of above Isometrics in AFC Status (Non Metallic Piping)	L				L/K*100
12	Number of Isometrics with all materials in BOM at stores (Non Metallic Piping)	M				M/K*100
13	Piping Front Avaialbility					C/A*100
14	Piping Engineering Progress					B/A*100=❶

B	Piping Spool Fabrication	Inch Diameter				
		As per Drg.	Achieved			
1	Weld Joint Fitup	A=A1+A2	a=a1+a2			a/A*100=S1
1(a)	Butt Weld	A1	a1			a1/A1*100
1(b)	Socket Weld	A2	a2			a2/A2*100
2	Welding	B=B1+B2	b=b1+b2			b/B*100=S2
2(a)	Butt Weld	B1	b1			b1/B1*100
2(b)	Socket Weld	B2	b2			b2/B2*100
3	Non Destructive Testing --(S3=S4*0.4+S5*0.3+S6*0.2+S7*0.1)					S3
3(a)	Radiography	D=D1+D2	d=d1+d2			d/D*100=S4
3(a)(i)	Butt Weld	D1	d1			
3(a)(ii)	Socket Weld	D2	d2			
3(b)	Ultrasonic Testing	E=E1+E2	e=e1+e2			e/E*100=S5
3(b)(i)	Butt Weld	E1	e1			
3(b)(ii)	Socket Weld	E2	e2			
3(c)	Magnetic Particle Testing	F=F1+F2	f=f1+f2			f/F*100=S6
3(c)(i)	Butt Weld	F1	f1			
3(c)(ii)	Socket Weld	F2	f2			
3(d)	Liquid Penetrant Testing	G=G1+G2	g=g1+g2			g/G*100=S7
3(d)(i)	Butt Weld	G1	g1			
3(d)(ii)	Socket Weld	G2	g2			
4	Post Weld Heat Treatment	H=H1+H2	h=h1=h2			h/H*100=S8
4(a)	Butt Weld	H1	h1			
4(b)	Socket Weld	H2	h2			
5	Hardness Survey of Welds	J	j			j/J*100=S9
5(a)	Butt & Socket Welds	J	j			
6	Hydrostatic Testing of Pipe Spool	K	k			k/K*100=S10
6(a)	Butt & Socket Welds	K	k			
	Piping Spool Fabrication Progress--(S1*0.04+S2*0.06+S3*0.05+S8*0.05+S9*0.02+S10*0.08)					❷

C	Piping Installation	Weight Kg		Inch Diameter		Inch Meter		
		Estimated	Erected	As per Drg.	Achieved	As per Drg.	Achieved	
1	Piping Spool Installation					A	a	a/A*100=E1
2	Valve & other inline equipment Installation	B	b					b/B*100=E2
3	Field Weld Joint fitup			C=C1+C2	c=c1+c2			c/C*100=E3
3(a)	Butt Weld			C1	c1			
3(b)	Socket Weld			C2	c2			
4	Field Welding			D=D1+D2	d=d1+d2			d/D*100=E4
4(a)	Butt Weld			D1	d1			
4(b)	Socket Weld			D2	d2			
5	NDT of Field Welds			E=E1+E2	e=e1+e2			e/E*100=E5
5(a)	Butt Weld			E1	e1			
5(b)	Socket Weld			E2	e2			
6	PWHT of Field Welds			F=F1+F2	f=f1+f2			f/F*100=E6
6(a)	Butt Weld			F1	f1			
6(b)	Socket Weld			F2	f2			
7	Hardness Survey of Field Welds			G	g			g/G*100=E7
7(a)	Butt & Socket Welds			G	g			
8	Installation of Pipe Supports Hangers etc.,					H	h	h/H*100=E8
8	Hydrostatic Test of Field Welds/ Golden Joints and Piping after completion					J	J	J/J*100=E9
9	Punch Listing & Clearing of Punches			K	k			k/K*100=S10
	Piping Installation Progress--(E1*0.10+E2*0.05+E3*0.02+E4*0.03+E5*0.02+E6*0.03+E7*0.01+E8*0.04+E9*0.05+E10*0.05)							❸
4	Overall Progress---- ❹ = ❶ X 0.30 + ❷ X 0.30 + ❸ X 0.40							

ANSI Standard Pipe Chart with SPF agaist Schedule							
US	Metric	OD		Schedule Designations	Wall Thickness		Standard Point Factor
		(Inches)	mm	(ANSI/ASME)	(Inches)	mm	(SPF)
1/8	6	0.405	10.3	10/10S	0.049	1.24	0.72
				STD/40/40S	0.068	1.73	1.00
				XS/80/80S	0.095	2.41	1.40
1/4	8	0.54	13.7	10/10S	0.065	1.65	0.74
				STD/40/40S	0.088	2.24	1.00
				XS/80/80S	0.119	3.02	1.35
3/8	10	0.675	17.1	10/10S	0.065	1.65	0.71
				STD/40/40S	0.091	2.31	1.00
				XS/80/80S	0.126	3.20	1.38
1/2	15	0.84	21.3	5/5S	0.065	1.65	0.55
				10/10S	0.083	2.11	0.70
				STD/40/40S	0.119	3.02	1.00
				XS/80/80S	0.147	3.73	1.24
				160	0.188	4.78	1.58
				XX	0.294	7.47	2.47
3/4	20	1.05	26.7	5/5S	0.065	1.65	0.58
				10/10S	0.083	2.11	0.73
				STD/40/40S	0.113	2.87	1.00
				XS/80/80S	0.154	3.91	1.36
				160	0.219	5.56	1.94
				XX	0.308	7.82	2.73
1	25	1.315	33.4	5/5S	0.065	1.65	0.49
				10/10S	0.109	2.77	0.82
				STD/40/40S	0.133	3.38	1.00
				XS/80/80S	0.179	4.55	1.35
				160	0.250	6.35	1.88
				XX	0.358	9.09	2.69
1 1/4	32	1.66	42.2	5/5S	0.065	1.65	0.46
				10/10S	0.109	2.77	0.78
				STD/40/40S	0.140	3.56	1.00
				XS/80/80S	0.191	4.85	1.36
				160	0.250	6.35	1.79
				XX	0.382	9.70	2.73
1 1/2	40	1.9	48.3	5/5S	0.065	1.65	0.45
				10/10S	0.109	2.77	0.75
				STD/40/40S	0.145	3.68	1.00
				XS/80/80S	0.200	5.08	1.38
				160	0.281	7.14	1.94
				XX	0.400	10.16	2.76
2	50	2.375	60.3	5/5S	0.065	1.65	0.42
				10/10S	0.109	2.77	0.71
				STD/40/40S	0.154	3.91	1.00
				XS/80/80S	0.218	5.54	1.42
				160	0.344	8.74	2.23
				XX	0.436	11.07	2.83

2 1/2	65	2.875	73.0	5/5S	0.083	2.11	0.41
				10/10S	0.120	3.05	0.59
				STD/40/40S	0.203	5.16	1.00
				XS/80/80S	0.276	7.01	1.36
				160	0.375	9.53	1.85
				XX	0.552	14.02	2.72
3	80	3.5	88.9	5/5S	0.083	2.11	0.38
				10/10S	0.120	3.05	0.56
				STD/40/40S	0.216	5.49	1.00
				XS/80/80S	0.300	7.62	1.39
				160	0.438	11.13	2.03
				XX	0.600	15.24	2.78
3 1/2	90	4	101.6	5/5S	0.083	2.11	0.37
				10/10S	0.120	3.05	0.53
				STD/40/40S	0.226	5.74	1.00
				XS/80/80S	0.318	8.08	1.41
				XX	0.636	16.15	2.81
4	100	4.5	114.3	5/5S	0.083	2.11	0.35
				10/10S	0.120	3.05	0.51
					0.156	3.96	0.66
					0.188	4.78	0.79
				STD/40/40S	0.237	6.02	1.00
				XS/80/80S	0.337	8.56	1.42
				120	0.438	11.13	1.85
				160	0.531	13.49	2.24
				XX	0.674	17.12	2.84
4 1/2	115	5	127.0	STD/40/40S	0.247	6.27	1.00
				XS/80/80S	0.355	9.02	1.44
				XX	0.710	18.03	2.87
5	125	5.563	141.3	5/5S	0.109	2.77	0.42
				10/10S	0.134	3.40	0.52
				STD/40/40S	0.258	6.55	1.00
				XS/80/80S	0.375	9.53	1.45
				120	0.500	12.70	1.94
				160	0.625	15.88	2.42
				XX	0.750	19.05	2.91
6	150	6.625	168.3	5/5S	0.109	2.77	0.39
				10/10S	0.134	3.40	0.48
					0.188	4.78	0.67
				STD/40/40S	0.280	7.11	1.00
				XS/80/80S	0.432	10.97	1.54
				120	0.562	14.27	2.01
				160	0.719	18.26	2.57
				XX	0.864	21.95	3.09
7	175	7.625	193.7	STD/40/40S	0.301	7.65	1.00
				XS/80/80S	0.500	12.70	1.66
				XX	0.875	22.23	2.91

				5S	0.109	2.77	0.34
				10/10S	0.148	3.76	0.46
				20	0.250	6.35	0.78
				30	0.277	7.04	0.86
				STD/40/40S	0.322	8.18	1.00
8	200	8.625	219.1	60	0.406	10.31	1.26
				XS/80/80S	0.500	12.70	1.55
				100	0.594	15.09	1.84
				120	0.719	18.26	2.23
				140	0.812	20.62	2.52
				XX	0.875	22.23	2.72
				160	0.906	23.01	2.81
9	225	9.625	35.0	STD/40/40S	0.342	8.69	1.00
				XS/80/80S	0.500	12.70	1.46
				XX	0.875	22.23	2.56
				5S	0.134	3.40	0.37
				10S	0.165	4.19	0.45
					0.188	4.78	0.52
				20	0.250	6.35	0.68
				30	0.307	7.80	0.84
10	250	10.75	273.1	STD/40/40S	0.365	9.27	1.00
				XS/60/80S	0.500	12.70	1.37
				80	0.594	15.09	1.63
				100	0.719	18.26	1.97
				120	0.844	21.44	2.31
				140/XX	1.000	25.40	2.74
				160	1.125	28.58	3.08
11	275	11.75	298.5	STD/40/40S	0.375	9.53	1.00
				XS/80/80S	0.500	12.70	1.33
				XX	0.875	22.23	2.33
				5S	0.165	4.19	0.44
				10S	0.180	4.57	0.48
					0.188	4.78	0.50
				20	0.250	6.35	0.67
				30	0.330	8.38	0.88
				STD/40S	0.375	9.53	1.00
12	300	12.75	323.9	40	0.406	10.31	1.08
				XS/80S	0.500	12.70	1.33
				60	0.562	14.27	1.50
				80	0.688	17.48	1.83
				100	0.844	21.44	2.25
				120/XX	1.000	25.40	2.67
				140	1.125	28.58	3.00
				160	1.312	33.32	3.50

				10S	0.188	4.78	0.50
				10	0.250	6.35	0.67
				20	0.312	7.92	0.83
				STD/30/40S	0.375	9.53	1.00
				40	0.438	11.13	1.17
14	400	14	355.6	XS/80S	0.500	12.70	1.33
				60	0.594	15.09	1.58
				80	0.750	19.05	2.00
				100	0.938	23.83	2.50
				120	1.094	27.79	2.92
				140	1.250	31.75	3.33
				160	1.406	35.71	3.75
				10S	0.188	4.78	0.50
				10	0.250	6.35	0.67
				20	0.312	7.92	0.83
				STD/30/40S	0.375	9.53	1.00
				XS/40/80S	0.500	12.70	1.33
16	400	16	406.4	60	0.656	16.66	1.75
				80	0.844	21.44	2.25
				100	1.031	26.19	2.75
				120	1.219	30.96	3.25
				140	1.438	36.53	3.83
				160	1.594	40.49	4.25
				10S	0.188	4.78	0.50
				10	0.250	6.35	0.67
				20	0.312	7.92	0.83
				STD/40S	0.375	9.53	1.00
				30	0.438	11.13	1.17
				XS/80S	0.500	12.70	1.33
18	450	18	457.2	40	0.562	14.27	1.50
				60	0.750	19.05	2.00
				80	0.938	23.83	2.50
				100	1.156	29.36	3.08
				120	1.375	34.93	3.67
				140	1.562	39.67	4.17
				160	1.781	45.24	4.75
				10S	0.218	5.54	0.58
				10	0.250	6.35	0.67
				STD/20/40S	0.375	9.53	1.00
				XS/30/80S	0.500	12.70	1.33
				40	0.594	15.09	1.58
20	500	20	508.0	60	0.812	20.62	2.17
				80	1.031	26.19	2.75
				100	1.281	32.54	3.42
				120	1.500	38.10	4.00
				140	1.750	44.45	4.67
				160	1.969	50.01	5.25

				10S	0.218	5.54	0.58
				10	0.250	6.35	0.67
				STD/20/40S	0.375	9.53	1.00
				XS/30/80S	0.500	12.70	1.33
22	550	22	558.8	60	0.875	22.23	2.33
				80	1.125	28.58	3.00
				100	1.375	34.93	3.67
				120	1.625	41.28	4.33
				140	1.875	47.63	5.00
				160	2.125	53.98	5.67
				10S	0.218	5.54	0.58
				10/10S	0.250	6.35	0.67
				STD/20/40S	0.375	9.53	1.00
				XS/80S	0.500	12.70	1.33
				30	0.562	14.27	1.50
24	600	24	609.6	40	0.688	17.48	1.83
				60	0.969	24.61	2.58
				80	1.219	30.96	3.25
				100	1.531	38.89	4.08
				120	1.812	46.02	4.83
				140	2.062	52.37	5.50
				160	2.344	59.54	6.25
				10	0.312	7.92	0.83
26	650	26	660.4	STD/40S	0.375	9.53	1.00
				XS/80S	0.500	12.70	1.33
				10	0.312	7.92	0.83
28	700	28	711.2	STD/40S	0.375	9.53	1.00
				XS/20/80S	0.500	12.70	1.33
				30	0.625	15.88	1.67

				10	0.312	7.92	0.83
30	750	30	762.0	STD/40S	0.375	9.53	1.00
				XS/20/80S	0.500	12.70	1.33
				30	0.625	15.88	1.67
				10	0.312	7.92	0.83
				STD.	0.375	9.53	1.00
32	800	32	812.8	XS/20	0.500	12.70	1.33
				30	0.625	15.88	1.67
				40	0.688	17.48	1.83
				10	0.312	7.92	0.83
				STD	0.375	9.53	1.00
34	850	34	863.6	XS/20	0.500	12.70	1.33
				30	0.625	15.88	1.67
				40	0.688	17.48	1.83
				10	0.312	7.92	0.83
36	900	36	914.4	STD/40S	0.375	9.53	1.00
				XS/80S	0.500	12.70	1.33
				STD/40S	0.375	9.53	1.00
42	1050	42	1066.8	XS/80S	0.500	12.70	1.33
				30	0.625	15.88	1.67
				40	0.750	19.05	2.00
48	1200	48	1219.2	STD/40S	0.375	9.53	1.00
				XS/80S	0.500	12.70	1.33

Appendix D
Radiographic Acceptance Criteria for Rounded Indication

As mentioned in Chapter 11, Table 341.3.2 of ASME B 31.3 provides acceptance criteria for all other types of weld defects other than rounded indications (porosity). Instead, ASME B 31.3 refers it to ASME Section VIII Appendix 4.

D.1 TERMINOLOGY AS PER APPENDIX 4 OF ASME SECTION VIII DIV (1)

Rounded Indications

Indications with a maximum length of 3× width or less on the radiograph are defined as rounded indications.

Aligned Indications

A sequence of four or more rounded indications when they touch a line parallel to length of the weld drawn through the centre of two outer rounded indications.

Source: Reproduced from ASME with permission ASME Section VIII Div 1 Appendix 4.

D.2   ACCEPTANCE CRITERIA AS PER APPENDIX 4 OF ASME SECTION VIII DIV (1)

Image Density

Density within the image of the indication may vary and is not a criterion for acceptance or rejection.

Relevant Indications

Rounded indications which exceed the following dimensions:

$1/10t$	for t less than 3 mm (1/8″)
0.4 mm (1/64″)	for t from 3 to 6 mm (1/8″–1/4″), incl.
0.8 mm (1/32″)	for t greater than 6–50 mm (1/4″–2″), incl.
1.6 mm (1/16″).	for t greater than 50 mm (2″)

Maximum Size of Rounded Indications

Isolated random	$1/4t$ or 4 mm (5/32″), whichever is less
When separated by 25 mm (1″) or more	$1/3t$ or 6 mm (1/4″), whichever is less
For t greater than 50 mm (2″)	10 mm (3/8″)

Aligned Rounded Indications

Acceptable when summation of diameters of indications is less than t in a length of $12t$

Rounded Indications Charts

The rounded indications characterized as imperfections shall not exceed that shown in the charts.

Source: Reproduced from ASME with permission ASME Section VIII Div (1) Appendix 4.

Aligned Rounded Indications

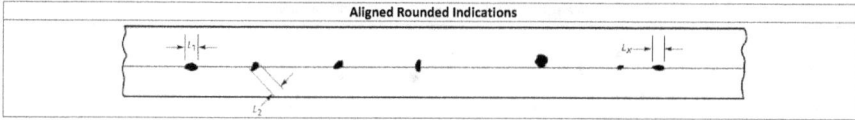

Groups of Aligned Rounded Indications

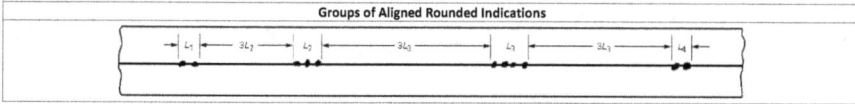

Charts for *t* equal to 3mm to 6mm (1⁄8" to 1⁄4"), inclusive

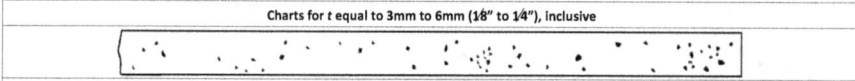

Random Rounded Indications [See Note (1)]

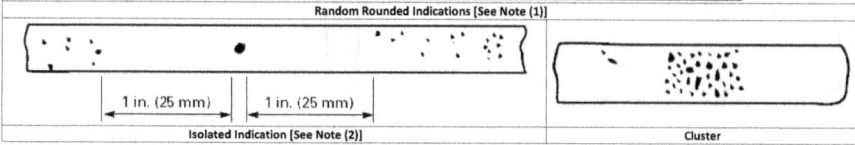

| Isolated Indication [See Note (2)] | Cluster |

Charts for *t* over 6mm to 10mm (1⁄4" to 3⁄8"), inclusive

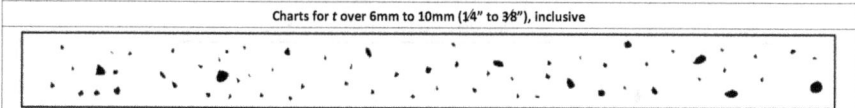

Random Rounded Indications [See Note (1)]

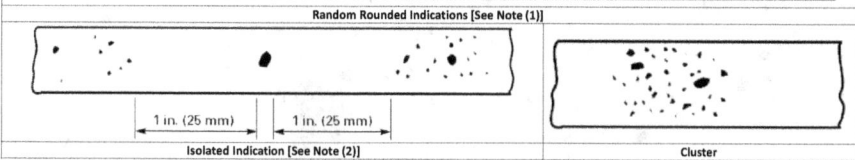

| Isolated Indication [See Note (2)] | Cluster |

Charts for *t* over 10mm to 19 mm (3⁄8" to 3⁄4"), inclusive

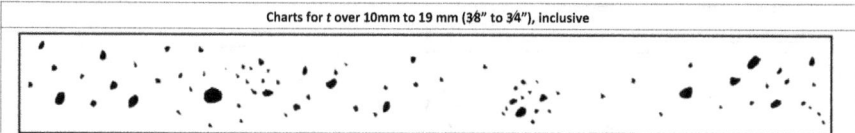

Random Rounded Indications [See Note (1)]

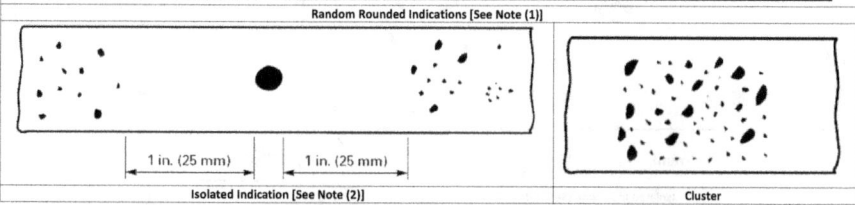

| Isolated Indication [See Note (2)] | Cluster |

Charts for *t* over 19mm to 50mm (3⁄4" to 2"), inclusive

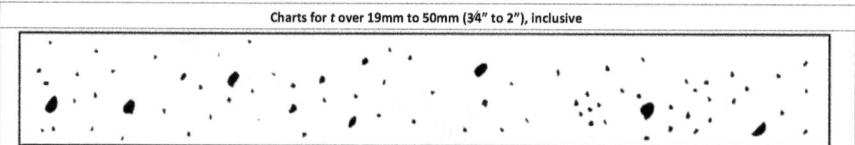

Random Rounded Indications [See Note (1)]

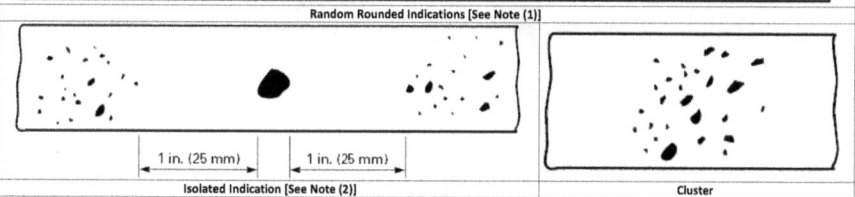

| Isolated Indication [See Note (2)] | Cluster |

Charts for *t* over 50mm to 100mm (4" to 8"), inclusive

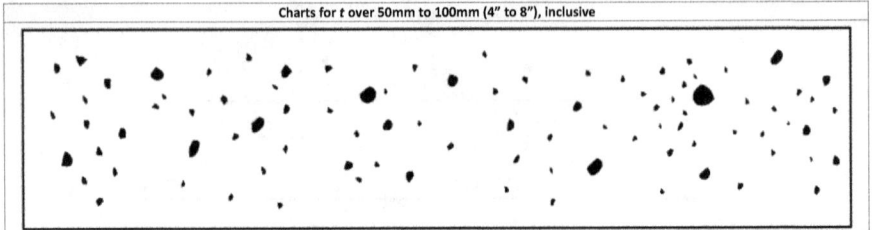

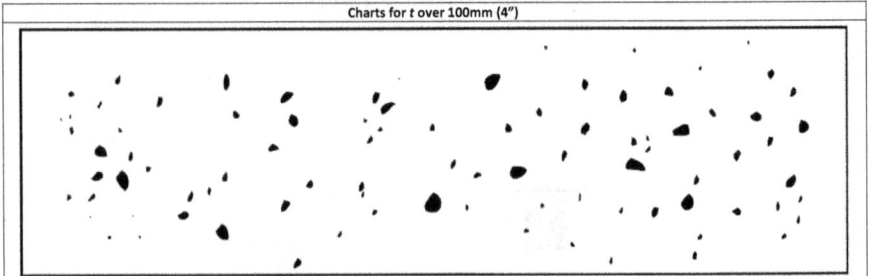

Random Rounded Indications [See Note (1)]

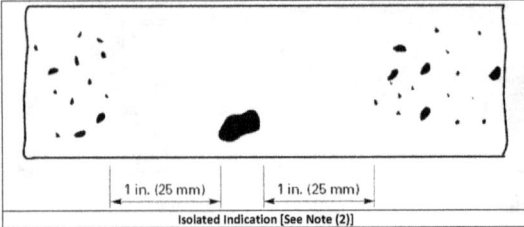

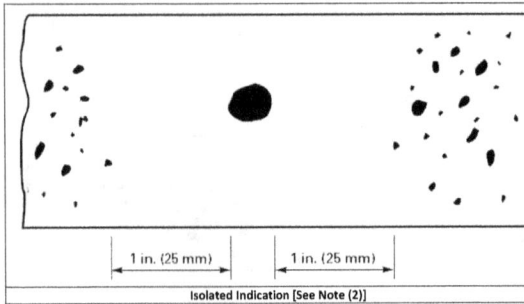

1 in. (25 mm)   1 in. (25 mm)

Isolated Indication [See Note (2)]   Cluster

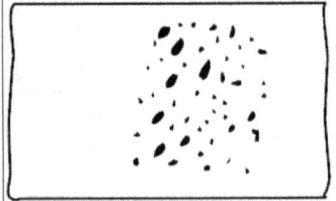

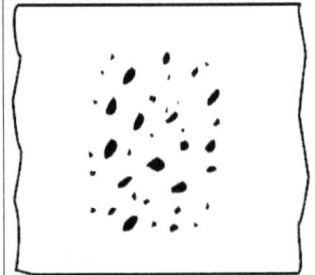

Charts for *t* over 100mm (4")

Random Rounded Indications [See Note (1)]

1 in. (25 mm)   1 in. (25 mm)

Isolated Indication [See Note (2)]   Cluster

Appendix E
Preheat and Post-Weld Heat Treatment Requirements as per ASME B 31.3

Required and recommended minimum preheat temperatures for materials of various P-Numbers are given in Table 330.1.1. If the ambient temperature is below 0°C (32°F), the recommendations in Table 330.1.1 become requirements. The thickness intended in Table 330.1.1 is that of the thicker component measured at the joint. How to arrive at these governing thickness is mentioned in Chapter 12.

E.1. REQUIREMENTS FOR PREHEATING

Preheat Temperatures (Table 330.1.1)

E.2. REQUIREMENTS FOR POST-WELD HEAT TREATMENT

Requirements for Heat Treatment (Table 331.1.1)

Base Metal P-No. or S-No. [Note (1)]	Weld Metal Analysis A No. [Note (2)]	Base Metal Group	Nominal Wall Thickness (mm)	(in.)	Specified Min Tensile Strength, Base Metal (MPa)	(ksi)	Minimum Temperature Required (°C)	Required (°F)	Recommended (°C)	Recommended (°F)
1	1	Carbon steel	<25	<1	≤490	≤71	-	-	10	50
			≥25	≥1	All	All	-	-	79	175
			All	All	>490	>71	-	-	79	175
3	2, 11	Alloy steels, Cr ≤½%	<13	<1/2	≤490	≤71	-	-	10	50
			≥13	≥½	All	All	-	-	79	175
			All	All	>490	>71	-	-	79	175
4	3	Alloy steels, ½% ≤ Cr ≤2%	All	All	All	All	149	300	-	-
5A, 5B, 5C	4, 5	Alloy steels, 2¼% ≤ Cr ≤10%	All	All	All	All	177	350	-	-
6	6	High alloy steels, martensitic	All	All	All	All	-	-	149[4]	300[4]
7	7	High alloy steels, ferritic	All	All	All	All	-	-	10	50
8	8, 9	High alloy steels, austenitic	All	All	All	All	-	-	10	50
9A, 9B	10	Nickel alloy steels	All	All	All	All	-	-	93	200
10	-	Cr–Cu steel	All	All	All	All	149–204	300–400	-	-
10I	-	27Cr steel	All	All	All	All	149[3]	300[3]	-	-
11A SG 1	-	8 Ni, 9 Ni steel	All	All	All	All	-	-	10	50
11A SG 2	-	5 Ni steel	All	All	All	All	10	50	-	-
15E	5	Alloy steels	All	All	All	All	-	-	-	-
9 Cr-1 Mo-V	All	All	All	All	204	400	-	-	10	50
21–52	-	-	All	All	All	All	-	-	10	50

Notes:

(1) P-Number or S-Number from BPV Code, Section IX, QW/QB-422.

(2) A-Number from Section IX, QW-442.

(3) Maintain interpass temperature between 177°C and 232°C (350°F and 450°F).

(4) Maximum interpass temperature of 316°C (600°F).

Source: Reproduced from Table 330.1.1 of ASME B 31.3 with permission from ASME.

Base Metal P-No. or S-No. [Note (1)]	Weld Metal Analysis A No. [Note (2)]	Base Metal Group	Nominal Wall Thickness (mm)	(in.)	Specified Min Tensile Strength, Base Metal (MPa)	(ksi)	Metal Temperature Range (°C)	(°F)	Holding Time Nominal Wall [Note (3)] (min/mm)	(hr/in.)	Min Time (hr)	Brinell Hardness, Max [Note (4)]
1	1	Carbon steel	≤20	≤¾	All	All	None	None	-	-	-	-
			>20	>¾	All	All	593–649	1,100–1,200	2.4	1	1	-
3	2, 11	Alloy steels Cr ≤½%	≤20	≤¾	≤490	≤71	None	None	-	-	-	-
			>20	>¾	All	All	593–718	1,100–1,325	2.4	1	1	225
			All	All	>490	>71	593–718	1,100–1,325	2.4	1	1	225
4 [Note (5)]	3	Alloy steels ½% <Cr ≤2%	≤13	≤½	≤490	≤71	None	None	-	-	-	-
			>13	>½	All	All	704–746	1,300–1,375	2.4	1	2	225
			All	All	>490	>71	704–746	1,300–1,375	2.4	1	2	225
5A, 5B, 5C [Note (5)]	4,5	Alloy steels, 2¼% ≤Cr ≤10%										
		≤3% Cr and ≤0.15% C	≤13	≤½	All	All	None	None	-	-	-	-
		≤3% Cr and ≤0.15% C	>13	>½	All	All	704–760	1,300–1,400	2.4	1	2	241
		>3% Cr or >0.15% C	All	All	All	All	704–760	1,300–1,400	2.4	1	2	241

(Continued)

Base Metal P-No. or S-No. [Note (1)]	Weld Metal Analysis A-No. [Note (2)]	Base Metal Group	Nominal Wall Thickness (mm)	(in.)	Specified Min Tensile Strength, Base Metal (MPa)	(ksi)	Metal Temperature Range (°C)	(°F)	Holding Time — Nominal Wall [Note (3)] (min/mm)	(hr/in.)	Min Time (hr)	Brinell Hardness, Max [Note (4)]
6	6	High alloy steels, martensitic	All	All	All	All	732–788	1,350–1,450	2.4	1	2	241
		A 240 Gr. 429	All	All	All	All	621–663	1,150–1,225	2.4	1	2	241
7	7	High alloy steels, ferritic	All	All	All	All	None	None	–	–	–	–
8	8, 9	High alloy steels, austenitic	All	All	All	All	None	None	–	–	–	–
9A, 9B	10	Nickel alloy steels	≤20	≤¾	All	All	593–635	1,100–1,175	–	–	–	–
			>20	>¾	All	All			1.2	½	1	
10	–	Cr–Cu steel	All	All	All	All	760–816	1,400–1,500	1.2	½	½	–
[Note (6)]			½	–	All	All						
[Note (6)]												
10H	–	Duplex stainless steel	All	All	All	All	Note (7)	Note (7)				
10I	–	27Cr steel	All	All	All	–	663–704	1,225–1,300				
[Note (8)]		[Note (8)]	1	1	All							
11A SG 1	–	8 Ni, 9 Ni steel	≤51	≤2	All	All	None	None				
			>51	>2	All	All	552–585	1,025–1,085	2.4	1	1	
[Note (9)]			1	–	All							
[Note (9)]									2.4			
11A SG 2	–	5 Ni steel	>51	>2	All	All	552–585	1,025–1,085	2.4	1	1	
[Note (9)]			1	–	All							
[Note (9)]									2.4			

(Continued)

Base Metal P-No. or S-No. [Note (1)]	Weld Metal Analysis A No. [Note (2)]	Base Metal Group	Nominal Wall Thickness		Specified Min Tensile Strength, Base Metal		Metal Temperature Range		Holding Time		Min Time	Brinell Hardness, Max
									Nominal Wall [Note (3)]			[Note (4)]
			(mm)	(in.)	(MPa)	(ksi)	(°C)	(°F)	(min/mm)	(hr/in.)	(hr)	
15E	5	Alloy steels 9 Cr-1 Mo-V	All	All	All	All	732–774	1,350–1,425	2.4	1	2	250
62	-	Zr R60705	All	All	All	All	538–593					
[Note (10)]	1,000–1,100											
[Note (10)]	Note (10)	Note (10)	1	-								

Notes:

(1) P-Number or S-Number from BPV Code, Section IX, QW/QB-422.

(2) A-Number from Section IX, QW-442.

(3) For holding time in SI metric units, use min/mm (minutes per mm thickness). For U.S. units, use hr/in. thickness.

(4) See para. 331.1.7.

(5) See Appendix F, para. F331.1.

(6) Cool as rapidly as possible after the hold period.

(7) Post-weld heat treatment is neither required nor prohibited, but any heat treatment applied shall be as required in the material specification.

(8) Cooling rate to 649°C (1,200°F) shall be less than 56°C (100°F)/hour; thereafter, the cooling rate shall be fast enough to prevent embrittlement.

(9) Cooling rate shall be >167°C (300°F)/hour to 316°C (600°F).

(10) Heat treat within 14 days after welding. Hold time shall be increased by 1/2 hour for each 25 mm (1″) over 25 mm thickness. Cool to 427°C (800°F) at a rate ≤278°C (500°F)/hour, per 25 mm (1″) nominal thickness, 278°C (500°F)/hour max. Cool in still air from 427°C (800°F).

Source: Reproduced from Table 331.1.1 of ASME B 31.3 with permission.

Appendix F
Quality Audit of Piping Girth Welds Based on Radiography

F.1 INTRODUCTION

In piping and pipeline construction projects, monitoring of weld quality and performance of welders and welding operators is a challenge to QC engineers on site. There are no clear-cut guidelines provided by applicable codes and standards. Moreover, in most of the occasions, client specifications also fail to provide a clear norm for assessment of weld quality. Therefore, various methodologies are adopted based on the experience of contractors, which are more favourable to interests of contracting firms. The appendix proposes a couple of methodologies for monitoring and analysing general weld quality, based on radiographic examination carried out during construction of piping and pipelines. Welding related to other equipment and structurals are excluded from the ambit of this appendix. The appendix is generally written based on ASME codes applicable for piping and pipelines, namely ASME B 31.3 (Process Plant Piping), ASME B 31.4 (Pipeline Transportation System for Liquid Hydrocarbons and Other Liquids) and ASME B 31.8 (Gas Transmission and Distribution Piping System), in view of oil and gas industry at large.

Before going into the details of weld quality monitoring, it is felt that a brief about certain salient aspects related to interpretation and evaluation of radiographs (often not described in specifications) would be helpful to the readers of this appendix, in the event of some disputes with regard to acceptance/rejection of radiographs.

F.2 SUMMARY OF BASIC STEPS IN INTERPRETING RADIOGRAPHS

1. Verify that radiograph corresponds to the component or part or weld being examined.
2. Verify that radiographic coverage is complete for the particular component or part or that of weld.
3. Verify that image quality indicators are correct and properly used and that proper image quality level was achieved.
4. Verify that film densities meet requirements of applicable codes/specifications.
5. Check for film artefacts and indications of surface phenomena and record the same if any found, on radiography report.
6. Retake radiographs having indications that cannot be resolved as an artefact or surface discontinuity.

7. Visually check surfaces of part radiographed for surface discontinuities or uneven contours or depressions matching in appearance of discontinuity revealed by radiograph.

8. Defects found on surface, matching with discontinuity in radiography, may be accepted as such with a remark on viewing record, provided the same falls within acceptance criteria for surface defects.

9. All remaining discontinuities are to be considered as internal discontinuities and to be evaluated to applicable standards and accept, reject or hold the part for further review. Record as "complies" or "does not comply" on the viewing record.

10. Decide on the extent of repair required for any non-complying discontinuities on film (for repair) and transfer to the part or component radiographed with maximum possible accuracy through tracings.

11. Prepare a test report indicating the nature, extent and disposition of all significant indications found on radiographs.

F.3 OTHER SALIENT FACTORS IN REVIEW OF RADIOGRAPHS

The usual acceptance criterion for density of films is in the range of 1.8–4.0. However, acceptance of radiographs with high densities (usually above 3.5) is purely based on illumination achievable by film viewer. Most of the applicable codes and client specifications are silent on this aspect and given below are some such tips.

a. Uniformly Illuminated Diffusing Screen

Examination of radiographs shall be carried out in "diffused light in a darkened room". Most illuminators also include a rheostat that enables brightness to be adjusted to accommodate radiographs of varying densities. In addition, film viewer shall be equipped with a sliding mask, so as to prevent direct bright light hitting the eyes of the reviewer, especially when shorter radiographs are to be reviewed.

b. Intensity of Light Source

For effective and meaningful interpretation of radiographs, ISO 5580 requires a minimum intensity of light, 30 candela per square metre (cd/m^2) transmitted through a radiograph under examination. In order to achieve this, the viewing facility shall have a minimum brightness as provided in Table F.1 below against specified density of radiographs.

Table F1 provides the brightness requirements against the upper value of density which may be used for densities below stipulated value. Brightness values provided above are the minimum brightness to view film, based on $30\,cd/m^2$ intensity of transmitted light through radiograph in review. The standard also suggests that brightness of $100\,cd/m^2$ as a more reasonable value.

c. Verification of Screen Brightness

Brightness of an illuminator can be checked with a photographic light meter by following these steps:

1. Set film speed indicator to 100 ASA or 200 ASA.

2. Place sensor element of light meter close to screen of the illuminator.

TABLE F.1
Density of Radiograph and Minimum Screen Luminance (cd/m²)

Sl. No.	Density of Radiograph	Minimum Screen Luminance (cd/m²)
1	1.0	300
2	1.5	1,000
3	2.0	3,000
4	2.5	10,000
5	3.0	10,000
6	3.5	30,000
7	4.0	100,000
8	4.5	300,000

TABLE F.2
Relationship between Screen Brightness and Exposure Reading

Sl. No.	F Number	Exposure (Seconds)	Screen Brightness	
			100 ASA	200 ASA
1	10	1/100	1,000	2,000
2	10	1/500	5,000	10,000
3	10	1/1,000	10,000	20,000
4	14.3	1/100	2,000	4,000
5	14.3	1/500	10,000	20,000
6	14.3	1/1,000	20,000	40,000
7	20	1/100	3,000	6,000
8	20	1/500	15,000	30,000
9	20	1/1,000	30,000	60,000
10	20	1/1,500	45,000	90,000
11	20	1/2,000	60,000	120,000

3. Record the "exposure" in hundredths of a second against a camera aperture setting of f 10, f 14.3 or f 20.

4. Use Table F.2 below to relate photographic exposure time to screen brightness.

d. Background Light

Background light shall only be sufficient enough to enable recording of details on viewing record (film flap and report), whereas too much background lighting might cause reflections off the film. This eventually reduces contrast, thereby making interpretation more difficult. Furthermore, room used for viewing shall be quiet and comfortable enough to avoid other unnecessary distractions to reviewer.

e. Breaks between Review

Frequent breaks shall be allowed to the reviewer to prevent eye strain and maximize concentration level. Though individual capacities of reviewers vary widely, it is recommended that no more than 5–10 minutes be spent viewing a radiograph. Moreover, before commencing a viewing session, the reviewer shall take sufficient time for his or her eyes to get accustomed to darkened conditions established for review of radiographs.

f. Dryness of Radiographs

Wash water on a radiograph has a significant effect on sensitivity, thereby making it difficult to detect fine discontinuities. Therefore, it shall be ensured that the radiographs are properly dried before viewing.

g. Quality of Radiographs

Prior to identification and evaluation of discontinuities, the radiograph shall be checked for processing/handling artefacts, film density and sensitivity. The reviewer of radiograph shall ensure that quality of radiograph is adequate to carry out a meaningful interpretation and also in accordance with applicable specifications (codes and client specifications). This is to ensure that relevant discontinuities can be detected without fail. The results of these preliminary checks and measurements should be recorded on the viewing record or report.

F.4   EVALUATION OF DISCONTINUITIES

The reviewer shall resist the temptation to have a quick glance to "spot the defect". Instead, a thorough examination by carefully scanning radiograph from one end to other, concentrating on a small area of the radiograph and moving progressively, is required. The following information would be helpful to the reviewer in offering an effective interpretation:

- Information about product radiographed like casting, forging, weld, etc.
- In case of welds, information like welding process, material type, groove design, etc.
- Manufacturing process in the case of castings/forgings, etc.
- Probable types of discontinuities likely to occur during manufacturing.
- Technique used in radiography in detail.

The other very important criteria that a reviewer shall clearly understand is the acceptance/rejection criteria for radiographs of component or part inspected. This requirement is generally contained within applicable codes and client specifications. Radiographic interpretation is a skill that can only be mastered through knowledge of material being tested and experience, with more weightage for experience. Hands-on experience during removal of reported defects would be of great advantage to reviewer to assess the accuracy of his or her evaluation. Many indications produce subtle low contrast or un-sharp images which would be very difficult to site and interpret.

F.5   WELD DISCONTINUITIES

There is a standard set of abbreviations used to describe most of the weld discontinuities. These abbreviations are listed in AS4749-2001, "Non-Destructive Testing – Terminology of and Abbreviations for Fusion Weld Imperfections as Revealed by

TABLE F.3
Standard Abbreviations for Weld Discontinuities

Sl. No.	Defect Description	Code	Sl. No.	Defect Description	Code
	Abbreviations for Surface Imperfections				
1	Excessive Penetration	SXP	8	Root Concavity	SRC
2	Incomplete Filled Groove	SGI	9	Shrinkage Groove	SGS
3	Under Cut	SUC	10	Excessive Dressing	SED
4	Grinding Mark	SMG	11	Tool Mark	SMT
5	Hammer Mark	SMH	12	Torn Surface	STS
6	Surface Pitting	SPT	13	Spatter	SSP
7	Linear Misalignment	HiLo			
	Abbreviations for Internal Imperfections				
14	Longitudinal Crack	KL	25	Transverse Crack	KT
15	Crater Crack	KC	26	Lack of Side Fusion	LS
16	Lack of Root Fusion	LR	27	Lack of Inter-Run Fusion	LI
17	Incomplete Root Penetration	LP	28	Inclusion	IN
18	Linear Inclusion	IL	29	Oxide Inclusion	IO
19	Tungsten Inclusion	IT	30	Copper Inclusion	IC
20	Gas Pore	GP	31	Worm Hole	WH
21	Crater Pipe	CP	32	Localized Porosity	PG
22	Linear Porosity	PL	33	Elongated Cavity	EC
23	Uniform Porosity	PU	34	Burn Through	BT
24	Diffraction Mottling	DM			

Radiography". Description of each discontinuity is provided with prints taken from an actual radiograph or a sketch to describe discontinuity. Basically, weld imperfections are categorized under two groups broadly, such as "surface imperfections" and "internal imperfections" (Table F.3).

While making a report of radiographic examination, the abbreviation (or the serial number) of objectionable imperfection shall be indicated, which would be helpful in applying filters to get statistics of a particular type of imperfection in a project, provided formats are made properly in Excel or other similar software to have an effective search function.

F.6 INTERPRETATION OF RADIOGRAPHS AND REPORTING

As mentioned earlier, the design and construction of piping and pipelines (within the realm of this appendix) is based on ASME codes like ASME B 31.3, B 31.4 and B 31.8, the radiographic acceptance criteria also shall be governed by above-mentioned codes. They address both random spot radiography and full radiography depending on criticality of service as envisaged by the designer. As per code for Process Plant Piping, ASME B 31.3, weld imperfections are categorized into the following types:

1. Crack
2. Lack of fusion

3. Incomplete penetration
4. Internal porosity
5. Internal slag inclusion, tungsten inclusion or elongated indication
6. Undercutting
7. Surface porosity or exposed slag inclusion
8. Surface finish
9. Concave root surface (suck up)
10. Weld reinforcement or internal protrusion

Of the imperfections listed above, the code does not tolerate presence of imperfections (1), (2) and (7) in welds. However, it provides acceptance limits for all other types of imperfection based on fluid and service conditions as categorized in ASME B 31.3. When the size of imperfection exceeds a specified limit, imperfection is required to be removed by a suitable process (either by gouging or gouging followed by grinding) and re-welded as per approved procedures. For both, spot and full radiography, the acceptance criteria mentioned in applicable codes remain the same.

Though the definitions of imperfections are given only in ASME B 31.3, very same terminology could be extended to codes for pipelines, namely ASME 31.4 and 31.8. However, there are marginal differences in acceptance norms in these codes (referred to API 1104 – Welding of Pipelines and Related Facilities). Based on the evaluation/acceptance criteria specified in applicable codes for each type of imperfections, they are categorized as either acceptable or non-acceptable imperfections. The radiographic segment containing such non-acceptable imperfection is considered as a defective segment or colloquially called as defective spot.

Later those unacceptable defects are marked on respective welds through tracings of defects taken from radiographs. Depending on type and nature of defects, repair is carried out either from outside or inside (based on proximity) by appropriate methods, and joints are re-welded using an approved weld repair procedure. The weld thus repaired shall be re-radiographed to confirm removal of unacceptable imperfections reported earlier. In the case of random spot radiography, each defective spot found will attract penal radiography as stipulated under progressive examination in applicable code or as specified in respective client specification if any provided.

F.7   WELD QUALITY ANALYSIS

For radiography consisting of both spot and full, there are two different ways in analysing defect percentage of any type of weld, both based on the outcome of radiography. Since the reckoning method is different, the general acceptance criteria for both modes have to be different; the reasons for the same are explained in ensuing sections.

a. Based on defective spots

The first and easy method is to express it as a percentage of defective spots over total spots radiographed.

For example, consider the case of radiographs taken from a piping girth weld of 12″ diameter.

OD of the pipe $= 324\,mm$

Outside circumference = 1,017 mm

Recommended film size = 240 mm

Number of films required = 5 Nos

Total length radiographed = 1,200 mm (for 100% coverage in radiography with overlap)

Film spot size (for calculation) = 120 mm

Therefore, number of spots for calculation 10 Nos.

Assume that girth weld so radiographed has a slag inclusion of 20 mm long (1 mm above acceptance maximum of 19 mm) which need to be repaired. Obviously, this will fall within one spot (considered for calculation).

Therefore, defect rate = 1/10 × 100 = 10%

b. Based on defect length

For the second method, the assessment of quality is based on the length of defect in comparison to total length of weld radiographed and interpreted. For the same defect size, considered in method (a) defect percentage calculation based on length of defect, the calculation shall be as below:

Percentage defect = Length of defect/length of weld radiographed × 100
$$= 20/1,017 \times 100 = 1.967\%$$

c. Comparison of methods (a) and (b)

In order to compare the reliability and consistency of the assessment methods, the following workout would be helpful.

As we all are aware, elongated defects like slag inclusions have a maximum acceptance criteria based on thickness of the weld. However, the maximum permissible size of any single elongated slag is 19 mm irrespective of thickness of weld. Defects like cracks are not acceptable even if it is only 1 mm long which is visible in the radiograph. Consider a case of weld wherein a crack of 3 mm length is found on a 12″ diameter pipe, the calculation of percentage defect goes as below:

According to method (a), the percentage defect based on spot size = 1/10 × 100 = 10% itself as in previous case.

Whereas defect percentage calculation based on length of defect {method (b)}, this works out to 3/1,017 × 100 = 0.295%, which is well below the percentage reported in the case of a slag, whose length is 20 mm.

In severity scale, cracks are considered much more detrimental to other kinds of inclusions. So in case of method (b) (based defect length), all types of defects are considered par and the only criterion considered is the length. But the methodology based on spot (though does not truly reflect the severity) is more sensitive towards this on account of its positive approximation (or rounding off). Therefore, from the above typical calculations, it can be inferred that the defect analysis based on spot size is more realistic compared to that based on real defect length.

As the industry does not have a norm indicating the acceptance criteria with regard to these percentages, as a practice, many in oil and gas/petrochemical/power sector project sites use 2%–2.5% as the threshold level, beyond which drastic actions are required to curb decline in quality. However, this percentage is based on percentage defective spots and not on percentage defect length. When the percentage is based on defect length over the length of weld radiographed, the threshold shall be kept at a much lower figure of 0.08%–0.10%, due to reasons stated above, and need not be relaxed any further as the figures are quite achievable.

F.8   OTHER BENEFITS

A continuous monitoring as mentioned above, at fixed intervals (say on a weekly basis), would reveal the trend of weld quality, so that corrective actions could be initiated well in advance so that hue and cry of the client on poor quality could be avoided.

The statistics so required to assess the quality could be utilized to make presentations as shown in the enclosed Excel file.

If made in Excel or any other similar software, the weld quality statistics against each welder/welding operator could be obtained at the flick of a button.

The statistics thus used against thresholds specified may be used as a yard stick in extending incentives to welders/welding operators.

The statistics (analysis based on type of defects) could also be considered as indirect pointer towards a few root causes of weld defects. For example, a seasonal increase in porosity could be pointing towards either improper baking of electrodes or adverse weather conditions. Similarly, abnormal increase in lack of root fusions could be pointing at improper initial fit-up of joint or edge preparation, or problems with clamping.

Acknowledgement: This article was previously published as Pullarcot, S. (2012) Reviewing Radiography. *World Pipelines, June 2012.*

Daily Status of Radiography

Sl. No.	Joints	Size (In)	No. of Spots	Defect Spots	Defective Spots 1	2	3	4	5	6	7	8	9	10	11	12	13	14	15	16	17	18	19	20	21	22	23	24	25	26	27	28	29	30	31	32	33	34
1	10	16	200	24	12							1						1		2	4															4			
2	8	14	160	17	2												3			8																		2	2
3	14	12	140	12	1		2					2								2	5																		
4	12	10	120	15			3															5			2			2											3
5	10	8	100	16	6		1					1									3		1					3				1							
6	8	6	80	23	4							9							10																			
7	14	4	70	13			2					3							2		4																	2
8	18	3	54	10								1								3		1			2							3						
9	20	2	40	15	1							3							1			3			3													4
10	15	1	30	8			2													3								2							1			
129			**994**	**153**	26	2	8	0	0	0	0	20	0	0	0	0	0	0	4	3	15	21	8	0	0	0	0	10	0	0	7	0	0	7	4	0	7	11

Notes

1	Numbers provided below Defective spots are the Serial numbers of defects identified in Table 3 of the write up
2	Spot size considered for calculation is 120 mm. Incase fim size 240 & 480 mm, it is considered as 2 or 4 spots respectively
3	If a defect of length (say) 17 mm is found in a film of 240 mm length consisting of two spots, then only one spot (of 120 mm) is considered defective
4	If there are two defects, separated considerably to fall in both portions of a 240 mm long film, both the spots are considerd defective
5	If the film size (length) is smaller than 120 mm, all such films shall be considered as single spots for defect analysis

Defect Analysis

Pie chart labels (defect serial number : percentage):
1: 17%, 2: 1%, 3: 5%, 4: 0%, 5: 0%, 6: 0%, 7: 0%, 8: 13%, 9: 0%, 10: 0%, 11: 0%, 12: 0%, 13: 0%, 14: 3%, 15: 2%, 16: 10%, 17: 14%, 18: 5%, 19: 0%, 20: 0%, 21: 0%, 22: 7%, 23: 0%, 24: 0%, 25: 5%, 26: 0%, 27: 0%, 28: 5%, 29: 0%, 30: 0%, 31: 3%, 32: 0%, 33: 5%, 34: 7%

References

CHAPTER 1

ASME (The American Society of Mechanical Engineers) (2022) *B31.3 Process Piping.* New York: ASME.

Ellenberger, P. (2010) *Piping and Pipeline Calculations Manual: Construction, Design, Fabrication, and Examination.* Oxford: Butterworth-Heinemann.

Nayyar, M.L. (1999) *Piping Handbook.* 7th Edition. New York: McGraw-Hill.

NPTEL IITM (n.d.) *National Programme on Technology Enhanced Learning.* Available at: https://nptel.ac.in/ (Accessed: February 6, 2023).

CHAPTER 2

Engineering Construction (2016) Learn Isometric Drawings (Piping Isometric), *Engineering Construction.* Available at: https://engineering-us.blogspot.com/2016/03/learn-isometric-drawings-piping.html (Accessed: February 10, 2023).

Sölken, W. (n.d.) *Welcome to Wermac, Explore the World of Piping.* Available at: https://www.wermac.org/ (Accessed: February 6, 2023).

Unitel Technologies (2021) *Unitel Technologies.* Available at: http://www.uniteltech.com/ (Accessed: February 6, 2023).

Walker, V. (2009) Designing a Process Flowsheet. *Chemical Engineering Progress, 105,* 15–21.

CHAPTER 3

CAD Standard (n.d.) *Drawing paper sizes A0, A1, A2, A3, A4.* Available at: https://www.cad-standard.com/technical-drawing-basics/drawing-paper-sizes (Accessed: February 14, 2023).

Sölken, W. (n.d.) *Welcome to Wermac, Explore the World of Piping.* Available at: https://www.wermac.org/ (Accessed: February 6, 2023).

CHAPTER 4

Matveev, A. (2022) *What Is Piping – A Blog to learn Piping, Mechanical, and Process Engineering Basics in a Simple Way.* Available at: http://www.whatispiping.com/ (Accessed: February 6, 2023).

CHAPTER 5

Envestis, S.A. (2018) *Learn Piping – Projectmaterials.com.* Available at: https://blog.projectmaterials.com/ (Accessed: February 6, 2023).

JFE 21st Century Foundation (1991) *Welcome to JFE 21st Century Foundation.* Available at: http://www.jfe-21st-cf.or.jp/ (Accessed: February 6, 2023).

CHAPTER 6

Sölken, W. (n.d.) *Welcome to Wermac, Explore the World of Piping*. Available at: https://www.wermac.org/ (Accessed: February 6, 2023).

CHAPTER 7

Ace Alpha International FZE. (2015) *Ace Alpha International FZE*. Available at: http://www.acealpha.ae/ (Accessed: February 6, 2023).

AMG Sealing (n.d.) *AMG Sealing: Metal, Graphite & Rubber Gaskets*. Available at: http://www.amgsealing.com/ (Accessed: February 6, 2023).

Coastal Flange (2022) *Carbon, Steel, & Alloy Pipe Flanges*. Available at: http://www.coastalflange.com/ (Accessed: February 6, 2023).

Excellence Engineering Corporation (2015) *Excellence Engineering Corporation*. Available at: http://www.excellenceengcorp.com/ (Accessed: February 6, 2023).

Ferguson (2023) *Ferguson: Plumbing Supplies, HVAC Parts, Pipe, Valves & Fittings*. Available at: http://www.ferguson.com/ (Accessed: February 6, 2023).

Garlock (n.d.) *Garlock: Leaders in Sealing Integrity*. Available at: http://www.garlock.com/ (Accessed: February 6, 2023).

Lamons (2023) *Lamons: Industrial Supply & Service – Gaskets, Fasteners, Hoses, Assemblies, and Expansion Joints*. Available at: http://www.lamons.com/ (Accessed: February 6, 2023).

Metal Gaskets (n.d.) *Metal Gaskets*. Available at: https://www.metalgaskets.net/ (Accessed: February 6, 2023).

Rits Int'l Korea LLC (n.d.) *Reducing Flange (id:3184682)*. Available at: https://radpey.en.ec21.com/Rits_Int'l_Korea_LLC–3184672_3184682.html (Accessed: February 6, 2023).

Seal & Design (2022) *Seal & Design, Inc*. Available at: http://www.sealanddesign.com/ (Accessed: February 6, 2023).

Sölken, W. (n.d.) *Welcome to Wermac, Explore the World of Piping*. Available at: https://www.wermac.org/ (Accessed: February 6, 2023).

CHAPTER 8

Fenghua Fly Automation Co. (n.d.) *Fenghua Fly Automation Co.: China Pulse Valves Manufacturer*. Available at: http://sino-pneumatic.en.hisupplier.com/ (Accessed: February 6, 2023).

Garlock (n.d.) *Valve Accessories*. Available at: https://legacy.garlock.com/fr/node/1351 (Accessed: February 14, 2023).

Kent Introl (n.d.) *Kent Introl: Industrial Valve Manufacturer UK*. Available at: https://kentintrol.com/ (Accessed: February 6, 2023).

NewsLiner (2023) *NewsLiner: Breaking News in Hindi*. Available at: https://newsliner.in/ (Accessed: February 6, 2023).

Piping Engineering (2013) *Piping Engineering: Knowledge Base*. Available at: http://www.piping-engineering.com/ (Accessed: February 6, 2023).

Quora (n.d.) *Quora – A Place to Share Knowledge and Better Understand the World*. Available at: https://www.quora.com/ (Accessed: February 6, 2023).

Sölken, W. (n.d.) *Welcome to Wermac, Explore the World of Piping*. Available at: https://www.wermac.org/ (Accessed: February 6, 2023).

Teksal Safety (2022) *Teksal Safety: Industrial Safety Solutions*. Available at: http://www.teksal.com.au/ (Accessed: February 6, 2023).

VirtualExpo Group (n.d.) *Direct Industry: The B2B Marketplace for Industrial Equipment*. Available at: https://www.directindustry.com/prod/rotex-automation-limited-61940.html (Accessed: February 6, 2023).

VseSdelki (n.d.) *VseSdelki*. Available at: http://www.vsesdelki.rv.ua/ (Accessed: February 6, 2023).

Waterworld (n.d.) *Waterworld*. Available at: https://www.waterworld.com/ (Accessed: February 6, 2023).

CHAPTER 10

ASME (The American Society of Mechanical Engineers) (2022a) *B31.3 Process Piping*. New York: ASME.

ASME (The American Society of Mechanical Engineers) (2022b) *B31.4 Pipeline Transportation Systems for Liquids and Slurries*. New York: ASME.

Fabricating & Metalworking (2020) *Fabricating and Metalworking: The Business of Metal Manufacturing*. Available at: http://www.fabricatingandmetalworking.com/ (Accessed: February 6, 2023).

GlobalSpec (2010) Miter Bend Drafting - CR4 Discussion Thread, *Engineering360*. Available at: https://cr4.globalspec.com/thread/58972/Miter-Bend-Drafting (Accessed: February 6, 2023).

KLM Technology Group (2011) *Project Standards and Specifications: Piping Fabrication, Installation, Flushing and Testing*. Available at: https://www.klmtechgroup.com/PDF/ess/PROJECT_STANDARDS_AND_SPECIFICATIONS_piping_frabrication_and_commissioning_Rev01.pdf (Accessed: February 6, 2023).

Kobe Steel, LTD. (n.d.) Welding Terminology, *KOBELCO*. Available at: http://www.kobelco-welding.jp/education-center/references/references03.html (Accessed: February 6, 2023).

Quora (n.d.) What is Leg Length in Fillet Welding if Plates of Unequal Thickness are Joining? *Quora*. Available at: https://www.quora.com/What-is-leg-length-in-fillet-welding-if-plates-of-unequal-thickness-are-joining (Accessed: February 6, 2023).

Sölken, W. (n.d.) *Welcome to Wermac, Explore the World of Piping*. Available at: https://www.wermac.org/ (Accessed: February 6, 2023).

CHAPTER 11

ASME (The American Society of Mechanical Engineers) (2022) *B31.3 Process Piping*. New York: ASME.

CHAPTER 12

ASME (The American Society of Mechanical Engineers) (2022) *B31.3 Process Piping*. New York: ASME.

CHAPTER 13

Bentley Communities (2013) *Double Trunnion Modeling in AutoPipe*. Available at: https://communities.bentley.com/products/pipe_stress_analysis/f/autopipe-forum/87342/double-trunnion-modeling-in-autopipe (Accessed: February 6, 2023).

Enerpac (2022) *Introduction to Torque Tightening*. Available at: https://www.enerpac.com/en-us/training/e/torque-tightening (Accessed: February 6, 2023).

Google Images (n.d.) Constant Load Pipe Support Mechanism. Available at: https://www.google.com/search?q=constant+load+pipe+support+mechanism&source=lnms&tbm=isch&sa=X&ved=0ahUKEwjr9b3Y2Y3gAhUIExQKHexCBLkQ_AUIDigB&biw=1600&bih=731 (Accessed: February 10, 2023).

HYTORC (n.d.) *HYTORC*. Available at: https://hytorc.com/ (Accessed: February 6, 2023).

Machine Design (n.d.) *A Primer on Pipe Supports*. Available at: https://www.machinedesign. com/mechanical-motion-systems/hydraulics/article/21833467/a-primer-on-pipe-supports (Accessed: February 6, 2023).

Piping Technology & Products, Inc. (2017) *Constant Support Hangers*. Available at: https:// pipingtech.com/wp-content/uploads/2017/07/New-Catalog-39-40.pdf (Accessed: February 6, 2023).

Piping Technology & Products, Inc. (2022a) *Insulated Supports (figures 4500 – 4800)*. Available at: https://pipingtech.com/products/pre-insulated-pipe-supports/insulated-supports/#spec-jump (Accessed: February 6, 2023).

Piping Technology & Products, Inc. (2022b) *Pipe Saddles & Supports*. Available at: https:// pipingtech.com/products/pipe-supports-hangers/pipe-saddles-coverings/#spec-jump (Accessed: February 6, 2023).

Sanger Metal (n.d.) *Recommended Torque Value Chart*. Available at: https://www.sanger-met.com/products/hi-force_hydraulic_tools/the_information_pages/recommended_torque_value_chart0727.html?sscid=26714&ssbid=3369 (Accessed: February 6, 2023).

Sölken, W. (n.d.) *Welcome to Wermac, Explore the World of Piping*. Available at: https://www. wermac.org/ (Accessed: February 6, 2023).

Terenzi, A. and Marcolini, M. (2017) Thermal Analysis of Vent Pipes and Clamped Saddles Interacting with Pipe Racks. *Journal of Pipeline Systems Engineering and Practice, 8(4)*. Figure 1 available at: https://ascelibrary.org/cms/attachment/668d6170-332b-44c0-8477-fca9b17362ce/figure1.jpg (Accessed: February 10, 2023).

Westorc (n.d.) *Bolting Patterns*. Available at: https://www.westorc.com.au/bolting-patterns/ (Accessed: February 14, 2023).

ANNEXURE B

ASME (The American Society of Mechanical Engineers) (2023) *BPVC Section IX: Welding, Brazing, and Fusing Qualifications*. New York: ASME.

ANNEXURE D

ASME (The American Society of Mechanical Engineers) (2023) *B BPVC Section VIII: Rules for Construction of Pressure Vessels Division 1*. New York: ASME.

ANNEXURE E

ASME (The American Society of Mechanical Engineers) (2022) *B31.3 Process Piping*. New York: ASME.

ANNEXURE F

Pullarcot, S. (2012) Reviewing Radiography. *World Pipelines, June 2012*.

Index

For Product Safety Concerns and Information please contact our EU
representative GPSR@taylorandfrancis.com
Taylor & Francis Verlag GmbH, Kaufingerstraße 24, 80331 München, Germany

www.ingramcontent.com/pod-product-compliance
Lightning Source LLC
Chambersburg PA
CBHW060802220326
41598CB00022B/2522